Thermal Management of Electronic Systems

Thermal Management of Electronic Systems

Proceedings of EUROTHERM Seminar 29,
14–16 June 1993, Delft, The Netherlands

Edited by

C. J. HOOGENDOORN,
R. A. W. M. HENKES
J. M. Burgers Centre for Fluid Mechanics,
Delft University of Technology,
Delft, The Netherlands

and

C. J. M. LASANCE
Philips B.V.
Eindhoven, The Netherlands

SPRINGER-SCIENCE+BUSINESS MEDIA, B.V.

A C.I.P. Catalogue record for this book is available from the Library of Congress.

ISBN 978-94-010-4472-1 ISBN 978-94-011-1082-2 (eBook)
DOI 10.1007/978-94-011-1082-2

Printed on acid-free paper

CONTENTS

PREFACE

The Eurotherm Committee has chosen Thermal Management of Electronic Systems as the subject of its 29th Seminar, at Delft University of Technology, the Netherlands, 14-16 June 1993. This volume constitutes the proceedings of the Seminar.

Thermal Management is but one of the several critical topics in the design of electronic systems. However, as a result of the combined effects of increasing heat fluxes, miniaturisation and the striving for zero defects, preferably in less time and at a lower cost than before, thermal management has become an increasingly tough challenge. Therefore, it is being increasingly recognised that cooling requirements could eventually hamper the technical progress in miniaturisation.

It might be argued that we are on the verge of a revolution in thermal management techniques. Previously, a packaging engineer had no way of predicting the temperatures of critical electronic parts with the required accuracy. He or she had to rely on full-scale experiments, doubtful design rules, or worst-case estimates. This situation is going to be changed in the foreseeable future. User-friendly software tools, the acquisition and integrity of input and output data, the badly needed training measures, the introduction into a concurrent engineering environment: all these items will exert a heavy toll on the flexibility of the electronics industries. Fortunately, this situation is being realised at the appropriate management levels, and the interest in this seminar and the pre-conference tutorials testifies to this assertion.

The Organising Committee succeeded in assembling an interesting programme embracing over 35 oral and poster presentations, four invited lectures, two pre- conference tutorials, a special session on European co-operation, and exhibits. It is a pleasure to look back and conclude that the seminar was a great success. About 100 people from 12 different countries attended the seminar, which was well above expectations. The same holds for the two tutorials, about 25 persons each participated. The subjects of the papers are equally divided between numerical and experimental analysis, which clearly shows that experimental work is by no means a dead end. The exhibition provided the opportunity to have a look at the newest versions of today's thermal analysis software and some very interesting products as well. A survey of the exhibitioners can be found under the Acknowledgements, where we take the opportunity to thank them for their financial support.

However, despite its success, other initiatives are needed as well. It is the objective of the Eurotherm Seminars to be a popular forum for high-level scientific and technical interchange of ideas on a wide range of special topics, while the primary aim is to stimulate discussion and liaison between specialists groups. Dr. Kaveh Azar, chairman of the 8th SEMI-THERM conference, Austin, 1992, said in his preface: "SEMI-THERM is perhaps the only conference that provides a medium to present current and near future development work. It also attempts to merge contributions from the universities and industrial communities."

We would be happy if the 'thermal management' community in Europe could establish the same tradition in communication as our colleagues in the USA, and one of the most appropriate ways is to organise an Eurotherm seminar like this on a regular basis. After the lively session on European coordination, it was decided to organise a Thermal Management seminar every two years. IMEC in Leuven, Belgium, volunteered to take the initiative for 1995.

We would like to acknowledge the contribution of the Scientific Committee for helping us to ensure the quality of the proceedings, and we would also like to thank the authors, invited speakers and tutorial lecturers for their valuable contributions to the success of this seminar. We owe many thanks to Mrs Sharmila Sewmar, who managed to stay in control of the many things that had to be done, and still remain smiling, besides her normal job as secretary to professor Hoogendoorn.

We are very grateful to the J.M. Burgers Centre for Fluid Mechanics, the host of the seminar, which got many compliments on the perfect organisation, the very nice facilities and the general ambiance of the old city of Delft.

C.J. Hoogendoorn, R.A.W.M. Henkes and C.J.M. Lasance, editors

ACKNOWLEDGEMENTS

exhibition participants

Cadence Design Systems BV
Bruistensingel 380
5232 AE Den Bosch
The Netherlands
Tel.: 31 73 429500
Fax: 31 73 429679

CHAM
Bakery House
40 High Street, Wimbledon Village
London SW19 5AU
ENGLAND
Tel.: 44 81 947 7651
Fax: 44 81 879 3497

FLOMERICS Ltd
Kingsgate Business Centre, 12-50
Kingsgate Road,
Kingston upon Thames
Surrey KT2 5AA
England
Tel.: 44 81 5473373
Fax: 44 81 5472682

Mentor Graphics
Benelux Regional Office
Haarlemmerstraatweg 133-137
1165 MK Halfweg
The Netherlands
Tel.: 31 2907 7115
Fax: 31 2907 6532

Powercool Europe BV
Echtenseweg 10
7963 PC Ruinen
The Netherlands
Tel.: 31 5221 2497
Fax: 31 5221 2675

Power Electronics Ireland
EOLAS
Glasnevin
Dublin 9
Ireland
Tel.: 353 61 333644
Fax: 353 61 330316

Sensor Partners
P.O. Box 270
5150 AG Drunen
The Netherlands
Tel.: 31 4163 78329
Fax: 31 4163 77439

THERMACORE Inc.
780 Eden Road
Lancaster, PA 17601
USA
Tel.: 1 717 569 6551
Fax: 1 717 569 4797

1. INVITED LECTURES

THERMAL MANAGEMENT OF AIR-COOLED ELECTRONIC SYSTEMS: NEW CHALLENGES FOR RESEARCH

C.J.M. LASANCE
Philips NV.
Centre for Manufacturing Technology
P.O.Box 218
5600MD Eindhoven The Netherlands

ABSTRACT. Thermal Management is but one of the many critical tasks in the design of electronic systems. However, in the past decade, its role has become increasingly important, resulting from the widespread introduction of micro-electronics in almost every branch of industry.
As a result of the combined effects of increasing heat fluxes, miniaturisation and the striving for zero defects, preferably in less time and at a lower cost than previous designs, the thermal management of generally complex electronic systems has become a growingly tough challenge.

If we accept, for the time being, the fact that quality and reliability are heavily dependent on temperature, it is obvious that the designer must be able to predict the critical temperatures with sufficient accuracy. Previously, there was no way to predict junction temperatures of components in an often complex system to within a couple of degrees C, and designers tended to rely upon full-scale experiments, doubtful design rules or worst-case estimates. Today, in a growing number of cases, the output of some computer tools appears reliable enough to use the results for quality prediction.

It is one thing to criticise the long-accepted design methodology, it is quite another thing to formulate adequate alternatives. Especially in the consumer electronics sector, another problem will arise in the future because the thermal analysis of new (or, unfortunately, often old) designs, which is still the task of the specialists, will be largely performed by the electronic and mechanical engineers. Transfer of knowledge, awareness increasing, and accessibility of input and output of thermal analysis tools in a concurrent engineering design environment is another area in which the support of research groups is needed.

The lecture will briefly address the current problems in fields such as the determination of the heat transfer coefficient, the component thermal resistance, the temperature-reliability relationship and the interpretation of numerical results, and possible directions of future research will be indicated. Special attention is being paid to the natural convection of horizontal plates.

Because a number of research fields suffer from lack of human, technical and financial resources, the need for co-operation within the European Community is steadily growing. The last chapter deals with this issue.

Nomenclature

c_p	[J/kgK]	specific heat	α	[W/m²K]	heat transfer coefficient
q	[W/m²]	heat flux	λ	[W/mK]	thermal conductivity
$\dot{m}$	[kg/s]	mass flow	ϕ	[W]	power dissipation
R_{th}	[K/W]	thermal resistance			
T	[K]	temperature			
Nu	[-]	Nusselt number = $\alpha \cdot L/\lambda = q/\Delta T \cdot L/\lambda$			
Ra	[-]	Rayleigh number= $g \cdot \beta \cdot \Delta T \cdot L^3 \cdot Pr/\upsilon^2$			

C. J. Hoogendoorn et al. (eds.), Thermal Management of Electronic Systems, 3–24.

Introduction

In Consumer Electronics, the functionality of electronic parts has increased dramatically during the past decades. In striking contrast, the reliability, expressed for example as a Field Call Rate (FCR), has not evolved along the same lines. The increased complexity resulting from the increased functionality seemed to provide a rationale for increased FCRs and, only half a decade ago, FCRs of 15% and more were no exception. It is only fair to remark that Japanese manufacturers should be acknowledged as being the first to realise that consumers, in the end, would buy products whose designers has placed more attention, if not emphasis, on their 'right' to buy reliable products. Needless to say that they have been very successful with such an approach, even to the extent that 'Made in Japan', which, for a long time, had been synonymous with low quality copies, has now been turned into the opposite. Besides, consumer satisfaction is not the only thing that is advantageously affected by paying more serious attention to quality and reliability. For example, failures in the later stages of a production process can be very costly indeed, thereby adversely affecting the bottom price of the product, and even more seriously, the timely introduction of the product on the market.

The future of many electronic companies will depend to a large extent on the ability to initiate techniques to bring scheduling, performance, testability, supportability, producibility, life-cycle costs and quality control into the earliest stages of the design cycle. Nowadays, consumer electronics manufacturers unanimously agree upon the fact that it pays off to develop methodologies to ensure that quality and reliability, among other things, are built into a new design right from the start. This means that the team of designers should have the tools available to predict items such as quality and reliability, and then obviously the next question is which parameters affect these items.

It has been realised for a long time that higher temperatures do affect reliability adversely, due to a variety of physical failure mechanisms. The exact relationship, however, is nowadays a matter of renewed dispute. For the time being, it is taken for granted that temperature is one of the important parameters which should be taken into account from the very beginning of a new design. The following chapters summarise the more elaborate discussions to be presented by the author in two papers titled 'Accurate Temperature Prediction in Consumer Electronics: a Must but still a Myth', and 'Pragmatic Methods for Determining the Parameters Required for the Thermal Analysis of Electronic Systems', further referred to as L_a and L_b respectively.

Chapter 1 answers the question why thermal management is much more important than it used to be, and why its importance is steadily growing. Chapter 2 deals with the definition of the heat transfer coefficient, and particular emphasis is laid upon the theoretical and practical problems with the heat transfer of horizontal plates, which play an important role in the consumer electronics industry. Also, some comments are given on the increasing need for a user-friendly handling and interpretation of the large sets of data resulting from the application of CFD codes. The problems with the thermal characterisation of components are summarised in chapter 3, while chapter 4 discusses the temperature reliability relationships. Finally, the paper concludes with a plea for speeding up the badly needed co-operation between the electronics industries.

It should be noted that, first and foremost, this lecture aims at the packaging engineers environment. Therefore, many subjects are not treated in-depth, and numerous papers that have significantly contributed to a more fundamental insight into the physics of the often complex phenomena have not been reviewed, for which the author offers his apology.

1. About the Need for Thermal Management of Electronic Systems

When trying to answer the question why, several reasons show up.
In the first place we can distinguish purely technical reasons. Industry is confronted on the one hand with a technological push, such as miniaturisation, the introduction of more features, more plastics etc., and on the other hand with a market pull, such as a better reliability, lower noise levels, zero defects, and increasing demands for Electro Magnetic Compatibility. Most technical trends cause the temperature to increase, while most market trends demand lower temperatures. For a more detailed discussion of electronic packaging in the nineties, see e.g. Tummala ('91), Ohsaki ('91), Wesseley et al. ('91) and Bar-Cohen ('92).

In the second place, one may distinguish a physical reason. Especially for natural convection driven systems, which encompass the majority of the consumer products, there exists an upper limit to what can be reached in so far as the maximum temperature is concerned, even if all the variables that influence said temperature are optimised. It is to be expected that these limits will be reached in the coming years. It is obvious that is very important from a management point of view to know exactly what these limits are, in order to decide timely if one should introduce, for example, controlled fans, which decision in many cases appears to be a psychological barrier rather than a rational one to overcome.

Because design rules live by their relatively wide margins, and because added complexity tends to widen these margins even more, their area of application will drastically decrease when the limits of what is physically possible come into reach. In any case, the old set of design rules should be replaced at some moment in the future.

In the third place, there are a couple of logistical reasons for paying more attention to thermal management. Historically, electronic and mechanical design methods are part of every design philosophy, mainly because the functional requirements should be incorporated from the very beginning. In many industries, thermal design is just an afterthought, to be addressed if measurements in a prototype reveal any temperature problems. Nowadays, time-to-market is a hot issue, and the first prototype should be right-first-time. In other words, thermal design should be approached in just the same way as electronic and mechanical design.

If a concurrent engineering approach is adapted as the only and ultimate design philosophy to remain a major player in the world-wide competitive market place, Thermal Management will be an important and necessary part of it, and must be recognised and accepted by all people involved.

Now, suppose everything is settled and concurrent engineering principles have been introduced everywhere. Suppose also that a system level packaging engineer who is responsible for an adequate thermal design of a new product is asked by the product manager to come up with an estimation of the critical temperatures, such as the die temperature of an active component, the coil temperature of a transformer, or the peak temperature inside an electrolytic capacitor.
Then, it is very important for both parties to agree upon the underlying objectives of the request. If the objectives are for example to have an idea about the worst case situation, or to support a feasibility study, or to have some Figure of Merit to compare different designs, the estimation of temperatures of critical components for the purpose of a parametrical study makes sense.
However, if the objectives are to give an estimation of the expected Field Call Rates, or the Life Cycle Costs, the packaging engineer is in deep trouble. Because of the supposedly strong dependency of the reliability on the temperature, it is crucial that it should be estimated within narrow tolerances, say within plus or minus 3°C.

From a technical point of view, we have to solve at least three big problems in order to meet this criterion: how to arrive at accurate values for the heat transfer coefficient, thermal resistances, and temperature-reliability relationships.

2. About the Heat Transfer Analysis of Complex Electronic Systems

2.1 ABOUT THE DEFINITION OF THE HEAT TRANSFER COEFFICIENT

The basic equation to describe the heat transfer at the interface between a solid surface at T_s and a fluid at T_0 by the use of a heat transfer coefficient is:

$$q = -\lambda dT/dy \equiv \alpha(T_s - T_0) \tag{2.1}$$

In most textbooks this law is referred to as Newton's cooling law, but it is really more a definition of α than a law. As Jakob ('53) stated, this magically simple equation was a grave handicap to progress in the field of heat transfer, because for a long time it had been assumed that α is a constant, which is clearly not the case. Now, forty years later, there is no doubt that very much progress has been made in heat transfer in general, but not so in the definition and the representation of the heat transfer coefficient. Apart from the problem with the non-constant α, the second and much more challenging problem is how to define T_0.

Let us first reformulate (2.1) by replacing T_0 by T_{ref}, which now refers to any reference temperature whatsoever. It is the merit of Moffat and coworkers at Stanford that renewed interest has emerged to attack this problem. A good overview of the principles can be found in Moffat and Ortega ('88).

Some canonical cases to define the reference temperature can be treated in a straightforward way:

a. The far field temperature: *isolated object with uniform temperature or flux in an infinite reservoir of fluid at a constant temperature, steady state.*

The definition is well suited for boundary layer type of flow over plates subjected to uniform boundary conditions.

b. The cup-mixed-mean temperature: *channel- or tube flow, uniform boundary conditions, 'well-behaved' flow.*

Well-behaved means in the first place well-mixed. If we are dealing with uniform boundary conditions and well-mixed flow, as is usually the case with heat exchangers, then the reference temperature at any cross-section can be calculated via $\phi_{upstream} = \dot{m}c_p\Delta T$.

However, this approach becomes questionable if the conditions of the flow are not so well-behaved, for example if the mixing is not perfect. Nakayama ('88) noted that the ratio of the actual temperature rise of the air in the immediate vicinity of a component to the rise of the mixed mean temperature, clearly indicates that differences of up to a factor four could be expected.

c. The inlet temperature: *practical systems.*

To define the inlet temperature as the reference temperature is purely based on practical considerations, because at least this temperature is easy to measure and mostly not subject to variations. The main flaw is that the definition has no physical significance.

d. An 'ambient' temperature: *some company, history, or nobody-knows-anymore-based reference temperature.*

Numerous official specifications for components state some reference temperature, mostly called ambient temperature, above which the component is not guaranteed anymore. Very often, designers and factory people, with no basic knowledge of heat transfer, measure such a temperature with some undefined, often rather robust, uncalibrated, preferably Copper-Constantan thermocouple, positioned at some unreproducible and undefined place inside a powered system, somewhere in the neighbourhood of the component in question. It is obvious that, inside a real system, it will be very difficult to define a reference temperature in a proper and manageable way, because the exact position of the thermocouple should be prescribed for every practical situation.

Considering the afore-mentioned items a-d, we can conclude that we have a problem, not only because a choice should be made between various alternatives, but also because the non-uniform boundary conditions and the poorly behaved flow patterns in most of the practical systems do not allow for a logical and sensible way of dealing with local air temperatures.

e. The 'adiabatic' temperature: *more generally applicable.*
A more attractive approach to define the reference temperature offers the so-called 'adiabatic' temperature, proposed by Arvizu and Moffat ('82) for forced convection and Ortega and Moffat ('86) for buoyancy induced flows, which is defined as the temperature a component or object reaches when its power is switched off. All experiments, which are based on heating one element at a time, or those that make use of techniques such as naphthalene sublimation or a swollen polymer technique, are the basis for the determination of the adiabatic heat transfer coefficient. T_{adiab} is relatively simple to measure, and is also very well suited for standardisation purposes. For example, a well-chosen location could be assigned to every component of interest, and also the type of thermocouple and how to fix it to the surface could be prescribed. However, maybe the most important property of the definition is that it makes sense from a physical point of view. Convective heat transfer, especially in electronic systems with their many discrete boundary conditions, is a very localised phenomenon. With a reference temperature assigned to the component itself, it stays that way. Finally, such an approach paves the way to superposition principles, which are nothing else but a very efficient way of separating important effects that can be treated subsequently.

It should be repeated here that the whole idea of superposition depends on the allowable separation of the temperature and flow fields. In principle, systems dominated by natural convection do not belong to this category. Fortunately, in most practical systems, the power dissipation of some critical component is only a fraction of the total dissipation, so locally we may consider the flow field as characterised by buoyancy-induced convection, as shown by Moffat and Ortega ('86).

To the author's best knowledge, only three conjugate heat transfer studies based on the adiabatic heat transfer coefficient concept have been published (Ortega and Kabir ('91), Ortega and Lall, ('92), Witzman et al.('92)). Some results of a fourth one, dealing with one component on one board in a small enclosure, are presented in L_b.

It is obvious that the term 'adiabatic' is only justified from a theoretical point of view if, in the equilibrium state, no energy is exchanged with the environment. However, in almost all practical cases, there is always some exchange of energy. Even when radiation and conduction do not play a role, a difference in local air temperature around a component gives rise to heat transport via the component. Because, historically, conjugate heat transfer effects were not taken into account in the pioneering work of Moffat and coworkers, the term 'adiabatic' has been the correct phrase to describe the superposition idea.

For more practical cases, where it becomes difficult to separate the various heat transfer modes, continuing the use of the word 'adiabatic' does not seem appropriate. Sometimes, the words 'unpowered' or 'zero-powered' heat transfer coefficient have been used. However, their use is strictly speaking only valid for those convection situations in which the flow field essentially does not change when switching off the power of the component. In all other cases, an incremental formulation should be used, and then the term 'unpowered' becomes meaningless.

Therefore, the author advocates the terms 'fluid-based' heat transfer coefficient (α_{fb}) to replace the 'standard' definition, and 'solid-based' heat transfer coefficient (α_{sb}) to replace the 'adiabatic' definition. 'Fluid-based' points at a reference temperature somewhere in the fluid, and 'solid-based' at the reference temperature defined at some location on the object under study.

Summarising, the use of α_{sb} has distinct advantages over α_{fb}. Unfortunately, life is not a bed of roses. As pointed out by Lasance ('91), computer applications in which measured heat transfer coefficients should be substituted for some reason pose a difficult problem. All computational fluid dynamics codes solve the Navier-Stokes equations directly, so, provided the grid is fine enough to model all the boundary layers correctly, there is no need to impose some user-defined heat transfer coefficient. However, this procedure would require an unrealistic number of mesh lines when modelling, for example, a TV set with its many critical components.

And even if this option does become a realistic proposition somewhere in the near future, the simulation of real flow, with its many instabilities that influence the magnitude of the laminar heat transfer coefficient, will not be feasible for a long time to come. Local meshing does not solve this problem. Then, the only option is to impose estimated heat transfer coefficients at every boundary of interest, and it is here that a problem is encountered.

Most values for the heat transfer coefficient are acquired via adiabatic methods. In short, in deriving such a value, <u>two</u> different measurements in time at <u>one</u> location in space are needed. On the other hand, the computer code needs a value which is dependent on some reference temperature, in other words: temperatures at <u>two</u> locations in space at <u>one</u> moment in time.

One way to overcome this rather fundamental discrepancy is to simulate the experiment, which means that the 'adiabatic' temperature should be calculated first before the user-defined value can be imposed. Because ''adiabatic'' is not 'adiabatic' in a strictly physical sense, some thinking will reveal that an appropriate solution is by no means trivial. Research could be directed to a mathematical correct way of tackling this problem.

2.2 ABOUT THE HEAT TRANSFER COEFFICIENT OF DISCRETE COMPONENTS

While in the past the research was almost exclusively directed towards the finding of Nu-Gr and Nu-Re correlations for uniform boundary conditions and smooth surfaces, the need to have data available for boards populated with electronic components did stimulate researchers to focus on the mutual influence of discrete elements on each other. Because this research is driven mainly by the computer and telecommunication industries, the majority of the papers deal with channel flow. Excellent reviews, focusing on the cooling of electronics, can be found in Kraus and Bar-Cohen ('83), Moffat and Anderson ('90), Peterson and Ortega ('90), Aung ('91), Pecht ('91). Relatively few studies incorporated the conjugate heat transfer problem, or the non-uniform heat release (Zinnes ('70), Graham and Witzman ('88), Culham ('88), Barozzi and Nobile ('90), Bertin and Saulnier ('90), Kishinami et al. ('90), Culham et al. ('91)).

A few things become abundantly clear from these studies: the heat transfer results for discrete components differ greatly from the uniform boundary conditions results to an extent that the 'familiar' handbook correlations cannot be used to predict the heat transfer coefficient of components (or even flat plates) in real systems, at least if an accurate estimation is what the designer wants. Turning again to real systems, still another difficulty arises. As indicated by Moffat and Ortega ('88), the heat transfer data obtained from the study of discrete systems with elements of a certain geometry cannot be translated, strictly speaking, to other geometries. No universal length scale has been found yet. Maybe a solution can be found by scaling all the dimensions which are manifest on a board populated with different 3D components at various distances from each other. A paper by Krane and Phillips ('86) on natural convection cooling of a horizontal board in a small enclosure, still a relatively simple configuration, uses dimensional analysis to find a correlation encompassing 32 dimensionless groups. Fortunately, 17 of these groups were kept constant, otherwise it is obvious that the resulting correlation will be a monstrosity, and the advantage of a similarity or dimensional analysis will be lost to a great extent.

The question may arise if nature itself prohibits such an approach. Without going into details, some arguments in favour of this notion will be presented. Almost all analytical correlations can be traced back to the well known boundary simplifications of the generalised Navier-Stokes equations. When applying textbook correlations derived for isolated plates or channel flows to solve electronics cooling problems, one tends to forget that the starting-point was the boundary layer concept. It can be argued, however, that this attack is doomed to fail under certain circumstances. For example, in order to apply boundary layer approximations, the boundary layer thickness δ should be small compared to the characteristic length L of the object under study. If we look at the heat transfer of a relatively small component, this criterion is likely to be violated. It is certainly violated at the leading edge, and in a real system with many dissimilar components there are many leading edges. Hence, theoretically, the application of the boundary layer theory is not justified, which means that the use of correlations is not warranted. Furthermore, if the distance between components is such that the layers are interrupted periodically, it also becomes clear that it makes no sense to use the pitch between the components, or the distance from the leading edge of the board, as a parameter in the correlation. The author would not be surprised if future research reveals that the concept of correlation based on characteristic lengths should be abandoned altogether for boards with discrete and non-uniform dissipation.

A promising alternative, certainly from an experimental point of view, could be the approach proposed by Maciejewski and Moffat ('90), who showed that it could be possible to correlate heat transfer coefficients with the degree of turbulence, which itself could be, to a large extent, a function of the local pressure only.

2.3. ABOUT THE NATURAL CONVECTION HEAT TRANSFER BEHAVIOUR OF HORIZONTAL PLATES

In all cases where miniaturisation with respect to volume is not of paramount importance, the packaging engineer prefers a vertical position of the PCBs, because of the supposedly inferior natural convective heat transfer capabilities of horizontal plates. This is often the case in the mainframe and telecommunication industries. The situation is entirely different in the consumer electronics sector. Think of car radios, mini-audio sets, LCD TV, notebook computers, electronic light bulbs. In all these cases, it is very difficult, if not impossible, to apply vertical boards. A literature survey reveals that much more attention has been paid in the past to the analysis of natural convection heat transfer in vertical channels and isolated plates, as compared to horizontal and inclined plates. For example, in a recent review by Aihara ('91) about natural convection cooling techniques for computers, the term 'horizontal' could not be traced. Restricting ourselves to horizontal plates only, it is also observed that many papers are devoted to plume formation, or differentially heated enclosures with water as the convective agent, or to large scales as observed in meteorology. Only a few papers focus on the cooling of electronics, and almost none discusses discrete boundary conditions. The following paragraph briefly reviews the existing literature.

One of the most striking observations when reviewing the literature is the large discrepancies between the reported results, not only between different research groups, but also between experimental and numerical analyses. Lewandowski and Kubski ('83) compiled the results of about 40 publications on the natural convection heat transfer behaviour of both vertical and upward facing heated plates. For the vertical plates they found a scatter of ± 25%. However, for horizontal plates, the situation is much worse. In the laminar region, differences of 45% are observed, and in the turbulent region even of ± 100%. Besides this, different values of the exponent for the Nu-GrPr power law have been reported, varying from 0.2 to 0.3, and also an additive constant has been proposed (e.g. Miyamoto, '85). On top of that, the critical Ra or Gr number determining the onset to transition and to turbulence is a matter of dispute.

Many authors have tried to formulate answers in searching for the 'truth', and many explanations have been suggested for the observed anomalies. Yousef et al. ('82), in performing a very well-designed experimental study on horizontal square plates, found random fluctuations in the laminar region that caused the average Nu number to vary about 25% around the mean value, and the local Nu number even 45%. Furthermore, they concluded that theoretical solutions considerably under-predicted the average heat transfer coefficient. They attributed the fluctuations to 3D disturbances which were ignored in all the theoretical analyses, and concluded that a correlation which yields a single value for Nu or Nu_x is unrealistic. Gebhart et al. ('88) pointed at a possible cause of the discrepancies between numerical and experimental analyses. Of course, the fluid motions should ideally be caused only by the effect of the heated surface. However, structural vibrations, flow disturbances in the ambient medium, fluid circulations and stratification caused by a finite ambient medium may influence the data. Another major question is how the thermal boundary conditions at the surface are precisely attained experimentally. Ishiguro et al. ('78) detected periodic, not random, oscillations in a time-dependent 2D numerical study, for Ra numbers well below the transition limit, justified by experiments with a heated strip in water. The 3D fluid lumps originating at the higher Ra numbers could be the reason behind the discrepancies between the reported 2D analytical and experimental results. Rotem and Claassen ('69), in studying semi-infinite plates, found that the flow is considerably less stable for a horizontal configuration than a flow over slightly inclined plates. Related to this, Gebhart et. al. ('88) remarked that horizontal flows are not a special case of inclined flows because the tangential buoyancy force vanishes. Clausing and Berton ('89) studied mixed convection, and reported that very small draughts can influence the natural convection results. Husar and Sparrow ('68) were the first to publish flow visualisation studies, studying square, rectangular, triangular and circular plates. Apart from the last shape, partitioning of the flow field along the diagonals became evident, as was the formation of a central plume. From the flow fields it could be concluded that generalisation to other geometries may involve more than simply using the short side as the characteristic dimension. Afzal and Hussain ('84) also studied mixed convection over isothermal plates, and, for a certain combination of parameters, dual solutions were found. Therefore, it seems plausible that, depending on the manner how a temperature field is imposed, different solutions could be obtained. Interestingly, Cohen and Reshotko ('56) also found dual solutions in a study concerning laminar boundary layers in compressible flow with non-zero pressure gradients, and, according to them, which of the two solutions that sets in an experiment is determined by the initial conditions which establish the pressure field acting on the developing boundary layer. Mikheyev ('68) concluded that the heat transfer from upward facing horizontal plates was 30% higher than for vertical plates under similar conditions, while for downward facing plates the heat transfer was 30% less. However, Gebhart et al. ('88) reported that horizontal flows are always weaker than vertical flows.

Another interesting field from the viewpoint of the cooling of electronics is the closed or partly open enclosure with its bottom at the higher temperature. Chu and Goldstein ('73), pursuing the pioneering work of Malkus ('54) in discussing turbulent convection in a horizontal layer of water, also discovered discrete transitions, showing themselves as abrupt changes in the Nu-Ra slope. Temperature reversals were observed for relatively low Ra numbers. The development of thermals (rising 'mushroom'-like fluid elements) played a decisive role in the heat transfer. It was also noted that the transition to turbulence occurred at an Ra number about four times lower for air than for water. Turbulent convection experiments in a cavity open at the top were conducted by Townsend ('59). He observed comparatively infrequent 'active' periods, during which the temperature oscillates wildly between wide limits with extended periods of quiescence, during which the temperature is low and varies very little. A satisfactory explanation could not be given, but he suggested that regions of hot rising air, though very different from the plumes rising above localised heat sources, were the cause of the measured chaotic behaviour. Pnueli ('82) addressed the mathematical side in a review paper dealing with natural and mixed convection. He argues that in many cases where forced convection in channel flow is thought to be dominant, in reality mixed convection occurs, the magnitude of which depends on the location of the temperature difference. Although the average flow field is not very often much affected, the temperature field is affected and thus the heat transfer. He proceeds by making a clear distinction between Boundary Value and Eigen-Value problems. External flows come close to the first category, while the second category encompasses most internal flows and also large horizontal plates. Boundary layer approximations are doubtful in the treatment of internal flows. An eigen-value problem permits the solution of zero velocity, thereby giving way to a critical Ra number. An implication is that even in slow flows the various possible patterns may vary in time, never quite repeating themselves. Unsteady flow cases have been reported in laminar situations quite removed from turbulence, and the prediction of heat transfer is still an open question. Then he makes a classification in boundary values, eigen-values, and mixed problems. He concludes with some engineering advice: only boundary value problems are reliable. Once a number of experimental results is known, they can be extrapolated, because the results are reproducible. Eigen-value problems are not; engineers cannot apply them with confidence to systems even minutely different from his or her original work. However, it should be stressed that he does not plea for giving preference to forced convection, but only for designing for localised boundary value solutions, which can also be acquired with buoyancy-induced forced convection around critical parts.

Stratification effects could possibly play a role in those applications in which for some reason the top of the enclosure is hotter than the bottom. Such a situation could, for example, occur in car radios, or CD players with an amplifier on top, or sun-irradiated systems. In such systems, the thermal flow can become 'stratification-trapped' (Gebhart, '79). If the stratification-induced density variation is of the same order as the density variation imposed by some relatively large object such as transformers, heat sinks, or electrolytic capacitors, heat transfer during the transient condition can be seriously hampered, resulting in much higher temperatures than expected. Eventually everything will be mixed, due to thermal diffusion, and the heat transfer coefficient will regains it 'old' value.

The most important conclusion emerging from this literature review is that natural convection above a hotter plate is not very well understood, neither from a physical, nor from a mathematical point of view. Another thing that becomes evident is that in many cases the heat transfer behaviour is not a steady phenomenon. Apparently more than one solution exists, as is often the case in situations far from equilibrium. From an engineering point of view, this behaviour could introduce quite a lot of problems, that can be collected in two different categories:

* A minor disturbance in an otherwise steady state operating system could cause a transition to a completely different flow field in an unpredictable way.
* If more than one power source contributes to the heat transfer, the resulting flow field could depend on the sequence of powering up. For example, take an ASIC on a horizontal board with some other components dissipating as well. There could be a difference in the final flow field, depending on whether the ASIC is turned on after a steady state for the rest of the board was

established, or whether the ASIC was first powered, and then the board. In other words, superposition principles cannot be applied in such cases.

If these phenomena indeed affect the heat transfer behaviour in electronic systems with predominantly horizontally placed boards, it could be envisaged that such behaviour adversely influences the predictability of the temperature and thus the reliability. However, nobody knows to what extent the effects arise, and if so, what the influence is on the local temperatures. Of course, flow fields could look very different, but what really matters is the consequence with respect to the local temperatures.

In 1978, Gebhart ('79) noted in his Freeman Scholar Lecture '.....that the study of many kinds of buoyancy-induced flows is still in its infancy and many areas remain largely or completely unexplored.' And he ends with: 'Looking back, from perhaps twenty years hence, the engineer will have the answers to problems we cannot even properly pose.' I am afraid he was a bit too optimistic.

From the point of view of the cooling of electronics, a few interesting research fields suggest themselves.

- Stratification influences: could they occur, and how serious could be the influence?
- Interaction of plumes arising above localised heat sources could generate a much stronger plume: what are the optimum locations as a function of the interaction?
- The influence of the experimentally observed random and periodic oscillations on the maximum attained component temperatures.
- Mixed convection will be increasingly important, especially in the consumer electronics industry, because the introduction of small fans could gradually take over the purely natural convection driven systems. The question is to what degree superposition principles still apply when the natural and forced convection flows are comparable.
- Another topic that remains to be exploited in a much wider area is 3D internal flows in an enclosure, be it completely closed or (partly) covered with grids.

Some examples resulting from numerical studies indicate clearly that problems are to be expected. These are presented below.

2.3.1. 2D numerical analysis of a source in a differentially heated small enclosure. Consider figure 2.2. The bottom is heated 50°C above the top plate temperature. The sides of the enclosure and the plate in the centre are adiabatic. A number of cases have been studied, differing in the power dissipation of the plate and the sequence of the initial conditions:

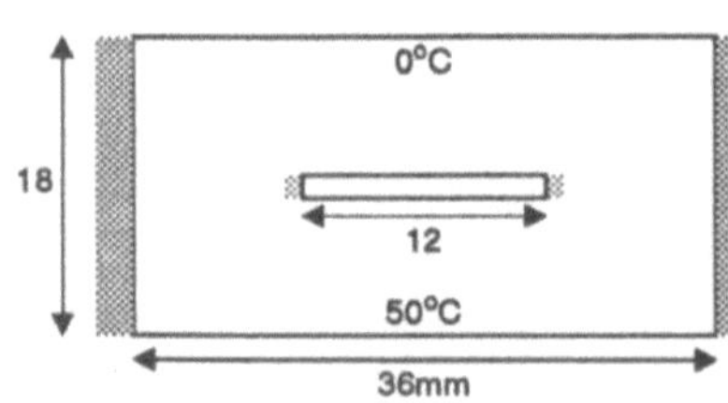

	Initial conditions First	Next
Fig.A	q_{plate} = 0 W/m²	ΔT = 50°C
Fig.B	q_{plate} = 0.46 W/m²	ΔT = 50°C
Fig.C	q_{plate} = 1.16 W/m²	ΔT = 50°C
Fig.D	ΔT = 50°C, q_{plate} = 1160 W/m²	q_{plate} = 0

Figure 2.2 Small enclosure data

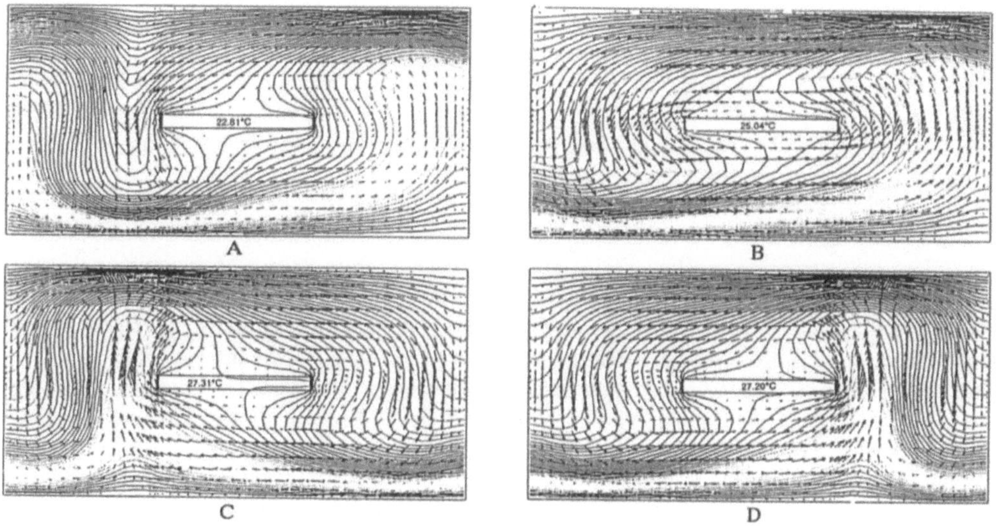

Figure 2.3 A-D Temperature and flow fields

Figures 2.3A-C show three cases in which first the power dissipation of the plate in the centre was set to 0, 0.46 and 1.16 W/m² respectively, and subsequently the temperature difference of 50 °C between bottom and top. Three very different flow fields can be distinguished (see figure 2.3A-C, figure D is discussed on the next page).

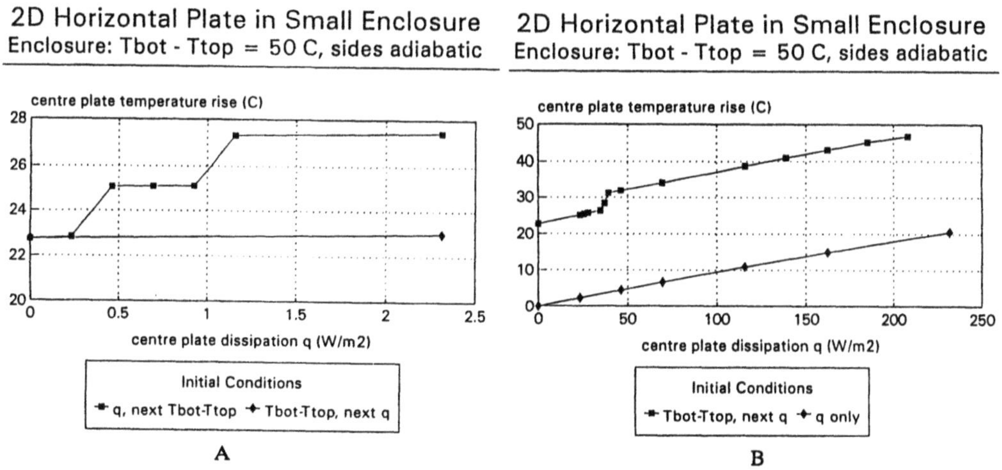

Figure 2.4 power dissipation vs. temperature rise

In figure 2.4A the temperature rise is plotted as a function of the plate dissipation. Notice the kinks in the slope. These kinks correspond with the transitions in the flow field. It is very remarkable that those minor power levels, compared to the total power dissipation that drives the flow field, are able to cause the transitions. When reversing the sequence, setting first the temperature difference and then the power dissipation, similar transitions have been observed, but at a much higher power level, see figure 2.4B. Figure 2.3D shows another phenomenon: a hysteresis effect.

In this case, with a fixed temperature difference, the plate dissipation was raised to 1160 W/m² before being switched off. Compare the resulting flow field with the one depicted in figure 2.3A. Note, except for the change in the flow field, also the difference in the acquired adiabatic temperature of the plate. Also, large differences were observed between the situations in which, with fixed ΔT, the power was raised immediately to a certain level, and in which the power was gradually increased to the same level.

The results make evident that the initial conditions should be taken into account. Of course, the problem that remains is how to validate the results, because experiments on such a small scale are nearly impossible to perform. However, in this fictitious case the author is of the opinion that the physics is correctly modelled through the conservation laws used in the analysis, because radiation effects are excluded, laminar flow is a fact, and environmental exchange is prevented by the enclosure. The question to be answered is whether the CFD code solves the equations properly. One way to address this issue is to compare the solutions of various CFD packages based on different solving techniques, e.g. Finite Difference, Finite Volume, Finite Element, Spectral Element or Boundary Element.

Interestingly, a similar behaviour has been observed on a much larger scale by studying the ventilation within a concert hall. With a fully populated hall, it turned out to be impossible to overrule the natural convection flow field raised by the audience by the low-velocity ventilation fans, while no problems arose after switching on the fans before the people filled the hall (TNO, '92).

2.3.2 About the Nu-correlations for upward and downward facing heated horizontal plates.
Textbooks often cite the following Nu-Gr correlations for heated or cooled horizontal plates:

$$Nu = 0.54 \cdot Ra^{0.25} \text{ (upward facing)}; \quad Nu = 0.27 \cdot Ra^{0.25} \text{ (downward facing)} \qquad (2)$$

Apparently, the conclusion should be that the heat transfer for upward heated plates is approximately twice as much as for downward heated plates. However, this is not what computer experiments show. For example, Goldstein and Lau ('83) presented 2D numerical results, studying the effects of horizontal extensions and vertical walls on the flow and temperature fields generated by upward-facing and downward-facing heated plates. 3D experiments with square plates confirmed the analyses to a sufficient degree. However, the boundary conditions chosen were those for a relatively large enclosure, and only one plate size was used in the numerical analysis. Therefore, it was decided to repeat the analysis with the CFD code FloTHERM (Flomerics), based on essentially the same set of equations and solution techniques, but with a larger solution domain with free boundaries and a much finer grid, in such a way that enlargement and refinement showed no further improvement. The plate size was varied between 12 and 160 mm, thereby extending the maximum Ra number, based on the half width of the plate, from $8 \cdot 10^3$ to $4 \cdot 10^6$. In this way the heat transfer behaviour of a horizontal plate in a virtually infinite ambient could be studied. Near the maximum Ra number, numerical instabilities pointed at the onset of transition. The deviations can be clearly observed in figure 2.5. Because only the power of the plate (with 'infinite' thermal conductivity, thus essentially imposing a constant temperature boundary condition), and the size were varied, and because the author is not in favour of using Nu-correlations in such a case (see L_b), the results are presented in the form shown in figure 2.5. However, for the purpose of comparison, Nu and Ra numbers were also calculated, resulting in $Nu = 0.64 \cdot Ra^{0.203}$ for upward and $Nu = 0.69 \cdot Ra^{0.186}$ for downward facing heated plates, using all data. The exponents in the correlations show very little discrepancy with the results of Goldstein and Lau ($Nu = 0.61 \cdot Ra^{0.203}$ and $Nu = 0.56 \cdot Ra^{0.190}$ respectively), thereby validating their numerical assumptions.

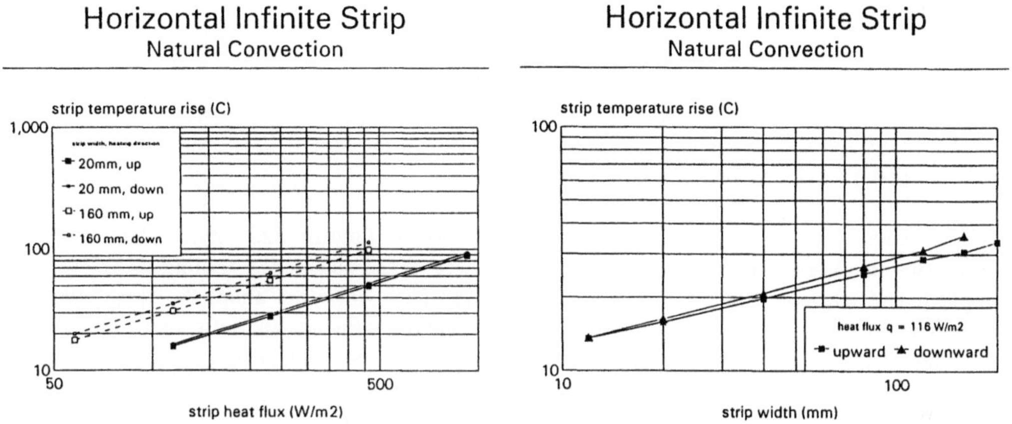

Figure 2.5 Horizontal flat plate, temperature rise vs. power dissipation

From these results, it is obvious that there is only a slight difference between the upward- and downward-facing plates. The reasons are not clear. Is the discrepancy caused by the unjustified extrapolation of large plates to smaller ones, or are the experiments not obeying the boundary conditions implicit in the numerical analysis? Is the 2D approximation not allowed, or are the boundary layer simplifications of the often analytical expressions not applicable? Further research should clarify these issues.

2.3.3. About possible flow reversals. Through a combined experimental/numerical analysis of some consumer product, a remarkable and worrying effect showed up. Figure 2.6 shows a schematic view of the most important objects. As an input for the code, the air resistance coefficient ξ of the outlet grid was imposed. By performing a sensitivity analysis on this parameter, it turned out that by changing ξ from 1 to 3, the flow field completely turned over. Figure 2.6 shows the top view of the set, with the z-velocity as the variable. Black is the negative direction, thus the airflow out of the system, and white represents the flow into the set.

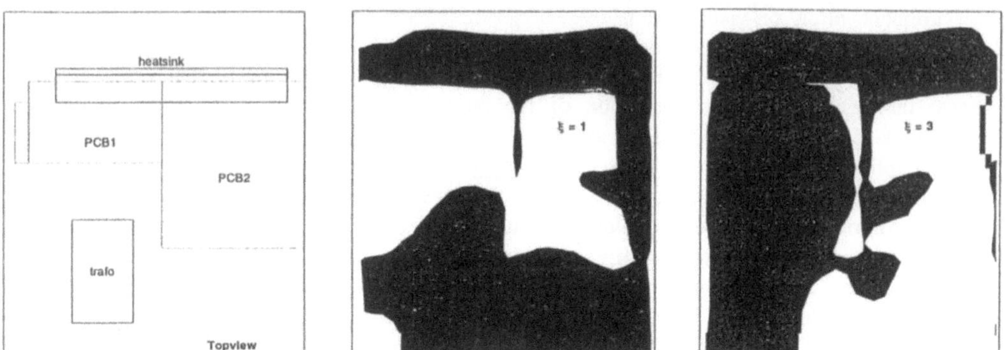

Figure 2.6 Observed flow reversal in electronic system (centre: ξ = 1, right: ξ = 3)

If this effect is really apparent in real applications, a serious problem is encountered, because a change in the average resistance coefficient, brought about by styling considerations, or worse, by the customer partly covering the outlet grid, can be the cause of an unexpected change in the overall thermal behaviour of the system. This example shows clearly one of the most important advantages of using CFD codes: the acquisition of physical insight by numerical flow visualisation.

Also, in this case, much more experimental and numerical research is needed to clarify the basic phenomena underlying the observed change in the flow field, the consequences on the critical temperatures, and how to design systems in such a way that the flow field is invariant with respect to influences that can be imposed by the customer.

2.3.4. About some observed mixed convection problems. Strange and unpredictable thermal behaviour also has been observed in those cases where mixed convection cooling techniques are applied, and when the forced and natural convection flow fields are about the same strength. Theoretically, the mixing of the flow fields and their mutual influence are not essentially different from the effects discussed in example 2.3.1. Practically, however, the designer has more degrees of freedom to overcome the problems in case they should become manifest. However, he or she should of course be aware of this. Consider the following example.

A few stacked horizontal boards, one of them containing a couple of highly dissipating ASICs, had to be cooled by small fans, the choice of which is constrained by noise level requirements. When measuring the temperature of the ASICs, it was found that one ASIC in particular showed an intolerably high temperature. After measuring the local velocities between the boards, it appeared that the velocity near the ASIC was nearly zero, despite the average velocity of 50 cm/s generated by the fans. Subsequent numerical analysis showed that the plume raised by the ASIC and spreading away exactly cancelled the air flow imposed by the fans, again showing the potential support of the CFD code in obtaining insight into the complex thermal behaviour of real systems.

2.4 ABOUT THE INTERPRETATION OF THE NUMERICAL RESULTS

Fluid flow phenomena occur in many engineering applications in the consumer electronics industry. A thorough understanding is often crucial for the proper design and operation of new products, and significant improvements could be the result of a better insight into the complex processes. Engineering education should therefore encompass the development of this insight. Most of the older and experienced designers with some background in heat transfer and fluid flow remember only the cumbersome differential equations, and the emphasis on laboratory work was usually on measurement, rarely on visualisation.

User-oriented CFD codes are nowadays commercially available, but the ever-increasing speed of the hardware will enable the engineer to simulate a design to such detail that a new problem arises: the handling and interpretation of the enormous amount of data. For example, 2D projections of a 3D velocity field are ambiguous and hard to interpret. To tackle this problem, a new discipline has emerged: scientific visualisation. Its main objective is to provide new insight through visual methods. Especially for CFD, the visualisation is far from trivial, due to the combination of vector and scalar fields, and the complexity of the resulting structures. Research should be directed to the mapping of physical quantities to optical and geometrical quantities, in such a way that an intuitively understandable image or animation of the flow field results. To demonstrate the potential of such an approach, the flow field inside a TV set, analyzed with the CFD code FloTHERM, was visualised using techniques developed by van Wijk ('92, '93), with very promising results.

3. About the Thermal Characterisation of Components

For decades, the manufacturers of electronic components have represented their thermal data by using some expression for a thermal resistance defined as:

$$R_{th} = (T - T_{ref})/\phi \qquad (3.1)$$

where T is some critical temperature, T_{ref} some reference temperature, and ϕ usually the steady state dissipation of the component.

However, we again find ourselves in roughly the same situation as in the former paragraph on heat transfer coefficient, because formula (3.1) is frequently misunderstood. At the same time, the need for more accurate junction temperatures becomes increasingly urgent, and the call for a precise definition of the various thermal resistances is heard by a growing number of researchers.

Furthermore, the author would be very surprised if future customers of the components will continue to accept the remarkable present day situation that it is A) very difficult to compare the thermal resistance data of different manufacturers, because the measurement methods are not sufficiently well standardised, and B), the data anyway cannot be translated from the measurement situation to the practical situation. Because of these reasons, a future redefinition of the standardisation methods is to be expected, which explains why the interest in the subject has risen sharply in the past few years.

The general opinion of the experts is that the thermal resistance values from junction to case should be treated with caution, but that the values for junction to ambient should not be used at all (e.g. Wills ('83), Andrews ('88), Dutta ('88)). This opinion originates from the following observations, when looking more closely at the definition of R_{th}:

A. What exactly is the junction temperature? Can this temperature be measured? If not, which temperature is measured? Is this temperature representative for reliability analyses?

B. What exactly is the reference temperature? Can this temperature be unambiguously defined? Is it possible to measure such a defined temperature? Does it make sense to define only one reference temperature for every component?

C. Is the equation really linear? Does R_{th} not depend on temperature level and power dissipation?

D. How accurate is the published data? What is for example the variance caused by the often unreproducible voids and contact resistances?

These questions are treated in more detail in L_a. Here, it suffices to remark that the posed questions still have not been answered satisfactorily.

The items mentioned under points A,C and D should raise concern among manufacturers and users, but these problems do not pose fundamental difficulties as far as the definition of the thermal resistance is concerned. Point B, however, is directly coupled to this definition and, as such, related to standardisation issues. From literature it becomes evident that a growing number of research laboratories are tackling this problem, and several alternatives have been formulated in the past decade. Paper L_a reviews most of the published proposals. It can be envisaged that those methods that fit into an integrated thermal design philosophy will ultimately survive.

In conclusion, in the opinion of the author, the focus with respect to standardisation should lie on those methods which describe the package itself, and nothing else, because the modelling of the real world around the package is the responsibility of the system level packaging engineer.

Two proposals that conform to this opinion are shortly surveyed here.

Bar-Cohen et al. ('89), and Krueger and Bar-Cohen ('91), restricted the analysis to the package itself. The authors realised that perhaps the best way to attack the surprising longevity of the error-prone use of the published $R_{thj\text{-}c}$ values was to retain the defined equation but to modify the meaning of one or more of the terms. The starting-point is the notion that it should be possible to characterise a package by a limited number of well-chosen thermal resistances, for example by placing nodes on all the surfaces that take part in the heat transfer and additionally on some lead groups. The resulting equation can be viewed as a generalised $R_{th\ j\text{-}c}$ based on an appropriately weighted average case temperature T_c.

The problem of course is to find suitable values for the weighting coefficients. This was performed in two steps. In the first step, a package was modelled using some FEM package, which provided, for example, the junction temperature as a function of the imposed temperature at every external surface of interest. In the second stage, the coefficients were obtained by minimising the mean square discrepancy between the temperature derived from the proposed definition and the FEM results. In this way, it appeared relatively easy to extract a set of thermal resistances that describe the package to a sufficient degree of accuracy. The paper applied this concept to a three-dimensional model of a PLCC, with excellent results.

The paper of Le Jannou ('91) is based on a similar philosophy. From 1985 onwards, the goal of this research group has been to describe the thermal behaviour of components in such a way that the outcome could be used in board or system level analysis software and databases, such as ATCOMP, CITHERE, PROMETHEE or COMPOSIT. Most papers related to the subject unfortunately have been presented in French, thereby hampering the distribution of their pioneering work throughout the scientific community. Le Jannou states that, for practical reasons, a thermal network cannot be drawn up for every component. This is why a classification is made according to type of the package, and within a type, according to die size as the most important parameter. For each of these categories, a simple topology with constant thermal resistances is provided and stored in a database. These rather simple networks replace the familiar thermal resistance in the data sheets. Finally, in order to be used in combination with board level analysis software, the network representing the package can be extended with a number of resistances corresponding to the external conditions imposed by convection and radiation. These values are not stored in the component database, but are the responsibility of the packaging engineer.

It is to be anticipated that the use of numerical analysis tools will become an increasingly important part of the thermal management of electronic systems, even more when taking into account concurrent engineering with its sharing of data bases and other software. In this philosophy, only certain methods for the thermal characterisation will survive. Both the methods discussed above, essentially similar, are limited to the analysis of the package itself, and not its environment, and are suitable for implementation in a database to be called by a board analysis software package. Also, from the viewpoint of the component manufacturer, the proposals appear to have a sound foundation, because the responsibility of the environmental influences is left to the packaging engineer.

To conclude this chapter, it is evident that the progress in acquiring meaningful data for reliability analysis, is seriously hampered by the non-standardised and often useless thermal information in the data sheets. Hence, the author pleas for concerted actions to arrive at an international consensus, shared by both vendors and customers, on the following topics:

- The data sheets should contain information about the methods used to arrive at the published values.
- A standardised definition for the package thermal resistances.
- A standardised experimental and numerical methodology to acquire these thermal resistances.
- A set of standard test chips, typical for a family of packages.

4. About the temperature-reliability relationships

From the viewpoint of a reliability engineer, the accurate determination of some temperature is not the ultimate goal of the analysis.

The current quality move causes reliability-requirements to drive new product designs. Consequently, accurate reliability predictions become increasingly important, since they can appreciably affect the decisions. And here we meet the last, and probably most difficult problem which still has to be solved: what exactly are the relationships between the calculated steady state temperature, and a reliability figure that determines important and often decisive issues such as Field Call Rates, Life Cycle Costs, Early Failures or Mean Time Between Failures? And is the steady state temperature indeed the dominant parameter? Micro-electronic device failures, after all, are not only a function of the absolute temperature. There is ample evidence that temperature gradients in space and time could be far more important than the absolute temperature, and miniaturisation will undoubtedly stress this point even further. To illustrate the problem, let us compare on the one hand often quoted design rules and specifications, and on the other hand the results that emerge from some recent research described in the open literature.

A very respectable and steadily growing amount of money is spent every time a new design reaches its final design stage in order to meet specifications such as: "The maximum ambient temperature should be 45°C". Timely introduction is frequently delayed, because the measured or calculated temperature is a couple of degrees C above the specifications. Even if we assume that the specifications are OK, one may wonder whether the measurement techniques and the calculation methods are reliable enough to firmly support decisions like saying yes or no to a new product. It is really very remarkable that very important and costly decisions are dependent on figures that seem to have the status of the Holy Bible and yet are based on quicksand.

However, even if we arrive at the situation where the temperatures that are measured are indeed the temperatures you wanted to be measured, and the calculation methods provide the right results, and the test protocols are improved, we are still confronted with the question about the appropriateness of the specifications. The reader who still believes in the blessings of MIL specs is advised to consult the papers by e.g. Feduccia ('88), Bar-Cohen '(88), Evans ('88), O'Connor ('88), Leonard ('88, '90), Wong ('89), Pecht ('91), Luthra ('90), Witzman ('90, '91), Hakim ('90,'91), Bowles ('92). A more detailed discussion of these papers can be found in paper L_a. Maybe the most important conclusions to quote here are that the mathematical models used for part failure rates are completely unsupportable by any physical or logical truth, and that very weak correlation has been found between Si maximum temperature and field failure probability.

How could this happen? One of the reasons could be that modern electronic equipment encompasses many more electronic components than in the past, and, purely for statistical reasons, according to the Arrhenius equation, temperatures should decrease in order to maintain at least the same reliability as before. This mere fact is the reason for the very conservative failure rate estimations that are based on the exponential temperature dependent relationships. The reliability prediction crisis has its roots in the professional sector, where meeting MIL specifications or not makes the difference between contract award or not. However, this argument does not play a role in the consumer electronics sector, as long at least as commercials and consumer tests are not based on a comparison between the MTBF rates of different vendors. Moreover, from a reliability engineering point of view, the two sectors act differently. In a large consumer electronics industry, a design team has to come up with a new design every year, and such a company houses hundreds of design teams. It is obvious that a few reliability experts cannot support all those teams all over the world. Transfer of up-to-date knowledge of the emerging reliability problems and the resulting change in culture then pose a much graver problem.

To conclude, it is a matter of great concern that, despite the overwhelming amount of evidence that failure rate based reliability prediction protocols do not fit into the current quality improvement philosophy, and despite the fact that in the USA as well as in Europe initiatives have been launched to attack this very unsatisfying situation, these facts are not realised at the appropriate management levels in the consumer electronics sector.

5. About the need for co-operation

As pointed out in the preceding paragraphs, numerous research fields lie fallow, waiting to be explored. In a number of cases, it is just a matter of preference or of the available expertise in a laboratory that governs the choice that could be made. In other cases, however, it would be very unwise for a single research group to pursue some pioneering work beyond the stage where standardisation issues are at stake. And, looking in particular at the unsolved problems related to the reliability in the field, or at the standardisation with respect to thermal characterisation, this kind of research can only be mobilised on an international scale, in view of human, technological and financial resources.

In the USA, this fact has been realised for a long time, and for more than a decade various initiatives have been launched to stimulate co-operation between industry, university and government, resulting in numerous industry/university consortia, various standardisation and research-oriented committees set up by the professional engineering societies, and very elaborate student exchange programs between the USA and Japan (see e.g. Pecht et al. ('89), Keys ('89), Akella et al. ('89), Hijikata and Peterson ('91)). In Europe, a couple of initiatives are just emerging. Some rough conclusions, based on discussions with some knowledgeable persons in the field, not on a profound study (so please do not quote them as a fact), are shown below.
The USA has, as compared to Europe:

- more thermal management experts per company of the same size (3:1)
- more designers trained in thermal techniques (10:1)
- more research contracts with universities (10:1)
- more conferences on thermal management related topics (5:1)
- more publications (50:1)

Of course, Europe is suffering from historical, lingual, political, religious and xenophobian barriers, but if the striving for co-operation cannot be founded on human or ethical arguments, maybe economical arguments are capable of encouraging co-operative networks.

A study conducted by the DG XIII committee of the European Community (EC, '91), related to the status quo of the European electronics and information technology industry, contains numerous graphs from independent sources like BIS-Mackintosh, Datamation, Dataquest, EIC. The study provides ample evidence of the alarming situation these industries are facing, with respect to market shares, trade deficits, net profits and global ranking.

From these and other studies, a couple of conclusions and recommendations could be extracted. For example, in the labour market, there are not enough engineers and researchers with recent training in the production, adaptation or use of the rapid technological advances. For the same basic population, Japan trains twice as many engineers a year, as compared with Germany and France together. This means that demand on the European market is less advanced and less receptive to innovations than in the USA or Japan. Therefore, the Community is willing to launch training measures to increase advanced skill levels of research workers and engineers.

Although there is plenty of co-operation on pre-competitive research in Europe, co-operation arrangements on the development of new products are all too few and far between, and the possibilities of co-operation with the Community and industrial partners have not been sufficiently exploited. What is needed here is the formation of networks of excellence which aims at the creation of durable links between established centres of complementary expertise in either academia or industry. In a later stage, perhaps an International Research Centre for Electronic Packaging could be founded at a strategic location. Furthermore, there is a lack of clearly defined medium- and long-term objectives set by the industry and on specific commitments from their side. If we restrict ourselves to topics related to thermal management, these observations are particularly true for reliability research, such as the development of advanced packaging technologies for single ICs and multi-chip modules, with an emphasis on the reliability/cost ratio.

Other fields that are open to international co-operation are standardisation on thermal characterisation of components, and the development of user-oriented software to embed within complex simulation and design packages, in order to support co-design and concurrent engineering.

To summarise, co-operation could be one of the essential keys to survival.

6. Conclusions

It is evident that the definition and use of important parameters governing accurate reliability prediction, such as the heat transfer coefficient, thermal resistance of devices, and temperature reliability relationships should be reviewed by the scientific community, in close co-operation with industry.

With the increasing and inevitable use of software tools by non-specialists, as driven by a concurrent engineering philosophy combined with a strategy to reduce the time-to-market, success or failure will depend on acquiring a science-based integrity of the data that result from applying these tools.

Even more serious perhaps, the lack of consistency in the definition of the parameters discussed will eventually lead to chaos if the expert's judgment is gradually replaced by computer predictions. Without co-operation on a large scale with respect to standardisation issues, the odds are that a sensible comparison between the performance of alternative designs within one industry, and of different products between industries, will be completely impossible.

The various initiatives, initiated by the professional societies in the USA and by the European Community, should deserve much more attention and support from the electronics industries.

In brief, it is hoped for that this paper offers enough material to stimulate the universities to partly reformulate their research objectives, and the electronics industries to initiate inter-company working parties to prepare proposals on standardisation, networks of excellence and training measures.

7. Epilogue

In continuation of this paper on New Challenges for Research, the reader is invited to read the book 'Lila, an Inquiry into Morals' by Pirsig ('91), a philosophical essay about the continuing, but often discontinuous evolution from Static to Dynamic Quality. Static quality can be characterised by traditions, memory, obedience to specifications and a firm belief in old design rules, while dynamic quality has something to do with inherent quality, creation, quality attitudes, quality moves. Dynamic quality is the source, not only the added value. Static quality rules the past, dynamic quality the future, and the present relates the two. However, both are essential for survival. Without dynamic quality, no future, but without static quality, no stability.

We are on the verge of a revolution in thermal management techniques. Our current problems are with Static Quality, and the ultimate challenge of research in this field is to develop tools and methods for the smooth introduction of Dynamic Quality concepts into the designer's environment.

Acknowledgments

The author acknowledges the assistance of all members of the group Heat and Mass Transfer of the CFT over the last couple of years, and especially the fruitful discussions with Dr. P. van Engen while preparing this lecture.

References

Afzal N., Hussain T., 1984, "Mixed Convection over a Horizontal Plate", Trans. ASME JHT, vol.106, pp.240-241

Aihara T., 1991, "Air Cooling Techniques by Natural Convection", chapter 1 in 'Cooling Techniques for Computers', Aung W., ed., Hemisphere Publ., USA

Akella R., Fox M., Smith S., 1989, "Center for Integrated Manufacturing Decision Systems-An Overview", IEEE/CHMT '89 IEMT Symp., pp. 169-173

Andrews J., 1988, "Package Thermal Resistance Model: Dependency on Equipment Design", IEEE Trans. CHMT, vol.11, pp.528-537

Arvizu D., Moffat R., 1982, "The Use of Superposition in Calculating Cooling Requirements for Circuit Board Mounted Electronic Components", 32th Proc. Electr. Comp. Conf., pp.133-144

Aung W., ed., 1991, "Cooling Techniques for Computers", Hemisphere Publ., USA

Bar-Cohen A., 1988, "Reliability Physics vs Reliability Prediction", IEEE Trans. Rel. vol.37, p.452

Bar-Cohen A., Elperin T., Eliasi R., 1989, "θ_{jc} Characterisation of Chip Packages-Justification, Limitations, and Future", IEEE Trans. CHMT, vol.12, pp. 724-731

Bar-Cohen A., 1992, "State-of-the-Art and Trends in the Thermal Packaging of Electronic Equipment", Trans. ASME J. Electr. Pack., vol. 114, pp. 257-270

Barozzi G., Nobile E., 1990, "Conjugate Heat Transfer in Forced Convection Cooling of Chip Arrays", Heat Transfer in Electronic and Microelectronic Equipment, Bergles A., ed., Hemisphere

Bertin Y., Saulnier J., 1990, "A Detailed Analysis of the Conductive/Convective Heat Transfer in a Single Electronic Board", Heat Transfer in Electronic and Microelectronic Equipment, Bergles A., ed., Hemisphere

Bowles J., 1992, "A Survey of Reliability-Prediction Procedures For Microelectronic Devices", IEEE Trans. Reliab., vol.41, pp. 2-12

Chu T., Goldstein R., 1973, "Turbulent Convection in a Horizontal Layer of Water", J. Fluid Mech. Vol.60, pp. 141-159

Clausing A., Berton J., 1989, "An Experimental Investigation of Natural Convection From an Isothermal Horizontal Plate", Trans. ASME, JHT, Vol.111, pp. 904-908

Cohen C., Reshotko E., 1956, "The Compressible Laminar Boundary Layer with Heat Transfer and Arbitrary Pressure Gradient", NACA Rep. 1294

Culham J., 1988, "Conjugate Heat Transfer From Surfaces with Discrete Thermal Sources", Thesis University of Waterloo, Ontario, Canada

Culham J., Lemczyk T., Lee S., Yovanovich M., 1991, "META-A Conjugate Heat Transfer Model for Air Cooling of Circuit Boards with Arbitrarily Located Heat Sources", ASME HTD-vol.171, 'Heat Transfer in Electronic Equipment', pp.117-126

Dutta V., 1988, "Junction-to-Case Thermal Resistance- Still a Myth?", 4th IEEE SEMI-THERM Symp., pp. 8-11

EC '91, anon., 1991, "The European Electronics and Information Technology Industry: State of Play, Issues at Stake and Proposals for Action", report from the DG XIII Committee

Evans R., 1988, "Superstition, Witchcraft, Prediction", IEEE Trans. Rel., August editorial

Feduccia A., 1988, "Reliability Prediction-Use it Wisely", IEEE Trans. Rel., vol.37, p.457

Gebhart B., 1979, "Buoyancy Induced Fluid Motions Characteristic of Applications in Technology", J. Fluids Eng., Vol.101, 5-28

Gebhart B., Jaluria Y., Mahajan R., Sammakia B., 1988, "Buoyancy Induced Flows and Transport", Hemisphere Publ. Corp., USA

Goldstein R., Lau K., 1983, "Laminar Natural Convection from a Horizontal Plate and the Influence of Plate-edge Extensions", J. Fluid Mech. Vol.129, pp.55-75

Graham K., Witzman S., 1987, "Analytical Correlations of Thermal Design of Electronic Packages", Cooling Technology for Electronic Equipment, Aung W., ed., pp.249-264

Hakim E., 1990, "Guest Editorial: Reliability Prediction; is Arrhenius Erroneous?", Sol.St.Techn., vol.33,nr.8

Hakim E., 1991, "Microelectronic Reliability/Temperature Independence", Qual. and Rel. Eng. Int., vol. 7, pp. 215-220

Hijikata K., Peterson P., 1991, "Panel:U.S./Japan Joint Research in the Thermal Sciences: Cooperative and Postdoctoral Programs", ASME/JSME Therm. Eng. Proc. Vol.1, pp. 513-524

Husar R., Sparrow E., 1968, "Patterns of Free Convection Flow Adjacent to Horizontal Heated Surfaces", Int. JHMT, Vol.11, pp. 1206-1208

Ishiguro R., Abe T., Nagase H., Nakanishi S., 1978, "Heat Transfer and Flow Instability of Natural Convection over Upward-facing Horizontal Plates", Proc. 6th Int. Heat Transfer Conf., Toronto, pp. 229-234

Jakob M., 1953, "Heat Transfer", vol.1, John Wiley Publ., USA

Keys L., 1989, "University/Industry (and Government) Interdisciplinary Centers, Programs, Consortia, Inter-relations, and Initiatives",Proc. IEEE/CHMT '89 IEMT Symp., pp. 165-168

Kishinami K., Saito H., Tokura I., 1990, "A Fundamental Investigation of Laminar Natural Convection Heat Transfer from a Vertical Plate with Discontinuous Surface Heating", Heat Transfer in Electronic and Microelectronic Equipment, Bergles A., ed., Hemisphere

Krane R., Phillips T., 1986, "Natural Convection Cooling of a Horizontally-Oriented Component Board Mounted in a Low-Aspect-Ratio Enclosure", ASME HTD-Vol. 57, Heat Transfer in Electronic Equipment, pp. 113-122

Kraus A., Bar-Cohen A., 1983, "Thermal Analysis and Control of Electronic Equipment", Hemisphere USA

Krueger W., Bar-Cohen A., 1992, "Thermal Characterisation of a PLCC - Expanded R_{jc} Methodology", Proc. 8th IEEE SEMI-THERM Symposium, Austin TX, pp. 263-270

Lasance, C., 1991, "Numerical Simulation of Natural Convection in a Television Set", ASME publ. HTD-Vol.171, pp. 101-108

L_a, Lasance C., 1993, "Accurate Temperature Prediction in Consumer Electronics: a Must but Still a Myth", NATO Advanced Study Institute on Cooling of Electronic Systems, June 21-July 2, Izmir, Turkey, to be published.

L_b, Lasance C., 1993, "Pragmatic Methods for Determining the Parameters Required for the Thermal Analysis of Electronic Systems", NATO Advanced Study Institute on Cooling of Electronic Systems, June 21-July 2, Izmir, Turkye, to be published.

Leonard C.,1988, "On US Mil-Hdbk-217 and Reliability Prediction",IEEE Trans.Rel., vol.37, pp.450-1, 461

Leonard C., 1990, "Mechanical Engineering Issues and Electronic Equipment Reliability: Incurred Costs Without Compensating Benefits", IEEE Trans. CHMT, vol.13, pp. 895-902

Le Jannou J.P., Huon Y., 1991, "Representation of Thermal Behaviour of Electronic Components for the Creation of a Databank", IEEE Trans. CHMT, vol.14, pp. 366-373

Lewandowski W., Kubski P., 1983, "Methodical Investigation of Free Convection from Vertical and Horizontal Plates", Warme-und Stoffubertragung vol.17, pp. 147-154

Luthra P., 1990, "Mil-Hdbk-217: What is Wrong with It?", IEEE Trans. Rel., vol.39, pp. 518, vol. A225,185

Maciejewski P., Moffat R., 1990, "A Correlation for Boundary Layer Heat Transfer Accounting for Free Stream Turbulence", Proc. 9th Int. Heat Transfer Conf., Jerusalem, pp.303-308

Malkus W., 1954, "Discrete Transitions in Turbulent Convection", Proc.Roy.Soc.

Mikheyev M., 1968, "Fundamentals of Heat Transfer", MIR Publishers, Moscow, pp.73-81

Miyamoto M., Katoh Y., Kurima J., Kurihara S., Yamashita K., 1985, "Free Convection Heat Transfer from Vertical and Horizontal Plates", Int. J. HMT, vol.28, pp.1733-1745

Moffat R., Ortega A., 1986, "Buoyancy Induced Forced Convection", ASME HTD-Vol. 57, Heat Transfer in

Electronic Equipment, pp. 135-144
Moffat R., Ortega A., 1988, "Direct Air Cooling of Electronic Components", chapter 3 in 'Advances in Thermal Modeling of Electronic Components and Systems, Vol.1, Bar-Cohen A. and Kraus A., eds.
Moffat, R. Anderson A., 1990, "Applying Heat Transfer Coefficient Data to Electronics Cooling", Trans. ASME, vol.112, pp. 882-890
O'Connor P., 1988, "Undue Faith in US Mil-Hdbk-217 for Reliability Prediction", IEEE Trans. Rel. vol.37, p. 468
Nakayama W., 1988, "Thermal Management of Electronic Equipment: A Review of Technology and Research Topics", chapter 1 in 'Advances in Thermal Modeling of Electronic Components and Systems, Vol.1, Bar-Cohen A. and Kraus A., eds.
Ohsaki T., 1991, "Electronic Packaging in the 1990's-A Perspective from Asia", IEEE Trans. CHMT, vol. 14, 254-261
Ortega A., Kabir H., 1991, "Substrate Conduction Mechanisms in Convectively Cooled Simulated Electronic Packages" Proc. 7th IEEE-SEMITHERM Conf., pp.1-6
Ortega A., Lall B., 1992, "A Clarification of the Adiabatic Heat Transfer Coefficient as Applied to Convective Cooling of Electronics", Proc. 9th IEEE-SEMITHERM Conf., Austin, pp.1-3
Ortega A., Moffat R., 1986, "Buoyancy Induced Convection in a Non-Uniformly Heated Array of Cubical Elements on a Vertical Channel Wall", ASME HTD-Vol. 57, Heat Transfer in Electr. Equipment, pp. 123-134
Pecht M., Barker D., Weiss D., 1989, "CALCE Research Opportunities at the University of Maryland", Proc. IEEE/CHMT '89 IEMT Symp., pp. 182-185
Pecht M., ed., 1991, "Handbook of Electronic Package Design", Marcel Dekker Publ., USA
Pecht M., 1991, "Advances in Reliable Design of Electronics", 7th IEEE-SEMITHERM symp., luncheon talk
Peterson G., Ortega A., 1990, "Thermal Control of Electronic Equipment and Devices", in 'Advances in Heat Transfer, vol.20, pp.181-314
Pirsig R., 1991, "Lila, an Inquiry into Morals",
Pnueli D., 1982, "Free Convection and its Applications", 7th Int. Heat Transf. Conf., Munchen, pp.253-259
Restrepo F., Glicksman L., 1974, "The Effect of Edge Conditions on Natural Convection from a Horizontal Plate", Int. J. HMT., vol.17, pp.135-142
Rotem Z., Claassen L., 1969, "Natural Convection above Unconfined Horizontal Surfaces", J. Fluid Mech. vol.38, pp. 173-192
TNO Apeldoorn, The Netherlands, 1992, Personal Communication
Townsend A., 1959, "Temperature Fluctuations over a Heated Surface", J. Fluid Mech., Vol.5, pp. 209-241
Tummula R., 1991, "Electronic Packaging in the 1990's-A Perspective from America", IEEE Trans. CHMT, vol. 14, 262-271
Wessely H., Fritz O., Horn M., Klimke P., Koschnik W., Schmidt K., 1991, "Electronic Packaging in the 1990's: The Perspective from Europe", IEEE Trans. CHMT, Vol.14, pp. 272-284
Wijk J. van, 1992, "Rendering Surface Particles", Proc. IEEE Conf. Visualisation '92, oct. 19-23, Boston USA, pp. 54-61
Wijk J. van, 1993, "Flow Visualisation with Surface Particles", IEEE Comp. Graphics and Applic., to be published
Wills M., 1983, "Thermal Analysis of Air-Cooled PCBs, Part 1", Electr. Prod., May, 11
Witzman S., Smith K., Metelski G., 1990, "Silicon Interconnect-A Critical Factor in Device Thermal Management", IEEE Trans. CHMT, vol.13, pp. 946-952
Witzman S., Giroux Y., 1991, "Mechanical Integrity of the IC Device Package- A Key Factor in Achieving Failure Free Product Performance", Proc. 41th IEEE ECTC conf., pp. 450-456
Witzman S., Nicoletta T., Mack B., 1992, "Enhanced Heat Transfer for Electronic Devices Operating inside Enclosures", 8th IEEE SEMI-THERM, Austin, pp. 95
Wong K., 1989, "The Bathtub and Flat Earth Society", IEEE Trans. Rel., vol. 38, pp. 403-404
Yousef W., Tarasuk J., McKeen W., 1982, "Free Convection Heat Transfer From Upward-Facing Isothermal Horizontal Surfaces", Trans. ASME JHT, vol.104, pp.493-500
Zinnes A., 1970, "The Coupling of Conduction with Laminar Natural Convection From a Vertical Plate", ASME JHT, vol.105, pp.528-535

ANALYSIS AND PREVENTION OF THERMALLY INDUCED FAILURES IN ELECTRONIC EQUIPMENT

AVRAM BAR-COHEN[1] & SORIN WITZMAN[2]

[1]Department of Mechanical Engineering, University of Minnesota, Minneapolis, Minnesota 55455, USA.

[2] Bell Northern Research, Ottawa, Ontario, K1Y-4H7, Canada

ABSTRACT.

This paper begins with a discussion of the physical phenomena associated with thermally-induced/accelerated failures, including diffusion-driven and thermo-mechanical failure mechanisms. Attention is then turned to the results of ongoing Northern Telecom "field-return" surveys and to the identification of the factors controlling device field reliability for newly designed equipment, mature technologies in mild environments, and equipment operated under harsh environmental conditions. It is concluded that new reliability models and figures-of-merit, based on the phenomenological understanding of the failure mechanisms, are needed in order to predict the thermal dependence of the probability of failure.

1. PHYSICS OF THERMALLY-INDUCED/ACCELERATED FAILURES

1.1 General Considerations

The useful life of electronic systems is constrained by a wide variety of electrical and mechanical failure mechanisms, sometimes occurring singley, but often appearing in bewildering combinations. The current flows encountered in electronic devices can induce mass transport by vacancy diffusion at the grain boundaries, leading to open circuit failures by void coalescence and short circuit failures by atomic accretion.

Loss of dielectric strength in the oxide layers is frequently cited as a failure mechanism in metal-gate MOS devices and, in non defective parts, is primarily affected by voltage, rather than temperature. Unfortunately, very seldom will a highly integrated device have "perfect" capacitors. Very small defects due to manufacturing variations can increase, locally, the electrical field by orders of magnitude. Some of the failure mechanisms for these defective capacitors can be temperature accelerated. (This phenomena is presented later on in this paper). Furthermore, highly energetic or "hot" electrons, encountered in high density MOS devices, may penetrate the oxide layers and lead to excessive current flow in the gates.

One of the most dramatic electrical failure mechanisms in microelectronics components is the "thermal run-away" in bipolar devices

C. J. Hoogendoorn et al. (eds.), Thermal Management of Electronic Systems, 25–42.

and "latch-up" in MOS technologies. These phenomena are associated with extreme heating of a small region within the silicon chip. Since the electrical resistivity of doped silicon typically peaks around 150°C, high current flows in a junction region can produce a thermal run-away condition, melting the silicon and at times producing electrical arcs. As the average chip temperature increases, there is a somewhat greater likelihood that local "hot-spots" will exceed this threshold temperature.

When a chip is exposed to human contact, dry air or a nearby charged surface, electrostatic discharges can damage the gate oxides and, at times, produce craters, micro-cracks, and open circuits. MOSFETs are thought to be especially vulnerable to this failure mechanism, at voltages as low as 100V. On-chip protection and use of other silicon devices can raise the damage threshold above 1kV.

Ionic and galvanic corrosion, as well as electrolyte and intermetallics formation, is often responsible for the chemical deterioration in the electrical characteristics of the signal paths.

Mechanical failure may involve fracture of the wire bond or sometimes of the chip itself. But, more often it occurs at the package level and includes cracking of the substrate and/or delamination of the encapsulant and loss of solder joint integrity. In most cases mechanical failure will precede and trigger an electrical failure.

The temperature dependence of some electrical properties, and electron as well as dopant mobility in the silicon, results in a link between temperature and the electrical failure mechanisms. Similarly, many of the chemical processes occurring in corrosion and intermetallic formation, are accelerated by elevated temperature. Differential thermal expansion between the various materials which are bonded together to create the chip package, is largely responsible for the observed mechanical failures. Consequently, as discussed in detail by Pecht et al. and shown in Table 2.5, pp. 126-133 [5] many failure mechanisms can be related to the local temperature and/or temperature gradients within the chip package. Unfortunately, however, no single expression has yet been found that can adequately describe the overall dependence of the component failure rate on temperature.

1.2 Diffusion Driven Failure Mechanisms

Most of the failure mechanisms which produce electrical and chemical failure at the device (chip) level are dominated by diffusion terms. These phenomena show a strong inverse relationship with the logarithm of the average temperature. It has become common to represent this dependence in the form of Arrhenius relations, with a unique, empirically determined activation energy for each failure mechanism and/or component type [8]. The relevant activation energy must be found empirically since it is rarely a "perfect" part that fails and it is extremely difficult to determine the relationship between the theoretical intrinsic activation energy and the value appropiate to the "weaker", defect-prone, segment of the component population.

As a general rule, the field effective activation energy is lower than the intrinsic activation energy for the phenomenon that generated the failure. Considering the actual manufacturing quality, the field effective activation energy varies between 0.5 - 0.8 eV. Using such values it is found that, in the normal operating range of microelectronic components, a 10-20°C increase in chip temperature can double the component failure rate and even a 1°C decrease may lower the failure rate by 2 to 4% [6].

On the other hand, if the quality of the parts is improved, the field effective activation energy rises, getting closer to the theoretical activation energy. For the relevant physical phenomenon at an effective field activation energy above 1 eV, the field failure rate will be practically independent of temperature, in the range of component temperature used today.

1.3 Thermo-mechanical Stress Related Failures

Thermally-induced mechanical failures in electronic packages occur at several distinct sites and are generally associated with thermal stresses produced by differential thermal expansion among materials constituting the chip package. Although steady-state operating temperatures that are substantially different than the fabrication or long-term storage temperature may be responsible for some failures, more often it is the temperature fluctuations that are most detrimental to the integrity of the chip package. Such variations may be produced by temperature cycling, associated with environmental changes or component testing. They may also result from variations in the power dissipation produced by functional requirements or by deliberate power cycling during component qualification.

If no compliant layers are present in the structure and the device has initial mechanical defects (micro cracks), a brittle-type failure (device cracking, passivation layer cracking and sliding, etc.) might occur. More often, the relatively fast and frequent temperature variations associated with power cycling, lead to fatigue-type structural failures.

Cyclic temperature variations can lead to continuous tensing and flexing of the wire and result in fatigue failure of the wire and/or bond pads. In the normal range of operating and qualification temperature, i.e. -55 to 125°C, the failure rates by these mechanisms are primarily determined by the amplitude of the temperature cycle. In most plastic packages, the fatigue of the wire bond is strongly accelerated when delamination of the chip inside the package occurs during temperature/power cycling.

Less frequent thermal transients, associated with environmental effects, often induce creep and structural deformation. The best known creep dependent failure is the failure of the solder joints in surface-mounted packages.

Although a mechanically failed package might pass undetected through the qualification tests, packaging failure constitutes a strong field failure hazard due to an increase in chip temperature associated with delamination along a critical thermal path. Most of the failures reported in harsh environments (temperature/power cycling, uncontrolled humidity, cool down periods) can be shown to result from an initial delamination, or other mechanical failure, of the chip package.

2. FIELD RETURN SURVEYS

Since laboratory tests cannot entirely reproduce field conditions, it is an absolute necessity that the starting point of any research in reliability be a comprehensive study of field returns and the associated field conditions. Unfortunately, only scant data can be found in the literature [8-11]. In most cases, field return data is reported for statistical purposes only [8,9], as the ratio of the returned parts to the entire field population, without any attempt to identify the failure mechanism and the root cause of the failure.

At Northern Telecom there is an on-going large-scale study on samples of field returned circuit packages. The study is aimed at determining the relative importance of the different degradation mechanisms and eliminating the causes of such failures. Failure analyses are performed at the mode, mechanism and cause level. The failures are then re-grouped by the root cause and the relative importance of the different causes calculated.

Mode level analysis is performed by capturing the data from the package repair process. In most cases, examination of the returned components will be sufficient to determine the failure mechanism(s). An example would be delamination of a small power device from a heat sink that allowed local heating and failure of the device.

Next, attention is turned to determining the cause that initiated the failure. In the previous example the cause might be an assembly defect that distorted the device-heat sink interface or it could be that load conditions over-stressed the circuit. This over-stress load could be due to a design error or it could be the expected response to an external fault condition that is acceptably rare.

From the example it can be seen that the same component damage can be caused by a design defect, a manufacturing defect or the expected response to an over-stress fault. Clearly any decision on corrective action should be based on knowing what fraction of the field failures are due to each root cause.

3. FIELD RETURN FAILURE ANALYSIS

Even though the failure analysis results tend to identify a relatively large number of failure modes that differ significantly from one application to another, studies of the types and history of the equipment as well as the environmental and electrical stress in which the device operates, allow us to group the equipment in distinct categories within which general trends can be identified.

As a general rule, design quality, part manufacturing quality and the environmental conditions (temperature and humidity) are the major factors that combine to determine the life of the part/system under field conditions. The distribution of the failures in the system can identify the relative importance of the specific factors in initiating the phenomena.

To explain the results, one should consider that very seldom are similar parts identical. Small variations in dimensions and material properties are always present. By design, these variations are limited to a certain domain, defined by the design tolerances. The part of the device population which meets these tolerances is the main or "good" device population. In spite of very tight and rigorous quality control procedures, there is a certain probability that some devices will show variations which exceed the design tolerances. This tail of the distribution represents only "marginally good" devices, i.e. the weak segment of the device population. The size of this population is determined by the quality of the manufacturing technologies and the quality control procedures. For a good manufacturer the order of magnitude of the size for this population is typically a hundred parts per million good devices. One of the most important stages of the field failure analysis is to find out if the failed parts belong to the main device population or if the failure mechanism only affects the weak segment or the defective population.

For the time being, by analyzing the type and the distribution of failures, we have concluded that electronic equipment can be divided into three main categories, based on reliability concerns only:

- Newly designed equipment.
- Mature design and manufacturing technologies; equipment operating in a mild environment.
- Mature design and manufacturing technologies; equipment operating in harsh environments.

3.1 Newly Designed Equipment

By our definition, this type of equipment represents new products. In some cases we might include established technology that is used in a different environment than that for which it was originally tested. In keeping with the experience of other companies [8-11,15], most of these failures were detected during the first year of the field trial period. It is important to mention here that the failure rate for the studied equipment was constant over time and usually lower or close to the overall equipment failure rates predicted by the MIL-HDBK-217.

For this category of products, the failure mechanisms vary wildly from one application to another. The one noticeable common failure is the mechanical damage of the interconnect, due to handling, testing or insertion which can account for as much as 40% of the total number of returns.

With respect to silicon devices, the failure mechanisms are numerous and very different from one product to another but electrical overstress tends to dominate. While the MIL-HDBK-217 predicts a relatively uniform distribution of failures in the system, in this study more than 70% of the hardware failures were located in the same place on/in a given product. Moreover, when a single product is analyzed, the failure mode for the damaged parts is similar 90% of the time. No "universal" correlation between the part temperature and the part failure rate was found.

The most plausible explanation for this distribution is that a "weak link" is present in the design. Electrical and mechanical stress levels are established and fixed at the design stage. If these levels of stress are high enough to produce aging or fatigue of the device structure during the useful life of the device, failure could occur.

Manufacturing errors and environmental conditions can modify the "built-in" failure mechanism, but, no significant reliability gains can be made without redesigning the product or parts of it. Interestingly, it seems that MIL-HDBK-217 tends to predict quite accurately the probability of making a design error in a complex system.

3.2 Mature Design and Technologies; Mild Environment

By incorporating the results of field return data, Northern Telecom is able to eliminate nearly all the design flaws in its circuit packages, leading to an actual return rate at least five times lower than the failure rate computed according to MIL-HDBK-217 methodologies and tables. Relatively low power swings (less than 0.5 W) and temperature swings (smaller than 10°C) may also play a significant role.

The participation of the silicon devices in the total number of returns is slightly higher (under 40%) than in the previous category, but much lower than the predictions made using standard prediction models (70% or more). The failures are also more evenly distributed among different parts of the assembly. The most important differences in the failure mechanism relative to the previous group, are that most failures can be associated with defective parts or manufacturing/assembly defects. The failure rate decreases steadily over time, reaching an asymptotically constant level after 5 or 6 years of operation in the field (see **Figure 1**).

These differences lead us to conclude that field failure is strongly related to the manufacturing variability (defectivity). Moreover, the variation of the failure rate in time can be easily associated with a Poisson distribution of the size of manufacturing errors, i.e. the parts with the largest variation from the ideal will fail first. Due to the use of micron-sized silicon features, even the very best manufacturing processes can result in significant variations among the devices. These variations are related to dimensions, chemical composition or the crystalline structure. Micro-defects related to handling and a certain degree of contamination are also present. If the part shows variations which are large enough to weaken the structure significantly, this part will fail in the field during the infant mortality period.

Consequently, for this category of equipment significant gains in reliability can be achieved by reducing the degree of defectivity at the parts and assembly level.

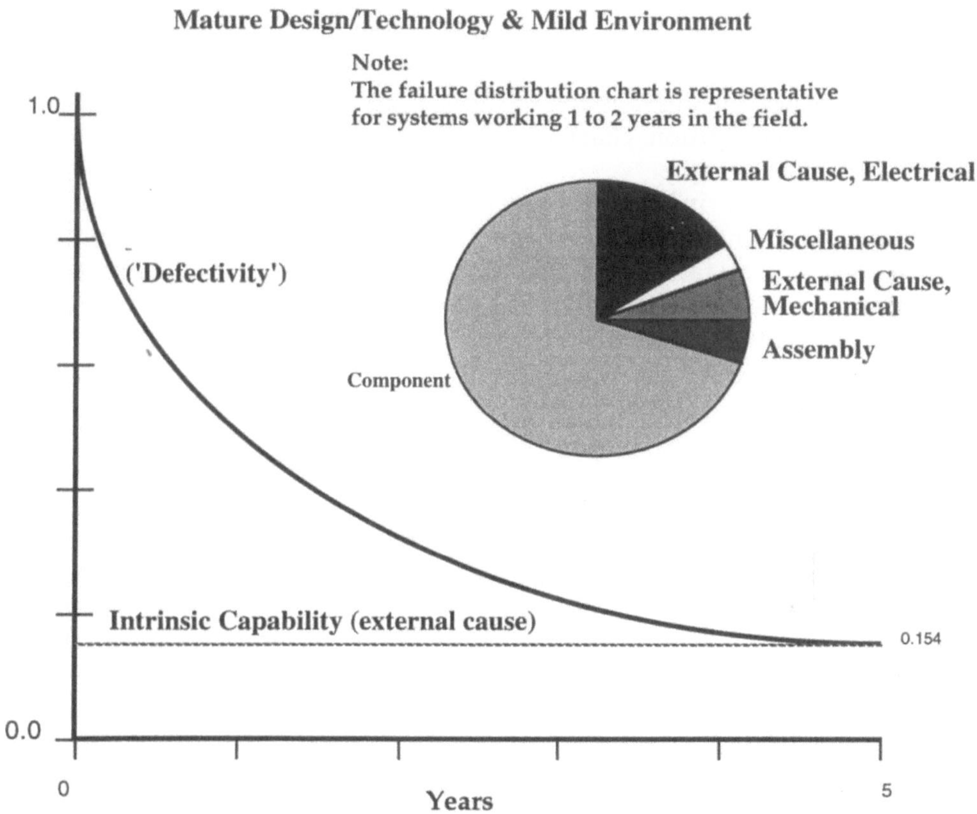

Figure 1

3.3 Mature Design and Technologies: Harsh Environmental Conditions

Even though harsh environmental conditions are usually associated with the military or the automobile industry, more and more communication equipment is installed in locations with very low or totally nonexistent environmental controls. The increase in power dissipation per chip associated with the use of the dormant state produces significant temperature swings at the device level (For example, for devices with power swings between the dormant stage and full power close to 1 Watt, the temperature swing at the silicon level is approximately 35 °C).

In the telecommunications industry, the environment of a silicon device is considered harsh when there is no control over humidity, temperature variations of large amplitude combined with power cycling are present, and the devices have extended cool-off periods in which the temperature drops below the dew point of the air inside the enclosure. Elevated static temperature levels alone is not enough to declare the environment as "harsh".

The most notable feature of the data for devices operated under harsh environmental conditions is the abundance of failures at the silicon interconnect level (wire bond). Even in cases where failure occurs at the silicon level (corrosion, die cracks, metal layer dislocation, etc.), it seems to be a consequence of the degradation of the packaging [15-18]. A "typical" failure mechanisms distribution chart for a silicon plastic encapsulated device is shown in **Figure 2** [15].

For equipment in this category, important gains in silicon device reliability can be made if the packaging of the device is improved to the point where mechanical fatigue of the silicon interconnect is eliminated. It is important to mention here that the level of humidity (both relative and absolute) constitutes a crucial factor in packaging degradation, along with temperature/power cycling.

3.4 Combined Failure Modes

In many circumstances, field failures are a result of a combination of design, manufacturing and environmental factors. Even though some factors might be considered as dominant for certain categories of equipment, the failure mechanism will involve factors from all three groups mentioned above. Whatever failure mechanism is being considered, the correlations for predicting the expected life of the device must contain the design stress levels (for normal and abnormal conditions), the manufacturing variations (in the form of a distribution of the defect size for a certain device population) and the environmental conditions that accelerate the degradation mechanism.

4. MODEL FOR DIELECTRIC BREAKDOWN

4.1 Introduction

Analysis of the NT field returns and qualification test results reveals that the most probable failure mechanism at the silicon level, for devices operating in mild environmental conditions, is time-dependent dielectric breakdown. This failure mode can be related to the very small thickness of the dielectric layer (100 to 300 Å) in VLSI circuits, making the oxide layer especially vulnerable to defect-related breakdown. Defects, which result in localized weak spots in the oxide, could be due to particulate contamination, crystalline defects in the substrate, localized thin regions or surface roughness at the weak spots. Such defects could lead to a higher field, a higher trap generation rate, or higher current density at these local areas. All these effects would shorten the time to breakdown, t_{BD}, of the oxide layer under a constant applied voltage.

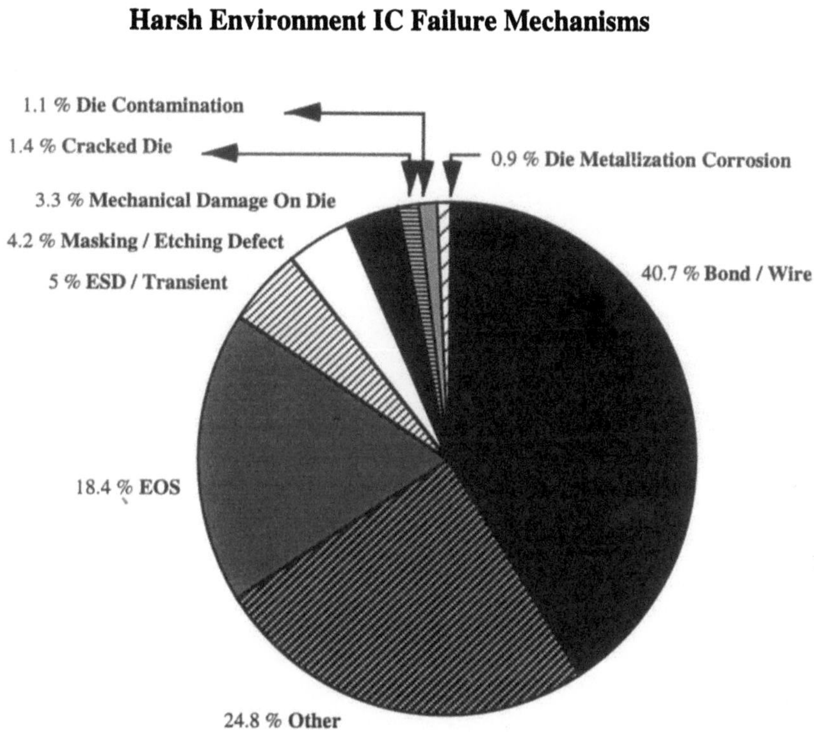

Figure 2

Even though a "perfect" dielectric layer has an expected life approaching the age of the universe, a microscopic defect can reduce its life to just a few minutes or seconds. In the data shown in **Figure 3** it may be observed that all the defects can be described by an "effective oxide thinning" model. "Unknown" means that the physical cause of the local perturbation in the electric field was not clearly identified.

4.2 Time Dependent Dielectric Breakdown Physical Models

In spite of extensive research in this field, there are not yet physical models for dielectric breakdown that are universally accepted. Furthermore, each type of oxide problem, whether it be silicon substrate defects, poly-silicon asperities, oxide thinning, point contaminating source, uniform contamination, etc. has a different voltage acceleration factor. The two principal models used in the industry today are commonly referred to as the E_{ox} and $1/E_{ox}$ models (proportional or inversely proportional to the electrical field) [13]. Research at UC-Berkeley has lead to the conclusion that $1/E_{ox}$

models can better accommodate the available experimental results and, it is the Berkeley Statistical Model of Oxide Breakdown that will be used [19].

Defect Distribution of Failed CMOS Devices (7 Volt EOS Test, 1 sec. duration)

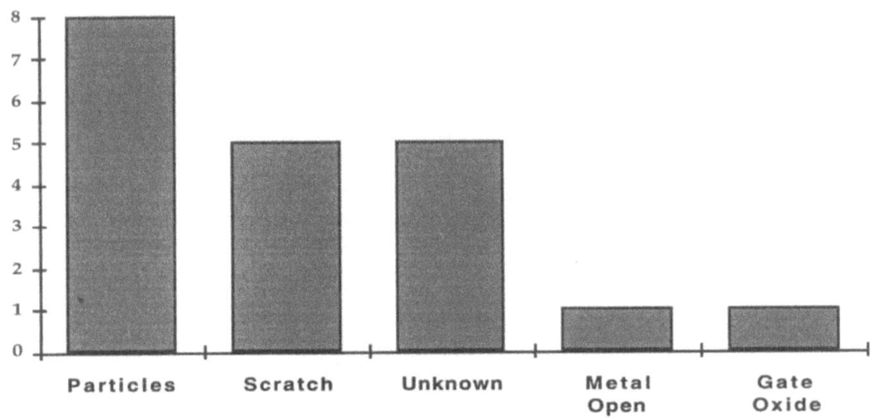

Figure 3

This statistical model assumes that oxide thinning is the one single physical cause for dielectric breakdown and that all the defects mentioned before will produce an "effective oxide thinning" effect (ΔX_{ox}) with statistical defect density ($D(\Delta X_{ox})$). It should be emphasized that the statistical model adopted here is independent of the physical breakdown mechanism. The input datum is not the actual thickness of the oxide layer, but the probability of having a certain number and distribution of capacitor defects.

Using this concept, the intrinsic time to breakdown, at constant temperature, is modified to account for the statistical variation according to:

$$t_{BD} = Ce^{(B+H)} (X_{ox} - \Delta X_{ox})/V_{ox} \quad \textbf{(1)}$$

where V_{ox} is the voltage across the oxide and C is determined by the intrinsic breakdown under the applied voltage of V_{ox}, B is the barrier height, and H the trap generation coefficient at 20°C. The effective oxide thinning, (ΔX_{ox}), can represent a "true" localized thin spot, surface asperities (higher E_{ox}), a lower barrier height (lower B), or an increased trap generation rate (lower H) at the localized defect spot.

Eq. **(1)** represents a "universally" valid correlation for predicting time-dependent dielectric breakdown as a function of the design factors, the electrical field, and the manufacturing defectivity contained in the sample. This expression is valid for all oxide thicknesses and typical processes. All the parameters in this equation are universal constants, independent of the type of the physical defects present in the sample. The value of the intrinsic breakdown time constant C is 1.1×10^{-11} sec. This expression agrees with the better known fact that Q_{BD} approximately equals 10 Coulomb/cm^2 for all typical oxides, at E_{ox}, equal to 10 MV/cm. [19]

Based on Eq. **(1)** it is possible to determine the lifetime of the oxide for arbitrary oxide areas if the cumulative defect density as a function of effective oxide thinning is known. Hence, the time to breakdown for a given oxide sample is determined by the ΔX_{ox} value of the weakest spot contained in the sample. Using the ΔX_{ox} model, one can also show that the field acceleration factor, β, is a function of the severity of the oxide defect:

$$\beta = \frac{-(B+H)\,(1-\frac{\Delta X_{ox}}{X_{ox}})}{2.30E_{ox}^2} = \beta_0\,(1-\frac{\Delta X_{ox}}{X_{ox}}) \qquad \textbf{(2)}$$

The acceleration factor decreases with increasing (ΔX_{ox}). This trend is supported by qualification test failures and it is similar to the variation of the failure rate during the infant mortality period in the field.

4.3 The Temperature Acceleration Factor

This statistical model can readily be extended to account for the temperature dependence of defect-related oxide breakdown. By considering the temperature dependence of the B and H factors in Eq. **(1)** as:

$$B+H = (B_0 + H_0\ )\,(c_1 + E_a/kT\ ) \qquad \textbf{(3)}$$

where B_0 and H_0 are the values of B and H at room temperature and c_1 and E_a

characterize the temperature dependence of (B+H). Incorporating this dependence and modifying Eq. (1) to account for changing temperature, the t_{BD} relation can be expressed as:

$$t_{BD} \propto \exp\left[\frac{(B_0+H_0)\,(c_1+\frac{E_a}{kT})\,(X_{ox}-\Delta X_{ox})}{V_{ox}} - \frac{E_b}{kT}\right] \qquad \textbf{(4)}$$

where E_b is the temperature dependence of the hole generation coefficient. The field acceleration factor becomes:

$$\beta \equiv \frac{d(\log(t_{BD}))}{dE_{ox}} = \frac{-\,(B_0+H_0)\,(1-\frac{\Delta X_{ox}}{X_{ox}})\,(c_1+\frac{E_a}{kT})}{2.30\,E_{ox}^2} \qquad \textbf{(5)}$$

Eq. 5 indicates that the field acceleration factor is inversely proportional to temperature, a fact which is supported by the experimental results published in the literature.

The effective activation energy for time-dependent dielectric breakdown is determined from a ln (t_{BD}) versus 1/T plot with the following definition:

$$Ea_{eff} \equiv k\frac{d(\ln(t_{BD}))}{d(1/T)} = \frac{(B_0 + H_0)\,(X_{ox} - \Delta X_{ox} E_a)}{V_{ox}} - E_b \quad \textbf{(6)}$$

This equation shows that as V_{ox} increases , Ea_{eff} decreases. This, again, is consistent with the field-dependent activation energy observations. Furthermore, as ΔX_{ox} increases, Ea_{eff} decreases. This implies that at a fixed E_{ox}, the dispersion in time to breakdown data decreases with increasing temperatures. In other words, the temperature variation modifies the distribution of the field failure without modifying the cumulative number of failures over a long period of time (10 years or more).

5. FIGURES-OF-MERIT (FOM) FOR PLASTIC PACKAGES

5.1 Introduction

For components operating in harsh environments, failure may be initiated by the mechanical degradation of the chip package, followed by the failure of the silicon interconnect or silicon interface to the package.

The significant temperature rise and internal temperature variations encountered in operating plastic IC packages can endanger the integrity of the die-bond, the chip, substrate, and/or encapsulant. It has been observed that separation of the die from the lead frame is generally the first mechanical fracture to appear in today's plastic packages, and often leads to a chain of degradations which end in an electrical failure of the device.

Although die-attach materials may fail by rupture or fracture, resulting from excessively high values of normal stress, the small thickness of typical bond layers suggests that shear stress failure, in the bulk or along the surfaces of the attachment, may be a more common failure mode.

Detailed analysis is required to determine the precise thermal stress and strain which develop within a package. The difficulty of finding an analytical or even numerical solution to this problem is primarily related to the existence of strong variations in the material properties with temperature and moisture absorption. However, more approximate, first order, relations can be used to guide the selection of adhesives as well as to identify promising directions, and target values, for the development of new, low-cost/high-reliability packaging materials.

5.2 Shear Stress Failures

Neglecting the stiffness of the adhesive and assuming perfect adhesion between the die-bond and the adherents, an estimate of the shear strain across the die-bond layer can be obtained by examining the thermal expansion of the material above and below the die-attach material.

Taking advantage of the observation that at steady-state the temperature difference between the chip and the encapsulant (or substrate) is relatively modest, an upper bound on the shear stress may be obtained by treating the package as an isothermal structure. The shear strain in the die-attach layer can then be determined from the difference between the unrestrained thermal expansion of the chip (above) and the encapsulant or substrate (below), divided by the thickness of the adhesive, h.

When multiplied by the Bulk Modulus, G, this approach yields an upper-bound value for the shear stress as:

$$\tau \leq (G/h)(T_f - T_o)(\alpha_{enc} - \alpha_c) \quad \textbf{(7)}$$

where T_f and T_o are the final and initial package temperatures, respectively, α_{enc} the expansion coefficient of the encapsulant and α_c that of the chip. Dividing this peak shear stress by the maximum allowable value, and selecting the relevant material properties, a shear-rupture or shear delamination FOM for die-attach materials can be obtained [25], as:

$$FOM\tau_{adh} = G / \tau_{adh} = E / 2(1+\nu)\tau_{adh} \quad \textbf{(8)}$$

When the thermal resistance of the die-bond leads to a significant temperature difference between the chip and the encapsulant, the thermal conductivity of the adhesives should be added to the denominator of the respective FOMs.

The values of these FOMs (based on the best available material property data) for seven candidates are shown in **Figure 4**. Use of the material with a low FOM will decrease the chance of a die-attach failure in the shear-stress mode.

It must be recalled that the FOMs presented above are based on first-order estimates of the thermal stress which develops in an operating plastic IC package. Nevertheless, the use of more sophisticated thermal stress formulations, as in [23, 24], appears to suggest that the property groups which have been incorporated into these FOMs are of greatest significance in material selection for minimum stress.

SHEAR STRESS FOM FOR CANDIDATE ADHESIVES

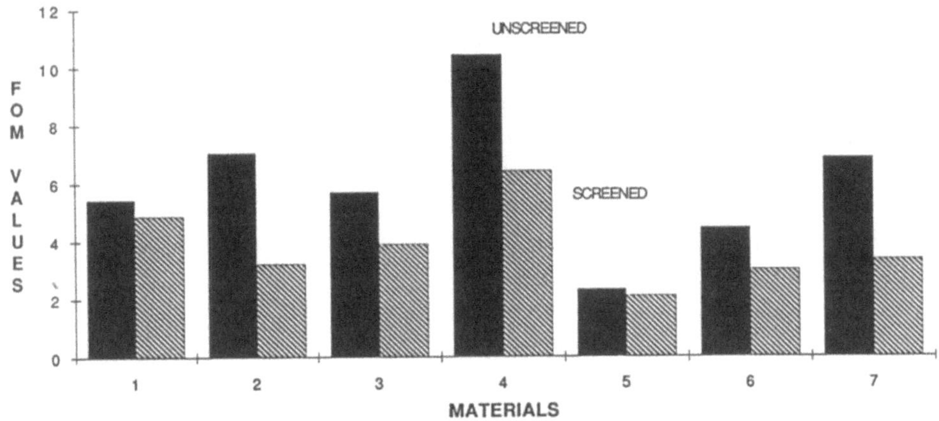

Materials:
1=Polyetheretherketone; 2=Polysulfone; 3=Polyetherosulfone; 4Polyetherimide; 5=Polybutyleneaterephthalate; 6=Polycarbonate; 7=Epoxy Thermoset

Figure 4

6. CONCLUDING REMARKS

Studies of device failure mechanisms show the dependency between the temperature level and the life of the device is, in most cases, a far more complex function than a simple Arrhenius correlation. Under moderate stress conditions, the temperature modifies the distribution of the field failure in time, without modifying the cumulative number of failures over a long period of time (10 years or more).

Cyclic temperature conditions are more damaging than elevated static temperature conditions. Cyclic thermo-mechanical stress creates fatigue micro-cracks in the packaging structure. Moisture penetrating into/through these micro-cracks produces further degradation of the packages. Mechanical failure of the packaging is the most probable cause of the device electrical failure under harsh environmental conditions such as temperature/power cycling, uncontrolled humidity, and long cool-down periods.

For new products, the failures are usually not randomly distributed, but concentrated in a very few locations. The failure analysis, in most cases, shows that the failed parts belong to the main device population, i.e. parts which meet all the design tolerances. In harsh environments (high temperature and humidity, cyclic power conditions), the failure occurs most often on the silicon interconnect level rather than the chip itself. Better design rules and more realistic qualification procedures, based on the understanding of the failure mechanisms, should be employed to eliminate the cause of the problem.

For properly designed equipment, the failure rate is not constant, but rather exponentially decreasing in time to an asymptotic value. To explain this behavior, one should consider that very seldom are similar parts identical. Small variations in dimensions and material properties are always present. The field failures are associated with the existence of a weak segment of the device population, representing the tail of the manufacturing tolerances distribution. Also, unexpected and unpredictable variations of the functional and environmental conditions will continue to produce damage of the least robust parts. The probability of encountering these conditions constitutes the reliability risk associated with specific applications.

The failure rate prediction methodologies in common use place an exaggerated importance on temperature as the primary factor that can be used in controlling or reducing the field failure rates. New models that show the physical correlations between the factors that initiate, control or accelerate the failure mechanisms should be developed. By replacing the existing statistical formulations by phenomenological ones, the designer will be able to identify the impact of the environmental or functional conditions, especially temperature, on the reliability (or probability of failure) of the product.

While in the past most of the efforts were directed towards predicting the steady state temperature in the system, more research is necessary today to predict the thermo-mechanical stresses generated in the electronic structures by variations of the functional and environmental conditions.

A great deal of effort was invested in the past to find new ways of extracting and transporting the heat generated in the system. As the level of integration in the system and the power dissipation rose, increasingly sophisticated and expensive cooling options were implemented. By understanding the impact of temperature on the functionality and the reliability of the part/system, new solutions may be found that will allow systems to operate at higher temperatures than are common today.

ACKNOWLEDGEMENTS

The authors would like to extend their thanks to John Moss from Bell Northern Research for his suggestions and insights, as well as for providing relevant statistical field data, and to Mr. Glen Showalter for his help in preparing this manuscript.

NOMENCLATURE

B	–	Barrier height
B_o	–	Barrier height at 20°C
C	–	Breakdown time constant
c_1	–	Adjustable fitting parameter (0.286)
E	–	Youngs modulus
E_a	–	Activation energy (0.019 eV)
E_b	–	Activation energy for holes generation
Ea_{eff}	–	Effective activation energy
E_{ox}	–	Oxide electric field
G	–	Bulk modulus
H	–	Trap generation rate
H_o	–	Trap generation rate at 20°C
h	–	Adhesive thickness
k	–	Stephan-Boltzman constant
T	–	Temperature
T_f	–	Final package temperature
T_o	–	Initial package temperature
t_{BD}	–	Time to breakdown
V_{ox}	–	Voltage across oxide layer
X_{ox}	–	Ideal oxide thickness
ΔX_{ox}	–	Effective oxide thinning
α_c	–	Chip expansion coefficient
α_{enc}	–	Enclosure expansion coefficient
β	–	Field acceleration factor
β_o	–	Field acceleration factor for ideal capacitors
τ	–	Shear stress
τ_{adh}	–	Adhesive shear-rupture strength

REFERENCES

[1] G.E. Moore, Progress in Digital Integrated Electronics, Proceedings IEEE International Electron Meeting, pp 11-13, 1975.

[2] J. R. Lloyd, M. Shatzkes, D.C. Challaner, Kinetic Study of Electromigration Failure in Cr/Al-cu Thin Film Conductors Covered with Polyimide and the Problem of Stress Dependent Activation Energy, Proceedings, IEEE International Reliability Physics Symposium, pp 216-225, 1988.

[3] F. C. Hsu, K. Y. Chiu, Temperature Dependence of Hot Electron Induced Degradation in MOSFETs, IEEE Electron Device Letters, Vol EDL-5, May 1984.

[4] R. Y. Moss, Caution-Electronic Discharge at Work, IEEE Transactions on CHMT, Vol 5, pp 512-515, 1982.

[5] M. Pecht, P. Lall, E. Hakin, Temperature Dependence on Integrated Circuit Failure Mechanisms, "Advances in Thermal Modeling III" Chapter 2, pp. 61-152, Edited by A. Bar-Cohen and A. Kraus, IEEE Press, Nov. 1992.

[6] G. N. Morrison, J. M. Kallis, L. A. Strattan, I. R. Jones, A. L. Lena, RADC Thermal Guide for Reliability Engineers, Report Number RADC-TR-82-172, Rome Air Development Center, Air Force Systems Command, Griffins Air Force Base, New York, 1982.

[7] D. A. Jeannotte, L. S. Goldman, R.T. Howard, Package Reliability, Chapter 5 in Microelectronics Packaging Handbook, Edited by Tummala and Rymaszewski, Van Nostrand Reinholds, New York 1989.

[8] S. F. Morris, MIL-HDBK-217 Use and Applications, Rome Air Development Center Technical Brief, April 1990.

[9] T. J. Green, A Review of IC Fabrication Design and Assembly Defects Manifested as Field Failure in Air Force Avionic Equipment, Proc. 1988 IEEE/IRPS, CH2508-0/88/0000-0226, 1988.

[10] C. T. Leonard, Mechanical Engineering Issues and Electronic Equipment Reliability: Incurred Cost Without Compensating Benefits, IEEE Transactions on CHMT, Number 4, Volume 13, pp 895-902, December 1990.

[11] C. Olson, Reliability of Plastic-Incapsulated Logic Circuits, Quality and Reliability International, Vol. 5, pp 53-72, 1989.

[12] T. R. Conrad, R. J. Mielnik, L.S. Musolino, A Test Methodology to Predict early Life Failure Mechanism, Proc. 1988 IEEE/IRPS CH2508-0/88/0000.

[13] D. L. Crook,Evolution of VLSI Reliability Engineering, Proc. 1990 IEEE/IRPS, CH2787-0/90/0000-0002.

[14] P. Kovwalczyk, Wafer Level Stress Testing: Containment of Infant Mortality Failures, 1990 WRR Final Report.

[15] R. J. Straub, Automotive Electronics IC Reliability, IEEE 1990 Custom Integrated Circuits Conference, CH2860-5/90/0000-0108.

[16] K. Van Doorselaer, K. de Zeeuw, Relations Between Delamination and Temperature Cycling Induced Failure in Plastic Packaged Devices, Proc. 1990 IEEE/ECTC 0569-5503/90/0000-813.

[17] S. Witzman, K. Smith, G. Metelsky, Silicon Interconnect- Acritical Factor in Device Thermal Management, IEEE Transactions on CHMT, Number 4, Volume 13, pp 946-952, 1990.

[18] S. Witzman, Y. Giroux, Mechanical Integrity of the Device Package-A Key factor in achieving Failure Free Product Performances, Proc. 1991 IEEE/ ECTC 0569-55503/91/0000-0450.

[19] J. Lee, I. Chen, C. Hu, Statistical Modeling of Silicon Dioxide Reliability, Proc. 1988 IEEE/IRPS, CH2508-0/88/0000-0131.

[20]. C.G.M. Van Kessel, S. Gee, J. Murphy, The Quality of Die-Attachment and Its Relationship to Stresses and Vertical Die- Cracking, IEEE Transactions on Components, Hybrids, and Manufacturing Technology Vol. CHMT-6, Number 4, December 1983.

[21]. D. Edwards, K. Heinen, S. Groothuis, Shear Stress Evaluation of Plastic Packages, IEEE Transaction on Components, Hybrids and Manufacturing Technology, Volume, CHMT-12, Number 4, December, 1987.

[22]. R. Thomas, Stress-Induced Deformation of Aluminum Metallization in Plastic Molded Semiconductor Devices, IEEE Transactions on Components, Hybrids and Manufacturing Technology, Volume CHMT-8, Number 4, December 1985.

[23]. E. Suhir, Thermal Stress in Electronic Components Chapter 5 "Advances in Thermal Modeling of Electronic Components and Systems", edited by Bar-Cohen, A., and Kraus, A., (Hemisphere Publishing Corporation, 1988, Volume 1), pp. 337-412.

[24]. H. Ewalds., R. Wanhill, Fracture Mechanics, Edward Arnold and Delftse Uitgevers Maatschappij (Publishers), 1985.

[25]. D. Mix, A. Bar-Cohen, Thermal Modeling of Plastic IC Packages, Material Research Society Symposium Proceedings, Volume 226, pp. 67-84, 1991.

2. NUMERICAL AND EXPERIMENTAL ANALYSIS OF SYSTEMS

Computational Analysis of Air and Heat Flow in Electronic Systems

S. Drake[1], M. White[2], A. Felthouse[3], C. Ganderton[3] and S. Glanfield[4]

[1] *Concentration Heat And Momentum Ltd., Bakery House, 40 High St., Wimbledon Village, London SW19 5AU.*
[2] *MOTOROLA ECID Ltd., 16 Euro Way, Blagrove Industrial Estate, Swindon, Wiltshire SN5 8YW.*
[3] *THORN Automation, P.O. Box 4, Rugely, Staffordshire WS15 1DR.*
[4] *Northern Telecom Europe Ltd., Brixham Road, Paignton, Devon TQ4 7BE.*

Abstract. The increasingly important contribution of Computational Fluid Dynamics to prototype testing by simulation of the thermal performance of modern electronic systems is described. A brief explaination of the physical and mathematical background is provided and advantages over alternative methods is discussed. Results are presented from four typical application types of a proprietary CFD code called Hotbox, and a case study is described in detail with some comparisons made with measured data.

1. Introduction

1.1 Continuous miniaturisation

Evolution of electronic systems has always been towards smaller more compact and multi-functional devices at low cost and with greatest reliability. The first substantial step towards miniaturisation of power and signal amplification was achieved when thermionic valves were replaced by discrete solid-state transistors, in Germanium for small low-power devices and later in Silicon to facilitate higher-power devices. Next, discrete transistors were superseded by integrated circuit boards where the much smaller multi-component configurations were mounted on ceramic bases.

With the advent of silicon-chip technology, all active and passive components of integrated circuits could be etched on to the single minute wafers of silicon which form the basis of todays highly compact modular electronic systems. The reduced spacing between components has led to far greater concentration of heat generation and fewer opportunities for its dissipation and removal. As a consequence, the designer is becoming increasingly aware of the importance of good thermal design of electronic systems, to avoid "hot-spots" and to retain reliability within increasingly miniaturised designs.

1.2 The role of thermofluid simulation

A thermal analysis can be made by a variety of methods. Often a working prototype is constructed and operated for measuring and recording a limited amount of thermal data, from which some understanding of the thermal performance can be reached. This method has some serious limitations such as

C. J. Hoogendoorn et al. (eds.), Thermal Management of Electronic Systems, 45–54.

the errors associated with experimental measurement and the high cost of construction and modification. Empirical analyses can be made by extrapolating or interpolating results from previous experimental tests; but these can be unreliable since empiricisms are usually unique to each system and can normally provide only global-performance parameters, not localised variations.

In this paper we describe a method of analysis based on the iterative solution of the non-linear equations which describe fluid flow and heat transfer. As many simultaneous equations are solved, the method relies heavily on the "number crunching" power of computers and with the advent of high-performance Personal Computers, the method is now within the economic reach of most electronics-design engineers.

1.3 Physical description

Modern sophisticated and delicate electronic equipment needs to be protected from moisture, dust and extremes of temperature. Hence, control of the immediate surrounding environment is necessary. This is usually achieved by housing the equipment in lightweight, sometimes insulated, cabinets. The passage of air through and/or within the cabinet is achieved through processes of natural convection and forced ventilation, often with filtration and sometimes with artificial cooling.

Inside a cabinet, the air which is to an adequate approximation incompressible, undergoes local heating and cooling by heat transfer from the various working components. Most cabinets possess electric fans which moderate the temperature of the components by forced convection. Where components make contact with other surrounding solid features such as the cabinet walls and other components, heat is transferred by conduction.

2. The numerical model

2.1 Equations solved

The numerical model to be described here is a finite-volume computer code called Hotbox, which is the electronics-simulation variant of the well known general-purpose fluid-flow simulator called PHOENICS. It incorporates the simultaneous solution of the three-dimensional Navier-Stokes equations of fluid motion, and the convection and diffusion of heat. The appropriate form of the generalised conservation equation for transient simulation as solved in Hotbox is:

$$\frac{\partial}{\partial t}\{\,\varrho.\varphi\,\} + div\,\{\,\varrho.\vec{v}.\varphi - \Gamma_{\varphi}.grad(\varphi)\,\} = S_{\varphi}$$

Where: φ represents the solved for variable.
ρ represents density.
$\vec{v}$ represents the velocity vector.
Γ_{φ} represents the diffusive flux vector.
S_{φ} represents the source of φ per unit volume.

For steady state simulation the first term is set to zero; for conservation of fluid mass, the φ term is set to unity; for conservation of each component of fluid momentum, the φ term is the respective fluid velocity; for conservation of energy, the φ term is the fluid or solid specific enthalpy, where for solid conduction of heat the first and second terms are set to zero. The solution procedure used by Hotbox is based on the pressure-correction method of Patankar and Spalding [1]. The differential equation above is replaced by an equivalent fully conservative finite domain equation and solved for the velocity components, pressure and energy in an iterative manner. Hotbox will deliver results in the form of the primary variables, fluid pressure, velocities and temperature at multiple locations throughout the calculation domain.

To represent turbulence, the code uses the two-equation k-ϵ model of Harlow and Nakayama [2], as developed by Launder and Spalding [3], and adapted for Low Reynolds number applications appropriate for mixed convection flows. Both the kinetic energy of turbulence and its rate of dissipation are scalar quantities which have their own conservation equations based on the general form above. The effect of turbulence on fluid momentum is then represented by a turbulent diffusivity derived from the values of k and ϵ.

In addition to the solved variables, various property laws and data can be invoked by Hotbox to represent the properties of both the fluid and solid(s).

2.2 Computing requirements

Simulation using numerical modelling is by definition a computationally-intensive process, since there are many equations to be solved, iteratively. Therefore, the accuracy achieved depends on the "number crunching" capability of computers. Prior to the advent of high performance PC's, numerical modelling was a technology available only to those who could afford mainframe computers and work-stations. Now, every office and laboratory has a PC, and with it comes the processing power to use a powerful numerical model.

The computing requirements of a numerical model can be quantified by the product of the number of variables to be solved, the number of time steps (if by transient simulation), the number of computational cells used, and the number of iterations required to achieve a converged solution. Typically, a demanding simulation using ten thousand computational cells, for say twenty time steps, might take about one hundred iterations per time step to converge. This means that 140 million separate calculations will be made, which on a PC 486 with a 50 MHz clock-speed would take about 9.25 hours cpu time.

Hence, a demanding simulation can be set to run over-night, although most applications will be less so and can be satisfactorily completed within one or two hours.

3. Application examples

3.1 Example 1: Inlet air contamination

An example of how numerical modelling has been used to test for possible design flaws before prototyping, is given by the simulation of two adjacent cooling apertures of a base-station control and computing device. The arrangement shown in schematic in Figure 1, consists of the exhaust aperture for the upper cage of the device situated immediately above the inlet aperture of its lower cage. It was envisaged that should warmer air from the exhaust aperture be drawn into the inlet aperture, then a reduction in cooling efficiency of the lower cage could result. A prompt evaluation of the proposed prototype configuration was needed to confirm that this would not happen and that the arrangement would be acceptable prior to building the prototype. The simulation showed that buoyancy of the warmer exhaust air was sufficient to prevent most of it from being drawn into the inlet aperture, but for a very small and insignificant amount. This resulted in a negligible increase in the average inlet air temperature and hence dispelled any doubt as to the risk of cross contamination, so that the prototype could be built and tested with a high level of confidence of success. This can be observed in the Colour Plate 1, showing contours of air temperature and velocity vectors on a plane traversing both the air inlet and exhaust.

3.2 Example 2: Product layout design

An example of how numerical modelling has been used to evaluate an alternative design, is given by an application of Hotbox to a cellular technology Analogue Base-station. The base-station which is a three rack system consisting of two radio bays and one power supply/communications bay, is designed for rapid and cost-effective enhancement of an existing analogue cellular network. It has a sealed enclosure with a heat exchanger designed for outdoor sites. The heat exchanger has an internal (clean) air circuit drawing hot air from a plenum above the racks. Heat transfer takes place between this and an external circuit within a cross flow matrix. Cooled air is then delivered to a base plenum before being drawn vertically up the racked equipment.

The model shown schematically in Figure 2, is a representation of one of the radio bays comprising power amplifiers and combiners from which about one third of the total heat dissipation takes place. Sets of amplifiers are grouped in three blocks. Each has a fan which forces air horizontally across the amplifier heat-sinks. This increases the complexity of the cooling scheme as elsewhere it is drawn vertically through the racks. A comparative analysis was required to evaluate two different arrangements of the amplifier blocks, one of which was preferable for the thermal design, and the other favourable for the cabling and assembly. The simulation was used to quantify the thermal penalty of opting for the latter arrangement.

The results in Colour Plate 2 clearly show that, with the blocks spaced out over the rack height, some air is drawn serially through amplifiers of successive blocks. Air entering the top group of amplifiers is therefore preheated by the lower ones, an unacceptable solution. The use of baffles to alter the flow was found, using Hotbox to be ineffective, unless many baffles were incorporated.

The thermally-preferred arrangement required close grouping of the blocks at the top of the rack, which allowed a parallel flow of air with no pre-heating and an acceptable air-temperature rise. The enhanced quality of the preferred solution can be seen immediately in the lower temperature contours achieved in the second arrangement shown in Colour Plate 2.

3.3 Example 3: Complex air movement

An example of how numerical modelling can enhance understanding of complex air movements within densely packed boxes, is provided by an application of Hotbox to a 'Remote Base-station Diagnostic System'. The RBDS is a diagnostic tool residing at a cellular base-station location. It provides additional fault detection and test facilities not otherwise possible at an operations and maintenance centre with which it communicates its results. The device is shown in Colour Plate 3 and consists of seven pcb's, a power supply unit, a radial cooling fan and a radio telephone, all contained in a small steel enclosure.

The results enabled a full visualisation to be made of the airflow and heat transfer within the RBDS. This would not have been possible with the existing method of analysis based on the estimation of pressure loss-coefficients over each feature and through each air-path in the equipment. Although validation was not possible due to the lack of experimental data, the model can attempt to isolate problems that would be missed by the existing method of analysis, such as the determination of a hotspot between pcb's, as shown in Colour Plates 4 and 5.

3.4 Example 4: Cooling with a heat bridge

In this case the heat source is a data-communications collection block which is situated at the end of an optical laser transmitter. Data is passed along an optical fibre through an infra-red medium causing an increase in the temperature of the block. This heat is dissipated via a heat-sink situated beside the block, but separated by an air gap. Introduction of a metal connecting-bridge between both the block and heat-sink is shown to conduct heat rapidly between both objects thus significantly reducing the block temperature, as can be seen in Colour Plates 6 and 7. A consequence of the much higher rate of heat transfer to the heat sink, is that the unventilated space behind the sink has warmed considerably from 3 ^{o}C to 41 ^{o}C, demonstrating its greater functional effectiveness.

4. Case study

4.1 Validation of the Hotbox code on a vehicle power converter

In this application Hotbox is used to simulate the steady state thermal performance of an Auxiliary Static Power Converter. This case was chosen for validation purposes, as the equipment had already been designed, built and tested, and detailed thermal measurements were available. The ASPC shown in Figure 3 is mounted underneath the chassis of a passenger-carrying light-rail vehicle which is part of an urban transit system in the tropical far east. The 31 kW rated ASPC uses thyristors to convert a traction power supply at 750 V d.c. to two outputs which are: a 3-phase, 380 V supply for air conditioning fans and other auxiliary equipment; and, a 28 V d.c. supply for battery charging.

The inverter enclosure is 1.5m long, 0.7m wide and 0.5m high. It is divided longitudinally into two equal compartments by an insulated barrier with cut-outs to allow heat-sinks of major assemblies to protrude between the compartments. The front compartment is totally sealed to provide protection for all electronic control components, and the rear compartment contains most of the electrical power equipment. The rear compartment is force cooled by extracting air from one end via a guide vaned axial fan, which is replaced with filtered ambient air at 40 °C and 100% relative humidity from the other end. Major components within the rear compartment are: ac filter chokes and heat-sink fingers protruding through the insulated divider from the sealed section. The front compartment contains: three inverting power switches which consist of gate turn-off thyristors, associated semiconductors, a pcb, power supply assembly and, three semiconductor diodes for the d.c. output, all mounted on large aluminium heat-sinks as complete sub-assemblies.

Underneath the chassis of a tram or train carriage the pocket of air surrounding the ASPC tends to move with the vehicle, thus only allowing heat dissipation by natural convection. The equipment configuration and method of cooling was chosen to enable the ASPC to generate its own air movement and hence not be limited by natural convection. The results of the validation are shown in Table 1 and it is clear that temperatures compare well in most positions.

Position	Measured Temperature (°C)	Simulated Temperature (°C)
Power switch 1 (internal face)	42	41
Power switch 2 (internal face)	50	42
D.C. input capacitors		
(middle)	60	57
(right-hand)	59	53
Air adjacent to control card	54	57
Air above gate drive cards (power switch 1)	64	64
Air between gate drive 4 and power switch 2	64	65
Air in centre of blast cooled section	27	28
Air out of fan	30	31

Table 1. *Comparison of Hotbox calculated temperatures with measured data for the ASPC.*

5. Conclusions

The versatility of numerical modelling as a tool to aid the thermal design of electronics systems has been demonstrated by way of illustrating four typical application examples. From this, it does not take much imagination to see that truly innovative designers could progress to using the method for design of sub-assemblies, which could be built into an assembly library, from which total systems can be designed. Thus utilising the power of numerical modelling to greatest effect.

A validation exercise of a leading commercially available numerical code, has shown that remarkable accuracy can be achieved by the method. The authors, all users of Hotbox, have benefited from quicker and more detailed analysis than was previously available to them. The ability to pre-solve, by locating hotspots or deficiencies in system performance and to investigate radical design changes has enabled them to increase design confidence, and enhance their perceived engineer capability in the eyes of potential customers.

As numerical modelling is used more widely, proper integration into CAD systems will be achieved by customising interfaces between the numerical model and CAD system. It is probable that the thermal analysis will also be interfaced with other models such as stress analysis and economic optimisation, or that the functionality of these models could become part of the evolution of the thermal analysis.

References

[1] Patankar S.V. and Spalding D.B. (1972)
"A Calculation Procedure for Heat, Mass and Momentum Transfer in Three Dimensional Parabolic Flows".
Int. J. of Heat and Mass Transfer, Vol. 15, pp 1787-1806.
Pergamon Press.

[2] Harlow F.H. and Nakayama P. (1968)
"Transport of Turbulent Energy Decay Rate".
Los Alamos Science Laboratory.
University of California, LA 3851.

[3] Launder B.E. and Spalding D.B. (1974)
"The Numerical Computation of Turbulent Flows".
Computer Methods in Applied Mechanics and Engineering, Vol. 3, pp 269-289.
Pergamon Press.

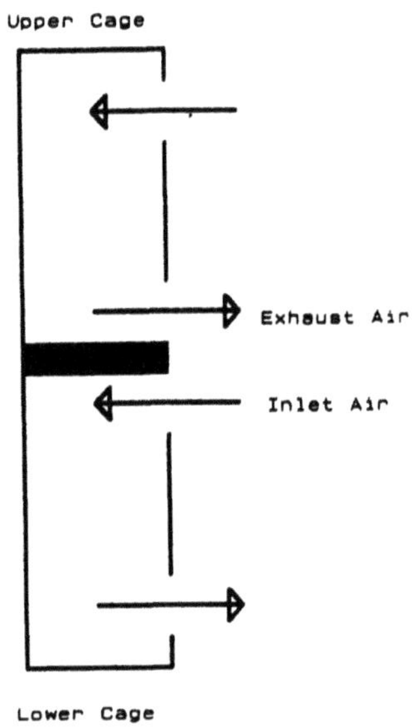

Figure 1. Schematic representation of inlet and exhaust flow paths for Example 1.

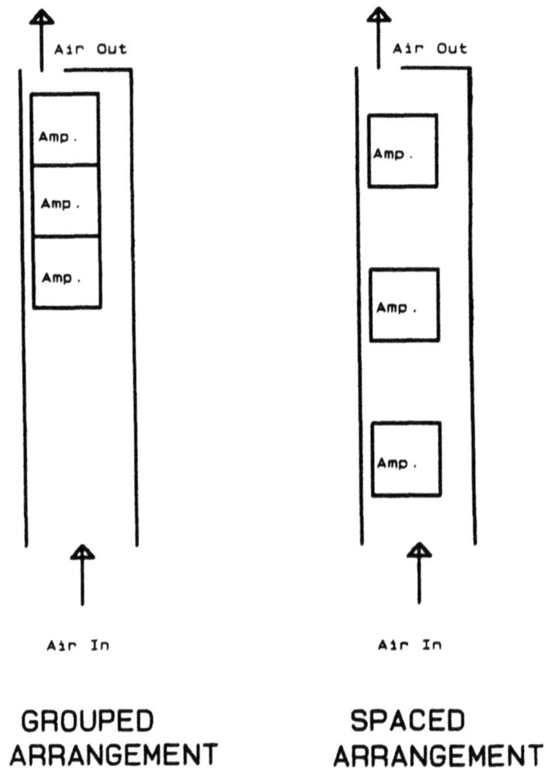

Figure 2. Schematic representation of amplifier group arrangements for Example 2.

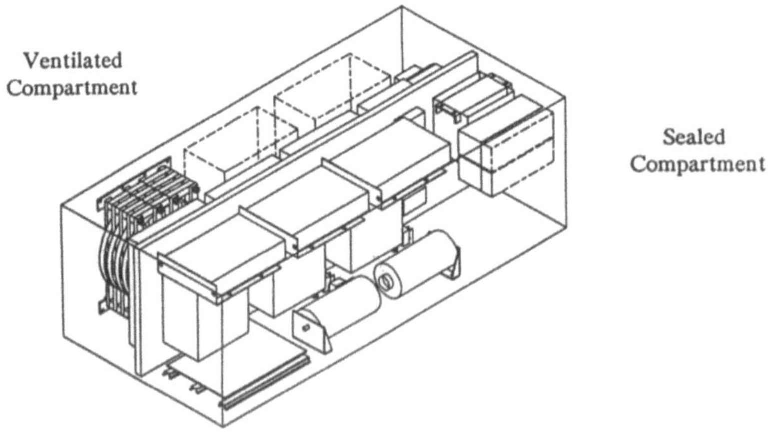

Figure 3. Isometric view of Auxiliary Static Power Converter.

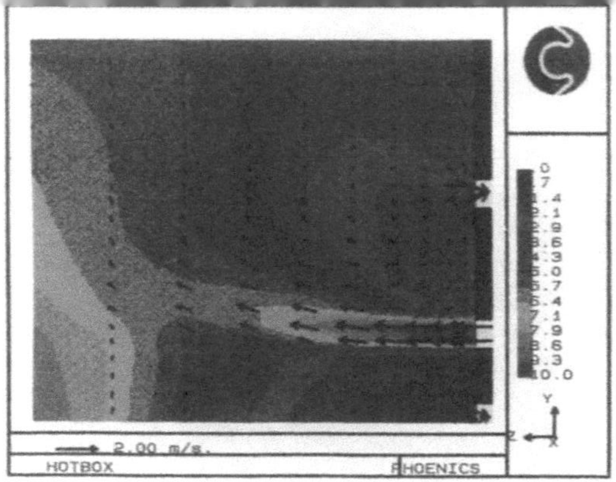

Plate 1. Air flow and temperature through apertures.

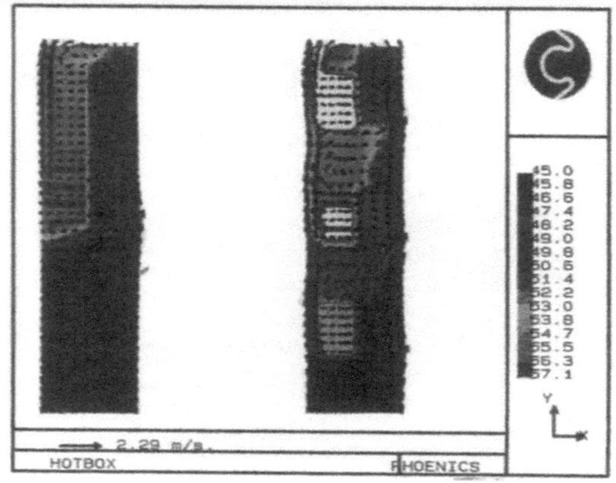

Plate 2. Air flow and temperature for the two alternative amplifier block arrangements.

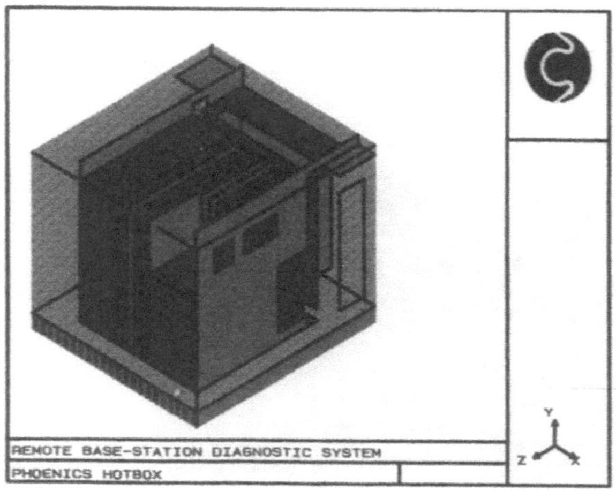

Plate 3. Remote Base-Station Diagnostic System.

Plate 4. Air flow and temperature on a plane traversing the pcbs.

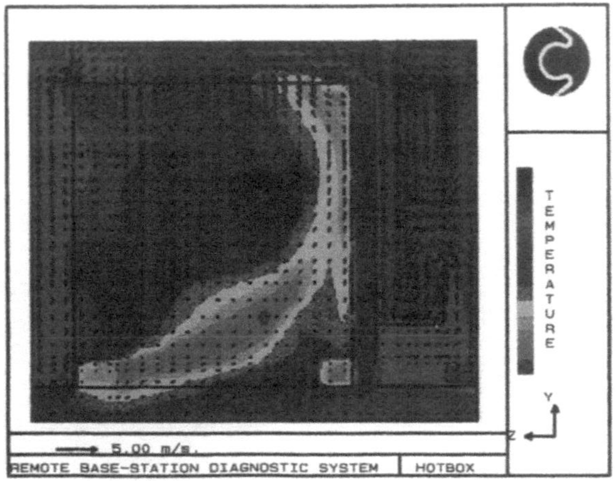

Plate 5. Air flow and temperature through hotspot shown in Colour Plate 4.

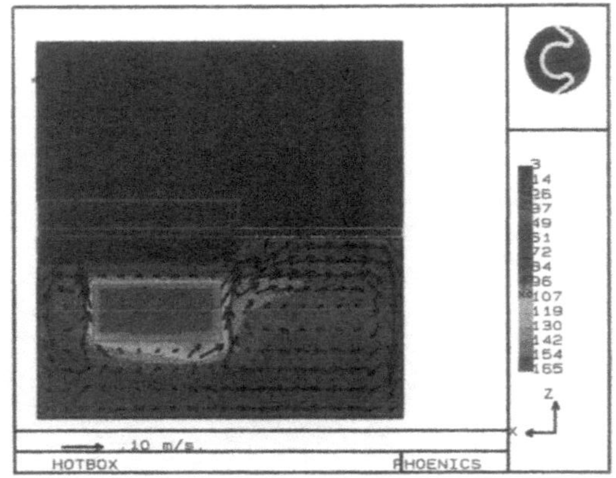

Plate 6. Temperature and flow vectors without heat bridge.

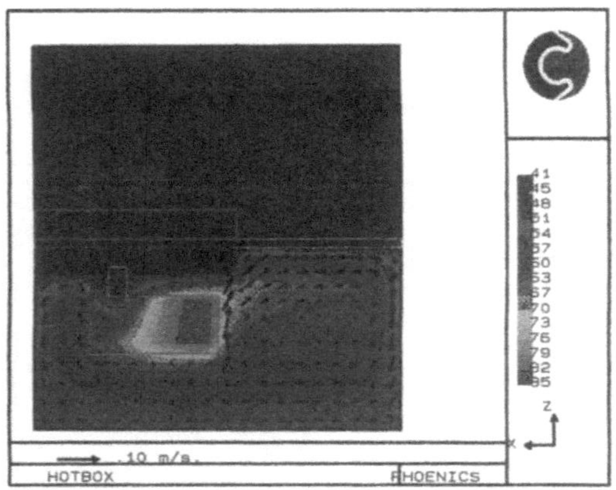

Plate 7. Temperature and flow vectors with heat bridge.

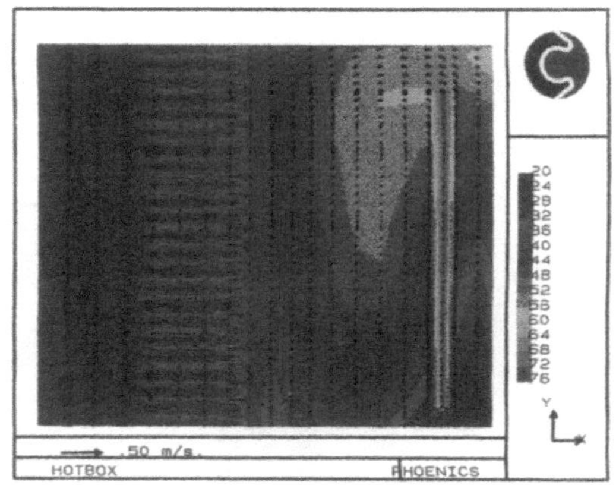

Plate 8. Air flow and temperature through fins traversing compartments.

Thermal Analysis of a Telecommunications Rack System

Drs. J. D. Parry & D. G. Tatchell
Flomerics Ltd, Kingsgate Business Centre, 12-50 Kingsgate Road, Kingston-upon-Thames, Surrey, England.

Abstract.

This paper describes a study of a prototype design of a transmission subrack under development by a major European manufacturer of telecommunications equipment. The paper details the thermal model, and presents the predicted thermal performance of the unit in three rack configurations. The study confirms the validity of the analysis by comparison with data for a single-subrack configuration; and the results for multi-subrack configurations provide insight into the thermal performance in practical operation.

1. INTRODUCTION

Thermal design of telecommunications equipment has become increasingly critical, due to:

- increased power densities, both in the electronic components, and in equipment,
- increased functionality (again both in components and equipment),
- preferences for natural convection cooling,
- increased user expectations for reliability,
- imposition of design and operating standards (eg ETSI) which include thermal performance,
- shorter design cycles - so less time for build-and-test to identify and solve thermal problems.

Consequently there is now a pressing need for thermal analysis to supplement practical test and measurement.

The objective of the reported study is to investigate the validity of applying thermal analysis techniques to the thermal design of a typical item of telecommunications equipment - a prototype design of transmission subrack - and to examine the thermal performance in a range of practical operating configurations.

2. PRACTICAL RELEVANCE

Transmission or switching equipment is generally made up of subracks, comprising a number of circuit packs each containing a vertically-mounted PCB. The subracks are mounted one above the other in vertical racks. The electronic components, mounted on the PCBs, are cooled by air flowing vertically between the PCBs, usually under natural convection.

Suppliers of equipment of this kind are required to conform to certain standards laid down by equipment operators, which define requirements in many areas including thermal design and operation. In Europe the relevant standards are those set down by

C. J. Hoogendoorn et al. (eds.), Thermal Management of Electronic Systems, 55–64.

the European Telecommunications Standards Institute (ETSI). For thermal design these state (Reference 1):

- Natural-convection cooling should be used except in special circumstances,
- Equipment should be able to operate in "miscellaneous racks" containing various types of subrack.
- Component junctions must not be subjected to temperatures which exceed maximum values quoted by the component supplier.

These requirements impose severe demands on the designer. He must ensure good thermal design (ie satisfaction of the third of the above requirements), without recourse to fans; and he must ensure this not just for a single subrack in isolation, but also for situations in which it operates in combination with other subracks of the same design, or of quite different design.

Conventionally, as a standard part of equipment design, single subracks are built and tested for thermal performance. However, using build-and-test methods alone, any kind of design optimisation can be a lengthy and expensive business; and, most crucially, it is difficult to extrapolate from single-subrack tests to actual performance in practical multiple-subrack configurations. In both aspects, thermal analysis is a vital tool to complement practical tests.

3. CASES CONSIDERED

The present work focuses on a typical transmission subrack. The subrack is analyzed first in the single-subrack configuration for which it was tested, and the analytical results are compared with the data. Then two typical mounting configurations are studied; three identical subracks in a 2.2m ETSI rack, and four subracks in a 2.6m ETSI rack.

The thermal model of the subrack is shown in Figure 1. The subrack contains 11 circuit packs. It is enclosed at the front by the front plates of the circuit packs, and, in front of that, by a plastic door. The sides and the back are enclosed. The space below the PCBs (the cable interface area) is usually occupied by bundles of cabling. As these are relatively sparse, and were not present at all in the measured configuration, they are neglected in the analysis. Above and below the PCBs, between the top and bottom horizontal rails used to support the card guides, are honeycomb screens, to contain electro-magnetic emissions. The overall dimensions of the subrack are: 0.532m high, 0.438m wide, and 0.274m deep. The total power dissipation is roughly 100W.

The subracks are mounted in a vertical rack enclosure, with an opening in the rack front below the lower subrack (to allow air to enter), and with the top open (to allow air to leave). The subracks are cooled by air flowing vertically through the slots between the PCBs. For most circuit packs the cooling is predominantly convective, the heat being transferred directly from the surfaces of the components and the board to the air. However, for the PSU circuit packs (located to the right of Figure 1) the designer has found it necessary to attach the power transistors to two C-section heat sinks, conducting heat via a heatsink extrusion to the fins on the front plate.

The thermal performance of a subrack of this type is affected by:

- The total air flow - which is affected by the geometry of the rack enclosure (height, and locations and sizes of vents), the number and power dissipations of the subracks, and the resistances to flow of the subracks (which can be strongly influenced by the presence of EMI shielding such as the honeycomb screens in the case considered here).
- The influence of the subrack geometry on the airflow distribution within the subrack - in particular the influences of the locations of the main devices and heat sources, and of the horizontal rails supporting the card guides.
- The air flow and temperature distribution entering the subrack from the subrack below.
- The effectiveness of any conduction planes and heat sinks, such as those in the power supply circuit packs in the current case.

All of these effects need to be included in any meaningful thermal analysis.

4. ANALYSIS APPROACH AND THERMAL MODEL

The analyses were carried out using the Flomerics FLOTHERM software (Reference 2), which is specifically designed for thermal modelling of all types of electronics equipment. It accounts simultaneously for heat transfer by:

- conduction in solid objects,
- convection in surrounding air, and
- radiation between surfaces of objects.

The air motion and associated heat transfer in forced, free or mixed convection are computed using computational fluid dynamic techniques to solve for the conservation of mass, heat and momentum. Heat interchange between air and solids is computed simultaneously with the air flow through locally calculated heat transfer coefficients taking account of the local level of turbulence.

FLOTHERM solves in 2-D or 3-D, steady-state or transient - the present study involves 3-D, steady calculations. Solution is obtained by an iterative, finite-volume method. The software is designed to run on PCs and workstations - the calculations reported here were run on an IBM RS6000/320H machine.

The thermal model of an item of equipment is created in FLOTHERM using generic "objects" or "building blocks", which can be combined, manipulated, and stored as required. In this case the thermal model of the subrack, as illustrated in Figure 1, is created as:

- Cuboid "Blocks" representing the subrack framework and cover plates - heat transfer by conduction is included in these items simultaneously with the air flow solution.

- "PCBs" representing the printed circuit boards, with "Blocks" (including conductive heat transfer) to represent major modules on the various boards and the heat sink assembly in the PSUs.
- "Planar Sources" of heat to represent the thermal footprints of the major power dissipating components on each PCB.
- An insulating "Plate" representing the plastic door on the front of the subrack.
- Flow "Resistances" to represent the pressure-drop effects of the honeycomb EMI screens just above and below the circuit packs.

The subrack model is then stored as a separate "assembly", and is copied and located within the appropriate model of the rack enclosure as required.

In the reported study, calculations were performed on the subrack mounted in three different configurations:

i)	Single subrack located at approx. mid-height in a TEP-1E rack.	grid: 40x30x37
ii)	Three subracks mounted in a 2.2m ETSI rack.	grid: 81x29x37
iii)	Four subracks mounted in a 2.6m ETSI rack.	grid: 87x29x37

In all cases the ambient temperature is 20°C.

5. RESULTS

5.1 Comparison with data - single subrack case

Figure 2 shows the predicted results in a vertical cross-section through the complete rack, cutting front to back (the left of the figure corresponds to the front). Air velocity vectors and temperatures are shown in a cross-section through the inner PSU. Air is drawn vertically up through the subrack. The highest air temperatures (of just over 46°C) occur in a central vertical band corresponding to the air gap above the board-mounted transformers in the PSU. The velocities in the subrack are highest in this region due to the high buoyancy effects. The velocities in the front and rear parts of the subrack are low due to the obstructions caused by the horizontal card guide support rails. In front of the front plates the flow is also relatively low. Above the subrack the Coanda effect causes the plume to attach to the back wall of the rack.

Figure 3 shows a front view of the predicted flow vectors and air temperatures in a cross-section through the PSU transformers. Clearly the flow and air temperatures in each part of the subrack are quite different.

Air temperatures at several locations within the honeycomb EMI screen at the top of the subrack and surface temperatures on the sides of the heatsink fins attached to the PSU front plates were measured by the equipment supplier using Type K thermocouples connected to a datalogger. The table below compares these measurements with the model predictions. For the air temperature measurements the predicted range of temperatures within the honeycomb above each card slot is compared with the measured range and shown in the table below. The comparison shows that the simulations predict temperatures to a reasonable level of realism - in general, the temperatures are correctly predicted to within one to two degrees of the measurements, compared with a total rise of up to 20 degrees. The surface temperatures on the heat sink fins are also within two degrees of the measurements.

Thermocouple Number	Location	$T_{local} - T_{ambient}$ (°C)	
		Measured	Predicted
Air Temperatures	Cardslot Number		
29	4	13.6 - 19.6	12.5 - 20.7
30	6	9.6 - 15.6	11.1 - 17.2
12	5	14.7 - 20.0	14.6 - 21.1
7	12	12.0 - 18.0	12.1 - 22.5
Surface Temperatures	PSU heatsink fin		
8	1	9.0 - 10.0	11.5
19	2	9.0 - 10.0	11.5

5.2 Practical mounting configurations - three and four subrack cases

Of practical interest is the operation of the subrack in three or four subrack configurations mounted within 2.2m and 2.6m ETSI racks.

Figure 4 is a plot of predicted air temperatures and air flow vectors for the four subrack case on the same cross-section as shown in Figure 1. The behaviour of each subrack is broadly similar, but differences arise from the flow profile at the entry to each subrack changing as the air passes up through the rack. As expected, air temperatures increase as air passes up through the rack. However, higher up the rack, the local high temperature region has spread across the surface of the boards. Comparison with Figure 1 shows differences in the air flow and temperature profiles between the tested configuration and those of practical interest.

The table below summarises the results by showing the maximum predicted temperature above each subrack for all three configurations. As expected, the temperature rise for each subrack is reduced as the number of subracks increases due to the increased air flow created by the increased chimney effect. However, the maximum temperature above the top subrack increases with the number of subracks, up to 49.4°C (29.4°C above ambient) for the four subrack case.

Subrack configuration:	Maximum predicted air temperature in top EMI screen (°C) vs. Subrack Number			
	1	2	3	4
i) Single subrack in TEP-1E rack	42.6	-	-	-
ii) Three subracks in 2.2m ETSI rack	39.7	44.2	47.8	-
iii) Four subracks in 2.6m ETSI rack	39.2	43.6	46.8	49.4

6. CONCLUSIONS

It can be concluded from the work performed that the thermal model provides a valid thermal representation of the prototype subrack under investigation.

Once a model has been created and verified, designers can use results of the kind reported here to:

- Investigate thermal performance of subrack designs, optimise heat sink locations, examine thermal effects of alternative types of EMI shielding, etc.
- Examine alternative designs of rack enclosure.
- Investigate alternative mounting configurations, including those involving "miscellaneous racks".
- Deduce whether the air temperature distribution leaving the rack is acceptable from the point of view of its effect on cabling or other equipment.

In addition, once the thermal model has been created and tested, it can be used to investigate other mounting configurations, including those in combination with other subrack types.

7. ACKNOWLEDGEMENTS

Flomerics would like to thank British Telecom plc who funded the study and the subrack manufacturer for permission to make use of design data and results.

8. REFERENCES

1. "Thermal Design Implications of the ETSI Equipment Engineering Standards"
Chris Jelly, British Telecom, UK.
Proceedings of the 1st FLOTHERM International User Conference, Sept. 1991

2. "FLOTHERM User's Guide" (Doc. No. FLOTHERM/UG/0492 Issue 1.0)
"FLOTHERM Reference Manual" (Doc. No. FLOTHERM/RM/1192 Issue 1.0)
Flomerics Limited, 1992.

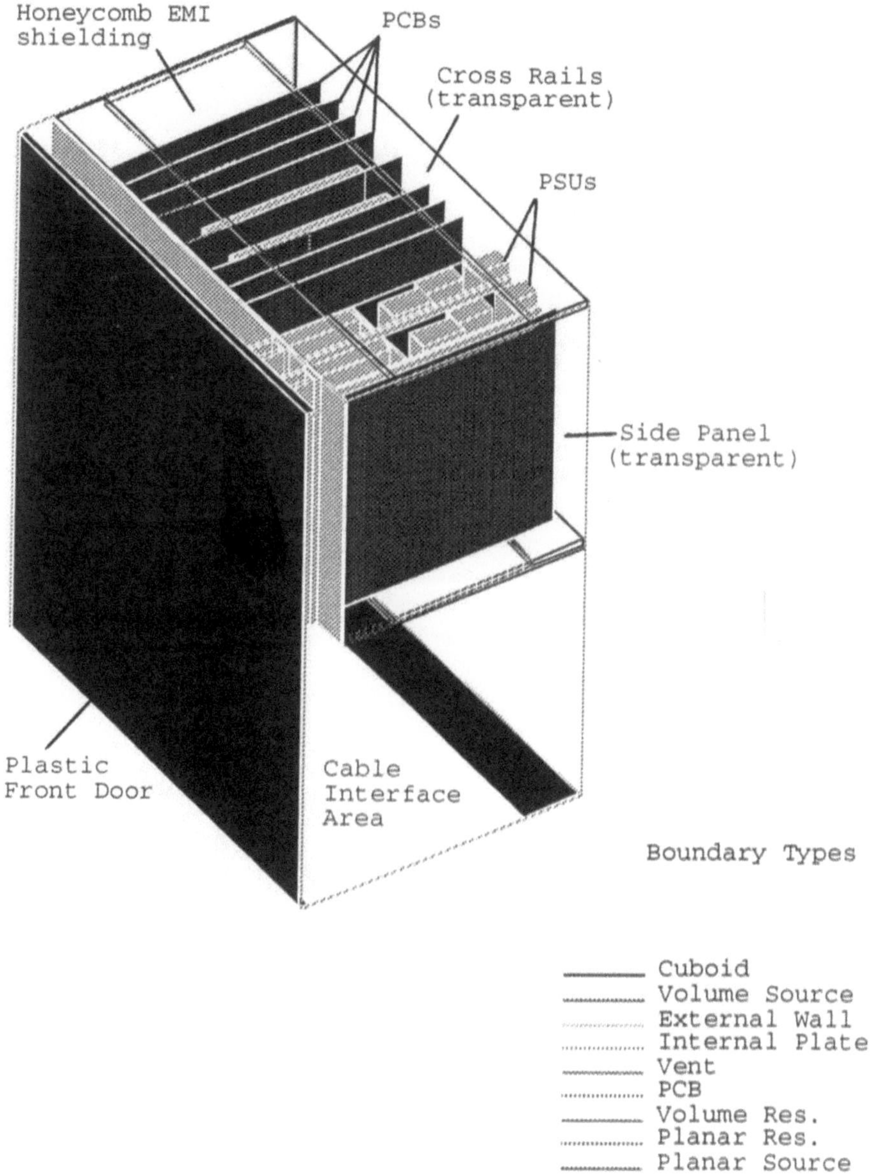

Figure 1. Subrack geometry with top and sides cut away showing details of the internals.

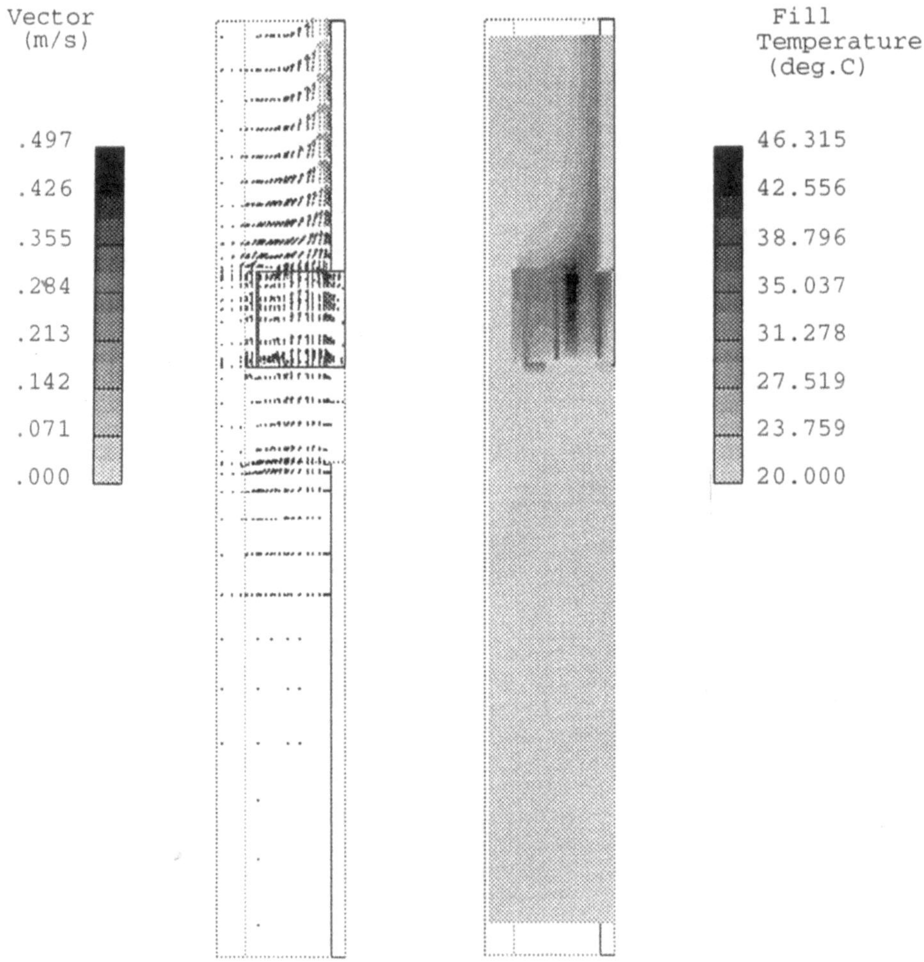

Figure 2. Side view of flow vectors and air temperatures above the PSU transformers (single subrack configuration)

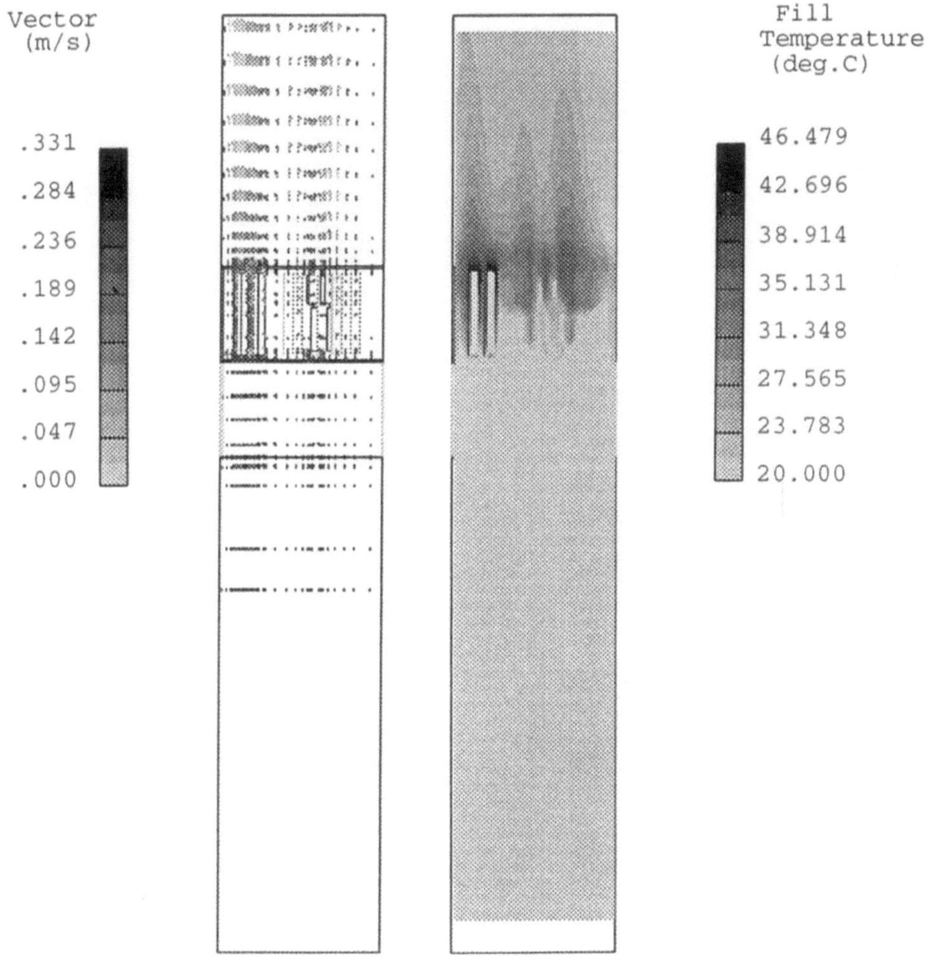

Figure 3. Front view of flow vectors and air temperatures through the PSU transformers (single subrack configuration)

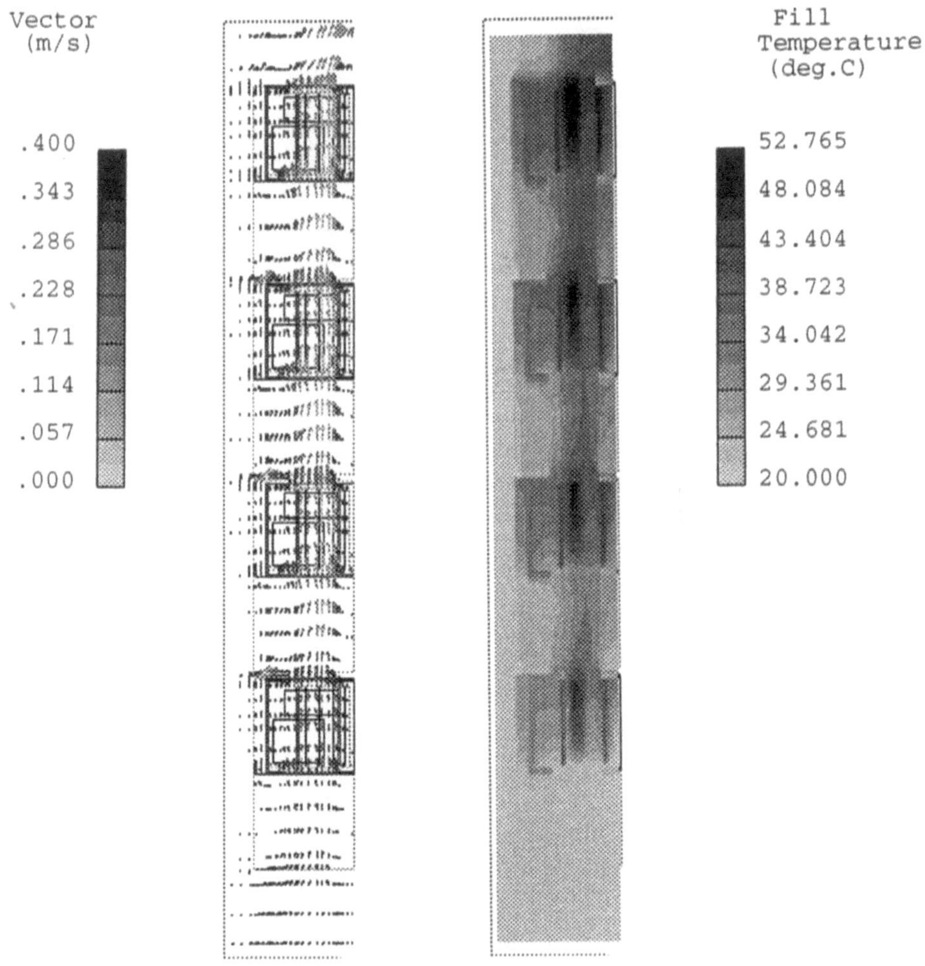

Figure 4. Side view of flow vectors and air temperatures above the inner PSU transformers (four subrack configuration)

Supporting experiments for CFD based thermal design of telecommunication equipment

H. BRUNEEL, B. BEERNAERT, G. MORTIER, J. DECLERCQ, B. BOESMANS [1]
W. TEMMERMAN, W. NELEMANS, E. LAUWERS [2]
[1] *K.U.Leuven, Dept. Werktuigkunde, Celestijnenlaan 300 A, B-3001 Heverlee (BELGIUM)*
[2] *Alcatel-Bell Telephone, F. Wellesplein 1, B-2018 Antwerpen (BELGIUM)*

Abstract. Experimental results are reported that are used to specify inlet and boundary conditions for computational fluid dynamics (CFD) simulation of forced convection cooling in racks containing telecommunication switching equipment. The flow field created by a fan-unit (inlet condition) is measured using Laser Doppler Velocimetry ; the thermal resistance and pressure drop of heat-sinks (boundary conditions) are measured in a low-speed windtunnel. The combined use of CFD and supporting experiments is shown to be indispensable for obtaining accurate simulation results.

1 Introduction

Thermal design of electronic equipment is becoming increasingly important because conflicting requirements, such as high reliability and small dimensions have to be fulfilled in a cost-efficient way.
Specific problems arise in the telecommunication industries : heat loads increase due to increasing switching frequencies and miniaturization, while convection air-cooling remains the preferred cooling scheme and international standards (e.g. ETSI) impose harsh environmental conditions. Therefore thermal design of telecommunication equipment is vital, both at system and at component level.
At present, in order to shorten product development time, there is a strong tendency to perform thermal design using computational fluid dynamics (CFD) tools instead of experiments. It is recognized that thermal experiments remain essential during the final design stages. CFD based modeling however has many advantages during preliminary design, because it is less time-consuming than experiments and because it allows greater flexibility. Early experience with CFD based modeling has shown that these computational tools should be used carefully. Any kind of CFD computation requires the specification of inlet and boundary conditions. Obviously these conditions determine the flow and temperature field resulting from the CFD computation. The specification of inlet and boundary conditions requires experimental information. Therefore supporting experiments are to be carried out before any attempt is made to obtain results from a CFD simulation.
In this paper the supporting experiments are presented that are performed for the thermal

C. J. Hoogendoorn et al. (eds.), Thermal Management of Electronic Systems, 65–74.

design of state-of-the-art switching electronics, mounted on vertical PCB's in a cabinet-type rack. This rack is equipped with axial fan units. It is possible to use heat-sinks or metal-core PCB's to ensure adequate cooling. In this case, the inlet and boundary conditions required for CFD simulation are :

- the velocity profile at the inlets of the channels formed by parallel PCB's , i.e. immediately behind each fan-unit ;
- the thermal resistance and pressure drop caused by the presence of heat-sinks in the flow channel

Neither of these conditions can be specified without supporting experiments. It is expected that the inlet velocity profile contains important horizontal velocity components. These contribute to the convective cooling and must therefore be taken into account. Moreover the vertical velocity component is not uniformly distributed, because the fans themselves block part of the inlet opening. Clearly an experimental investigation into the inlet velocity field is needed.
The detailed flow and heat transfer around a heat-sink can not be obtained from CFD simulations at rack-level. Instead, the thermal behaviour and the flow resistance of a heat-sink are specified as boundary conditions. It is expected that both thermal resistance and pressure drop depend on the velocity. A single, constant value (as often supplied by heat-sink manufacturers) is thus inappropriate for use in a non-uniform flow field. Here too, experiments are needed.

2 Inlet velocity profile

2.1 Experimental set-up

The flow field is measured using Laser Doppler Velocimetry (LDV) in an assembly consisting of 2 fan units and 2 subracks with 9 PCB's each, mounted conforming to ETSI standards. This assembly is mounted in a transparent cabinet that allows a laser beam to penetrate between the PCB's for velocity measurements. A front and side view of the test assembly is shown in figure 1. Each fan unit contains four fans. Figure 1 shows the position of the fans with respect to the PCB's. In the test assembly one thermal evaluation PCB is dissipating heat (80 W). This PCB contains thermal testchips whose power dissipation can be controlled and whose junction temperatures can be measured. Furthermore there are four heat-dissipating dummy cards. All other cards are non-heat dissipating dummies. All measurements are performed in the channels at both sides of the thermal evaluation PCB. This test assembly is a thermal mock-up of the real product environment influencing a single PCB.
The measurements are performed with a DANTEC 1D LDV system with a Flow Velocity Analyzer to determine the frequency shift and a 300 mW ILT Argon laser. This system measures the velocity in a small volume formed by the intersection of two laser beams, originating from the same source. The distance between the parallel beams is set at 60 mm and focussing optics are used with a focal distance of 300 mm. This short focal distance is required to obtain a small measuring volume. A description of the LDV technique and its applications can be found in [2].

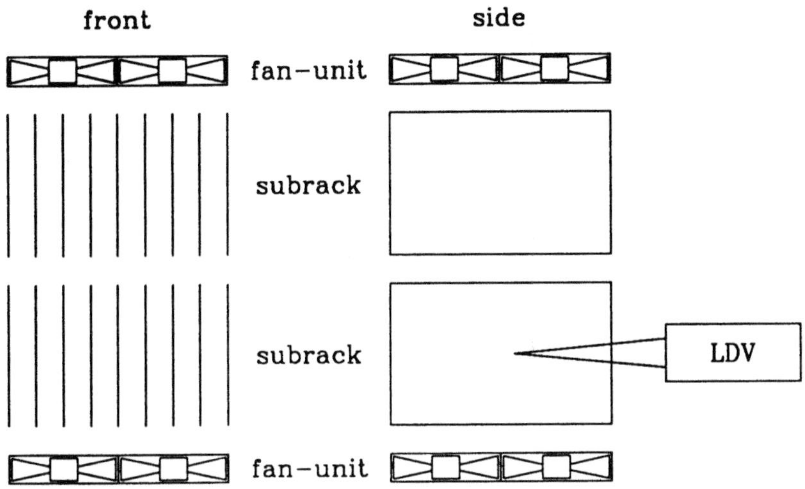

Figure 1: front and side view of test assembly

The LDV measurement technique is based on the Doppler frequency shift of the light beam reflected by moving particles. The LDV technique therefore requires the presence of tracer particles in the flow field. These tracer particles must be light enough to ensure that they follow the air flow. On the other hand they have to generate sufficient reflection of the laser beam. Incense particles proved to be the best solution for those two goals.
The LDV technique is well suited for the purpose of this investigation because

- it is a non-intrusive technique, i.e. the flow field is not disturbed
- the measuring volume results from the intersection of two 1.5 mm laser beams and is therefore small enough to record the flow pattern in detail

The LDV system allows point measurements of averaged velocity and turbulence intensity. Only both velocity components normal to the main axis of the LDV system can be measured. All three velocity components can therefore only be measured in points which are optically accessible in at least two perpendicular directions. This excludes points in-between the PCB's.

2.2 Results

Measurements are performed in four different configurations :

- in the normal configuration the PCB's are placed at a distance (pitch) of 30 mm. The fans are mounted 36 mm from the subrack.
- the second configuration includes additional free space between the fan-unit and the subrack. The distance between fan-unit and subrack is in this case 136 mm.

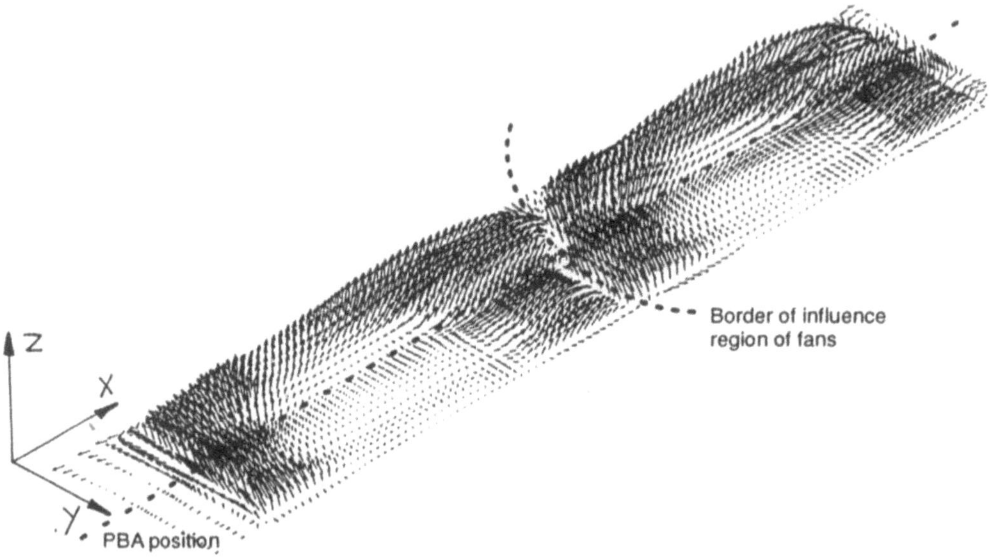

Figure 2: flow field at 21 mm above fan-unit

- the third configuration is similar to the normal configuration but has a 60 mm pitch between PCB's.
- the fourth configuration is the normal configuration but with one of the bottom fans switched off.

For each configuration measurements are performed using metal-core PCB's as well as standard PCB's with heat-sinks mounted on the IC's.

Horizontal and vertical velocity components and turbulence intensities are determined in 3 horizontal planes. One plane is located 21 mm above the bottom fan-unit. The second plane is located at the lower edge of the lower subrack ; the third plane is located at the top edge of the lower subrack.

Furthermore the vertical velocity component is measured in the vertical centerplane between two adjacent PCB's.

The density of the mesh which defines the measuring points varies from 5 x 5 mm to 20 x 30 mm, according to the plane.

The initial horizontal flow field is measured at 21 mm above the fan-unit. This flow field is shown in figure 2 for the normal configuration with a metal core PCB, covering the channels at both sides of the thermal evaluation PCB.

The results reveal the non-uniformity of the inlet flow field due to the fan shape. The vertical velocity drops from 6.5 m/s at the fan edge to -1 m/s at the fan centre (this region

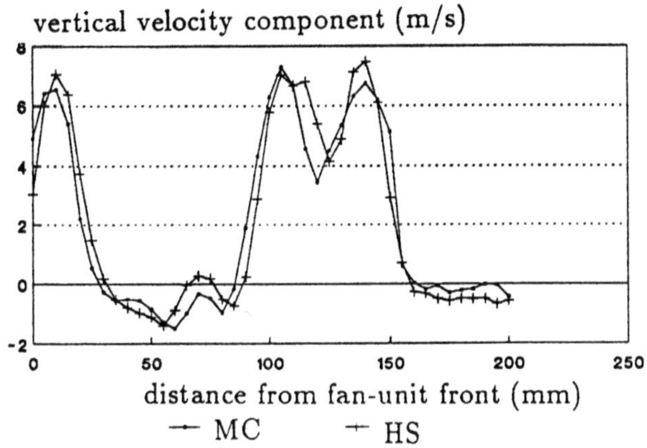

Figure 3: vertical velocity component for different downstream conditions

contains the fan motor). Evidently this could cause insufficient cooling at some locations. The influence of the PCB and the border between the region of influence of both fans are clearly visible.

The magnitude of the horizontal velocity component varies between 0 and 4 m/s. This is of the same order of magnitude as the vertical component, and thus can not be neglected in a CFD simulation of the cooling problem.

The different experiments show that the profiles of both velocity components are only slightly dependent on the downstream conditions. This is illustrated by figure 3 which shows vertical velocities for the normal configuration with metal core PCB's as well as PCB's with heat-sinks.

3 Heat-sink characterization

3.1 Experimental set-up

The heat-sink thermal resistance and pressure drop is measured in a low-speed wind tunnel at velocities ranging from 0.8 to 4.5 m/s and turbulence intensities between 3 % and 22 %. In the testsection of the wind tunnel a small channel is positioned which fits the heat-sink exactly (figure 4). The heat-sink is heated with a flat resistance heater connected to the heat-sink base surface. The pressure drop is recorded by means of an electronic differential pressure gauge connected to two static pressure taps on both sides of the heat-sink. The thermal resistance is defined here as :

$$R_{th} = \frac{T_b - T_0}{Q_{in}} \tag{1}$$

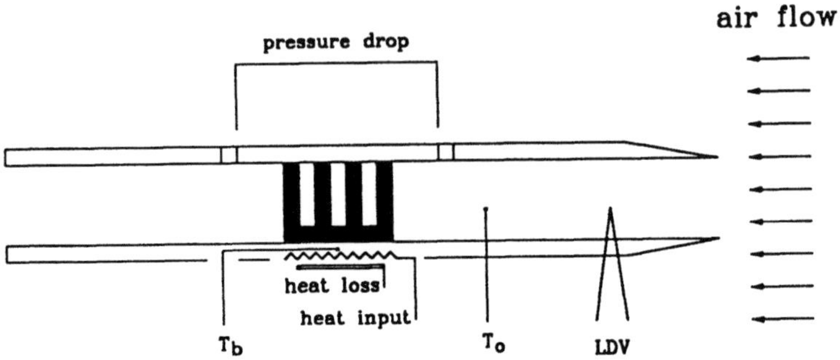

Figure 4: detail of set-up for heat-sink characterization

where T_0 is the free stream air temperature, T_b the average temperature at the heat-sink base surface, and Q_{in} the power supplied to the heat-sink by the resistance heater. Q_{in} is determined from the total dissipated power and the heat losses to the surrounding air (measured with a heat-flux sensor). All temperatures are measured with fine-gauge K-type thermocouples. Air velocity and turbulence level are measured with the same LDV system as used for the determination of the inlet flow profiles.

3.2 Results

Experiments are performed for two different types of heat-sinks :

1. a pin fin heat-sink (Thermalloy 2327B) with 7x7 pin fins and a size of 24.76 x 27.94 x 15.24 mm ;

2. an extruded plate fin heat-sink (SK 136) with 31 plate fins and a size of 50 x 150 x 15 mm.

Here only results are presented for the pin fin heat-sink, but the results for the plate fin heat-sink are similar.
Figures 5 and 6 show the dependence of thermal resistance and pressure drop on air velocity and turbulence level.
The results show that the heat sink thermal resistance decreases with increasing air velocity, but only up to a certain limiting velocity. Above this velocity the convective part of the overall thermal resistance becomes much smaller than the internal (conductive) part. As expected, the pressure drop increases as a quadratic function of velocity. The thermal resistance is quite insensitive to free flow turbulence but the pressure drop increases considerabely with increasing turbulence.

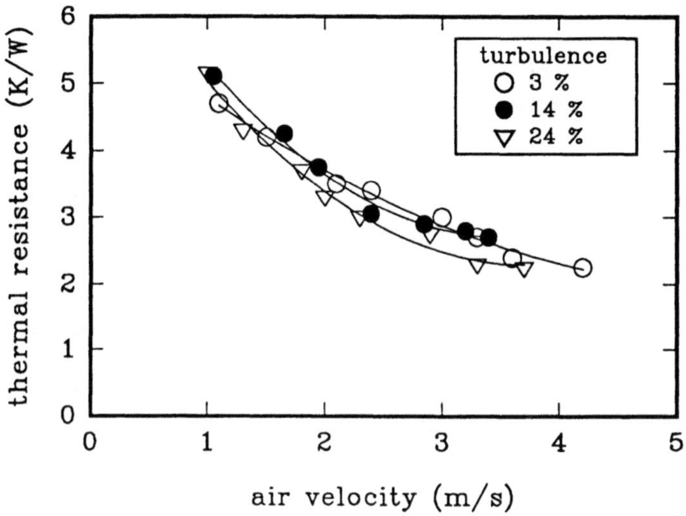

Figure 5: thermal resistance of pin-fin heat-sink

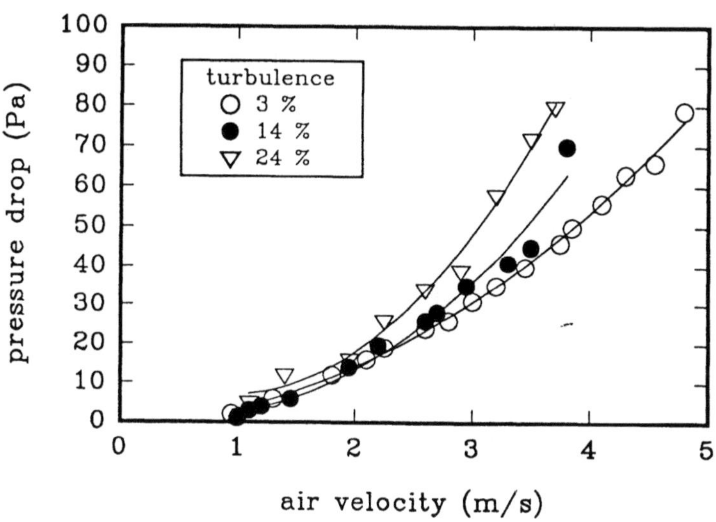

Figure 6: pressure drop of pin-fin heat-sink

4 Application to CFD simulation

The results of the supporting experiments are incorporated in a CFD based simulation of the flow and heat transfer in the channel between two parallel PCB's with heat dissipating IC's.

The fan-units are represented in the CFD model by means of the flow field at the inlet of the flow channel. As long as the shape of the velocity profile immediately behind a fan-unit does not depend much on the downstream conditions (i.e. on the pressure drop), it is possible to define a standard velocity profile shape for a specific kind of fan. In general however, the total volumetric flow rate produced by the fan depends on the total pressure drop in the channel. Therefore the velocity profiles behind the fan-unit is also expected to depend on pressure drop.

The heat-sinks are not modelled in detail, but are represented as porous blocks, through which air can flow. The pressure drop caused by these blocks is distributed uniformly over their volume. This way the exact flow rate of air that passes through the heat-sink can be calculated. The junction temperature is obtained from the value of the local air temperature, the internal thermal resistance and the heat-sink thermal resistance, the latter of which depends on the local air velocity.

The coupling between experimental results and CFD simulation is currently underway, and is discussed in more detail in ref. [3]. As a first step, the experimental velocity profile (fig. 2) and turbulence level above the fan-unit is used as input to a simulation of flow and heat transfer around a single metal-core PCB (without heat-sinks). The simulation is performed using Flotherm and the simulation volume comprises the two channels on both sides of the PCB. The computed velocity vectors (figure 7) show a shift of the air flow to the right, which corresponds with the experiments. Figure 7 also highlight the pronounced non-uniformity of the velocity field. Obviously, the non-uniform distribution of the air flow, causes a non-uniform distribution of the convective thermal resistance. Therefore the junction temperature in a particular IC depends on its location on the PCB, and computation results based on uniform air flow should be used carefully. It must however be emphasized that the local air velocity is not the only parameter influencting the junction temperature. The computation results ([3]) show that the lowest junction temperatures do not necessarily occur at locations with the highest air velocity. This can be attributed to the combined effect of internal and external heat transfer resistance. From the first CFD simulations it is concluded that the use of supporting experiments (concurrent with CFD) is promising, but more experimental information is still needed :

- A full threedimensional velocity and turbulence profile for a single fan should be determined experimentally. This should allow more flexibility in positioning the fans with respect to the PCB's. The dependence of velocity and turbulence profiles on the overall pressure drop should also be investigated in detail.

- The sensitivity of the computation results on the turbulence parameters should be investigated.

- The generation of turbulence by heat-sinks or by other porous blocks should be incorporated into the CFD model.

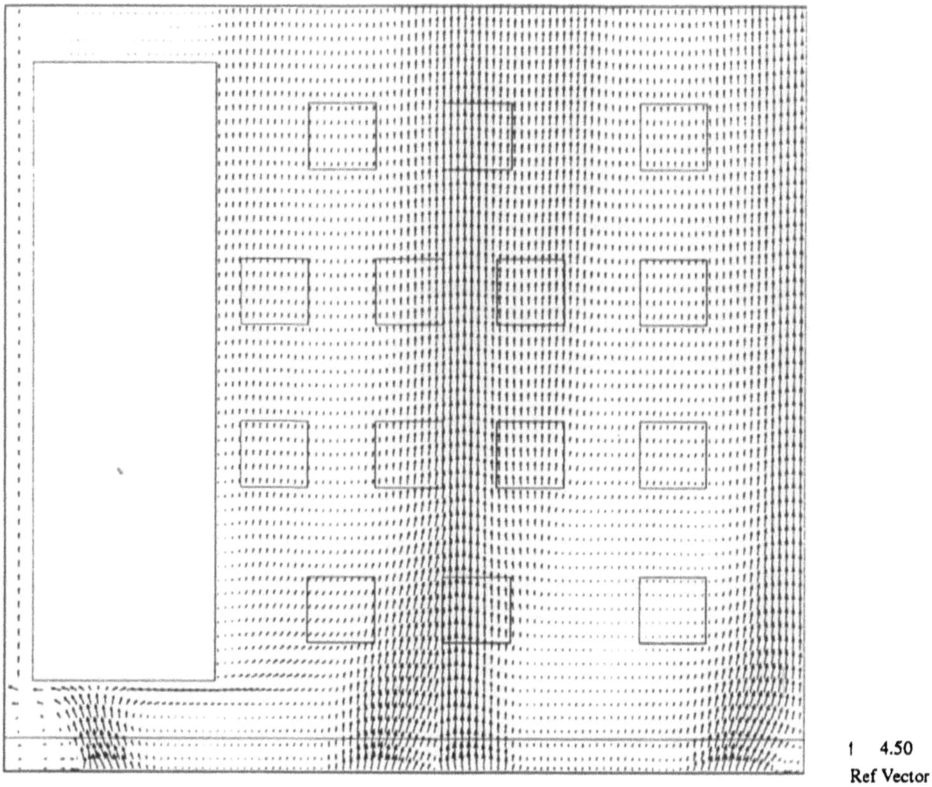

Figure 7: Computed flow in the middle of the channel between two PCB's

5 Conclusion

The quality of CFD results depends on the specification of the boundary conditions for the CFD model. Supporting experiments can be performed to obtain detailed information about these boundary conditions. The same type of experiments can be used for validating the CFD model.
In a convectively cooled rack, the velocity boundary conditions depend on the position and the shape of the fans. A non-intrusive technique, such as Laser Doppler Velocimetry, must be used to measure the flow field created by the fan. This flow field is highly non-uniform and causes a non-uniform distribution of thermal resistance throughout the rack.
The combined use of CFD and supporting experiments is considered as a prerequisite for obtaining accurate simulation results.

References

[1] Beernaert B., Bruneel B., Mortier G., Thermische metingen rack Alcatel-Bell, report KUL-Bell, 1992

[2] Boutier A., Introduction to Laser Velocimetry Techniques, VKI lecture series Laser Velocimetry, von Karman Institute, Brussels (B), 1991

[3] Temmerman W., Nelemans W., Goossens T., Air flow modelling requirements nearby fans, to be presented at Flotherm User Conference, 1993

Two complementary approaches for the thermal simulation of electronic equipment

J.L. BLANCHARD & J.M. MORELLE

Racal-Redac TAD. Parc Technopolis. Bâtiment Alpha.
ZAC de Courtabœuf. 3 avenue du Canada. 91966 Les ULIS cédex B. France

Abstract. This paper describes two consistent and complementary approaches for the thermal simulation of electronic equipment. They both consist of performing a PCB thermal analysis to compute the component junction temperatures, so as to prevent overheating problems, but they deeply differ in the context of their use and accordingly in their capabilities. Basically, the first approach integrates the thermal functions in one program to maximize ease of use and interactivity while the second one integrates the programs dedicated to these functions in one framework to maximize flexibility and accuracy. Both approaches share the same PCB geometrical description, as generated by the PCB design system.

1 Introduction

The continuing increase in the integration of electronic equipment requires the control of design constraints which are no longer purely electronic. Among these, thermal considerations play a crucial role, as demonstrated by the fact that most equipment failures are due to overheating problems, which shows that the use of thermal simulation is still uncommon.

This paper describes two complementary simulation approaches, through discussion of product features and usage, and their application within the complete design cycle. These two approaches are oriented identically towards board thermal simulation, and they can share the same PCB geometrical description, as generated by the PCB design system. However, they differ in the context of their use and in the way followed to perform the thermal functions.

For the first approach, these functions are packaged in the same program. In particular, the thermal simulation is completely integrated within the component placement function so as to visualize instantaneously the influence of a component moving on the board thermal map. Similarly, the product includes a tool to assist

C. J. Hoogendoorn et al. (eds.), Thermal Management of Electronic Systems, 75–84.

in the specification of the boundary conditions. The benefit of this approach is to provide a non-thermal specialist with an interactive and easy-to-use tool allowing the prevention of the most severe placement errors at an early stage of the design.

By contrast, the second approach relies on a framework ensuring the cooperation of separate programs to perform the thermal functions. In particular the PCB geometrical descriptions can be loaded in a CFD (Computational Fluid Dynamics) package to determine the global thermo-hydraulic behaviour of an electronic system. The thermal simulator can then use these results to model the local boundary conditions required by the thermal analysis of a particular PCB. The benefit of such a methodology lies in its high degree of flexibility and accuracy. This feature is reinforced by a dedicated library attached to the thermal simulator to model the component behaviour when reported on a board.

These two approaches are illustrated through the analysis of an instrumented PCB enclosed within a rack.

2 Thermal placement editor

The first approach involves a product named FloorPlanner 2000 whose basic thermal specifications are ease of use and the capability to compute interactively the board thermal maps. The computational algorithm of FloorPlanner 2000 uses an analytical approach based on a Fourier series expansion [1]. The main modelling assumptions are as follows:

- the temperature distribution $T(x, y)$ within the board $[0, a] \times [0, b]$ of conductivity k and thickness w is two-dimensional;
- the convective heat transfers on the upper and lower board faces are characterized by a unique pair (T_h, h) where the fluid temperature T_h and the heat transfer coefficient h are uniform;
- components are modelled by rectangular heat sources with uniform rate Q;
- steady-state is only considered.

If its edges are further assumed to be adiabatic, the board temperature is such that:

$$-\left(\frac{\partial^2 T}{\partial x^2} + \frac{\partial^2 T}{\partial y^2}\right) + \lambda^2 (T - T_h) = Q/k \tag{1}$$

$$\frac{\partial T(0,y)}{\partial x} = \frac{\partial T(a,y)}{\partial x} = \frac{\partial T(x,0)}{\partial y} = \frac{\partial T(x,b)}{\partial y} = 0 \tag{2}$$

with $\lambda^2 = h/kw$. The board temperature is then given by an expansion of the form:

$$T(x,y) = \sum_{i=0}^{n} \sum_{j=0}^{n} a_{ij} \cos\left(\frac{i\pi x}{a}\right) \cos\left(\frac{j\pi y}{b}\right) \tag{3}$$

The superposition principle allows to compute the board global thermal map, i.e. the map of the board equipped with its components, as the sum of the board map assumed to be free of components and of each component thermal map. Accordingly, when a component placement is modified, the operation cycle is as follows:

- the individual component map is substracted from the global map;
- the component map is computed at the new location;
- this new map is added to the global map.

From the computational viewpoint, the second sequence is obviously the critical one. The direct formulation (3) is actually inadequate for an interactive updating of the thermal maps. The algorithm optimisation, through a modified formulation and a careful implementation, whose details are beyond the scope of this article, is the main difficulty in meeting the "real-time" requirement. By contrast, seen from the user viewpoint, two parameters are sufficient to control the algorithm numerical behaviour. One is used to specify the computational "grid" density while the other allows control of the accuracy of the series convergence. The term "grid" here refers to an arrangement of points suited for computational optimization and isotherm plots.

Setting the problem also requires the specification of the heat transfer coefficients on the board faces. This constant difficulty of thermal simulations is here circumvented by providing an assisting tool. The user interface of this tool allows the designer to characterize rapidly the board cooling configuration: natural convection on a horizontal or vertical board, parallel or perpendicular flow on a board cooled by forced convection and fluid temperature. Internally, the program uses the physical properties of air and common correlations between the relevant non-dimensional numbers to derive and propose heat transfer coefficient values. The user remains free to impose his own values.

Eq. (3) or equivalent formulations allow the computation of the board thermal map and indication of the location of hot spots. However, junction temperatures are the ones the designer is the most interested in. For that purpose, the program simply computes the temperature rise due to the power flow through the Junction-to-Case resistance R_{jc}, assuming that the component case is at the same temperature as the board at the component location. The R_{jc} values, extracted from manufacturer's data sheets are stored in the PCB Design Database. Maximum junction temperatures are also included in this database to immediately point out the components with excessive junction temperatures.

The thermal algorithm is tighly integrated with the Expert Series PCB Design system, which provides the capability to modify or re-evaluate the placement at any point in the Place/Route cycle.

3 Integrated thermal suite

Due to the usage of an analytical method, FloorPlanner 2000 is potentially able to yield very high precision results, provided its thermal assumptions are strictly satisfied: actually, the method was originally developed for test purposes [2]. However, it is obvious that real configurations may significantly deviate from the restrictive assumptions used by FloorPlanner 2000. So this product cannot compete with the flexibility and related accuracy FD (Finite Difference) or FE (Finite Element) methods can bring. Among the critical points, the specification of boundary conditions and the component modelling play a crucial role.

3.1 Specification of Boundary Conditions by System Analysis

Ideally, the boundary conditions applied to a board must reflect its environment as realistically as possible. For the widespread cases, this environment is made by an air flow, whose characteristics are inherently complex. Firstly, this flow is influenced by the 3D geometry of the enclosure. Secondly, heat transfer within enclosures can be very hard to predict. The example of a system cooled by natural convection illustrates this feature: the power dissipated by each component is the cause of the flow in the enclosure while the local characteristics of this flow at the level of each component govern the way their power is dissipated. In short, it is deceptive to use a high-precision discretisation method if the specification of boundary conditions is not performed at the same level.

To tackle this problem, the approach proposed by Racal-Redac is to integrate the System Analysis into the design cycle. First, the user determines the component placement of the PCB — using for instance FloorPlanner 2000. With the component placement determined, the board placement in the system is performed. The system description is completed by adding geometrical elements other than boards (power supply unit, deflectors, internal obstacles, ...) and by specifying the system operating conditions (fan characteristics, fluid inlets and outlets, ...). Using Flotherm, the overall flow within the system is determined. Inspection of the result allows decisions to be taken about boards which may overheat if for instance a flow stagnation zone exists in their vicinity. Then the local environment of the selected board is exported to VTAT (Expert Series Thermal Analysis Tool).

This local environment is mainly described through the fluid temperature and heat transfer coefficients maps, for each board face. So, when the PCB thermal simulation is performed in conjunction with a CFD system analysis, the user can choose by default to use these maps without alteration. He may also prefer to employ his own correlations by taking into account the fluid velocity maps also recovered from the CFD package for that purpose. In short, the user has control capabilities on the way the CFD results are exploited at the PCB analysis level.

From the practical point of view, it is obvious that the cooperation between softwares dedicated to PCB design, System Thermal analysis and PCB Thermal Analysis must be performed in an appropriate framework to manage in a simple and consistent way the generation and transfer of data between programs. Accordingly, the data transfer between applications is done within the Vision framework using the *Controlled Directory Structure* while the applications themselves are driven from a System Interface named *Viewtool*.

The integration of System Analysis, Board Analysis and Layout CAD is a major step for a better design optimisation, by reducing size, noise, packaging costs, etc... The combination of these analysis techniques contributes to overcome the tendency to over engineer, using larger and therefore noisier fans for example, in the hope of avoiding potential problems.

3.2 Component Modelling

The modelling of the component behaviour is one of the other key-points of PCB thermal simulations. Seen from a structural point of view, the constitution of a component is too complex to be highly detailed during a PCB analysis: that would lead to an untractable number of unknowns. This means that the component behaviour has to be processed through macro-models whose basic requirement is to provide realistic junction temperatures.

R_{jc}-based macro-models, the only ones available for FloorPlanner 2000, are the simplest models VTAT can use. However, studies have revealed that such models may be inaccurate [3]. Accordingly, VTAT proposes more complex models based on local resistive networks. During the meshing process, this network is connected to the FD-board model, through a set of resistances modelling the mounting gap, and to the environment, through resistances modelling the radiative and convective heat transfer.

Up to now, there is no general agreement on standard component models and same remark obviously holds true on their creation, which can be based on experimental measurements or on software-aided approaches. When this last method is used to create VTAT models [4], the first step consists in a fine structural description of the component inner constitution. This description is then discretized as if the component was about to be simulated, but the conductance matrix is actually reduced by a *condensation* [5] procedure. The conductances contained in this reduced matrix are the local network data. For the end-user this complexity is however irrelevant since the models are delivered in ready-to-use libraries.

4 Simulation example

The two preceding approaches are applied to a rack (Fig. 1) whose dimensions are $428 \times 350 \times 444$ mm. This device cooled by natural convection with air at 24 °C encloses two single-sided CPU boards inserted in slides and in a mother board. One of these CPU boards is dissipating 5.2 W and is equipped with thermocouples allowing to measure the package top and pin temperatures of certain components, as well as the temperature of the board surface at ten locations labelled *L1* to *L10* (Fig. 2). The board is made of epoxy glass ($0.24\,\mathrm{Wm^{-1}{}^\circ C^{-1}}$) and its dimensions are $324 \times 292 \times 1.6$ mm. 135 components are mounted on it, mainly of DIP type, with a mounting air gap of 0.4 mm. The results obtained are given in table 1. Figs. 3 and 4 give the flow within the enclosure and the PCB isotherms respectively generated by Flotherm and VTAT. [6] gives another application example where the comparisons are based on the component junction temperatures.

5 Conclusion

The two proposed approaches differ in the context of their use and so in the range of functionality they provide. Their respective main features can be emphasized as follows:

First approach		*Second approach*
Interactive evaluation of PCB thermal maps	*Objective*	Accurate determination of PCB thermal maps
Simultaneous with placement	*Context of use*	Post-layout analysis
Estimation of heat transfer coefficients by an interactive assisting tool	*Boundary conditions*	Computation of flow features and heat transfer coefficient maps by a CFD package
Fourier series expansion	*Computational method*	FD method
R_{jc}-based	*Component models*	Local resistive network
Combination of the placement and thermal functions in a unique product	*Integration*	Combined usage of PCB design system, CFD package and PCB simulator – Links by files in a controlled structure

6 References

Acknowledgments. *We wish to acknowledge MM. Pathechi, Klausas, Polidoro and Zoccato (consultant) from NECSY Network Control Systems SpA for their contribution to this work.*

1. Ottavy, N., *Méthode de superposition de maillages en éléments finis*. Thèse n° 493. Université de Poitiers, ENSMA, pp. 802–803 (1989).
2. Blanchard, J.L., Schirar, S., *Environnement de simulation thermique des cartes électroniques en éléments finis*. Third conference on heat transfer, energy and environment applied to communication and computer fields, Perros-Guirrec (june 1989).

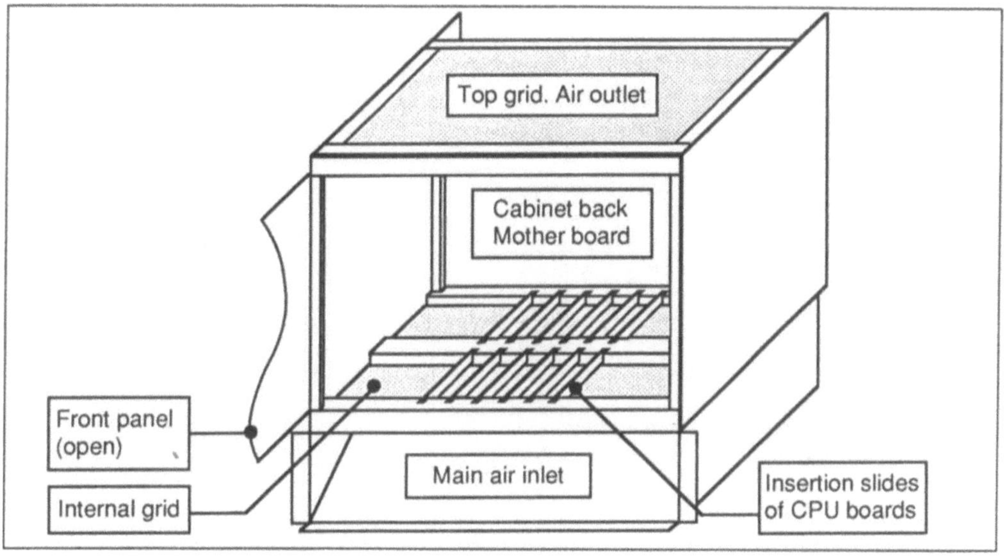

Fig. 1: Enclosure geometry

3. Le Jannou, J.P., Huon, Y., *Representation of Thermal Behavior of Electronic Components for the Creation of a Databank*. IEEE transactions on components, hybrids, and manufacturing technology, vol. 14, no. 2, pp. 366–373 (june 91).
4. Gauthier, T., *Construction and Validation of Thermal Model of Packages*, 7th annual IEEE Semitherm 91, pp. 74–79 (1991).
5. Huebner, K.H., *The Finite Element Method for Engineers*, Wiley&Sons, p. 155 (1975).
6. Blanchard, J.L., Morelle, J.M., *Two Thermal Simulation Approaches for Electronic Equipment*, 6th conf. on interconnection technologies for extreme integration, Milan, pp. 207–216, (sept. 1992).

Table 1a: Component comparative temperature (°C)

Location		Exp.	A1	A2
U3	Top	38.4	39.	36.8
U11	Top	39.3	34.	34.6
U14	Top	51.5	40.	52.3
	Pin	39.5		35.2
U20	Top	30.4	38.	28.1
	Pin	29.7		29.4
U44	Top	41.8	40.	44.1
	Pin	33.5		32.8
U45	Top	37.3	36.	39.7
	Pin	29.6		31.1
U136	Top	35.6	36.	36.0

Location: See Fig. 2
Exp.: Experimental results
A1: First approach (FloorPlanner 2000)
A2: Second approach (CFD system analysis & FD-based PCB thermal simulator)

Table 1b: Board comparative temperature (°C)

Location	Exp.	A1	A2
L1	34.9	37.7	32.2
L2	33.2	33.7	31.7
L3	29.7	33.5	30.2
L4	29.3	33.2	32.2
L5	29.1	33.1	31.9
L6	27.4	33.0	29.8
L7	26.9	33.1	29.5
L8	25.6	33.0	28.7
L9	25.4	32.8	28.1
L10	25.1	33.1	28.4

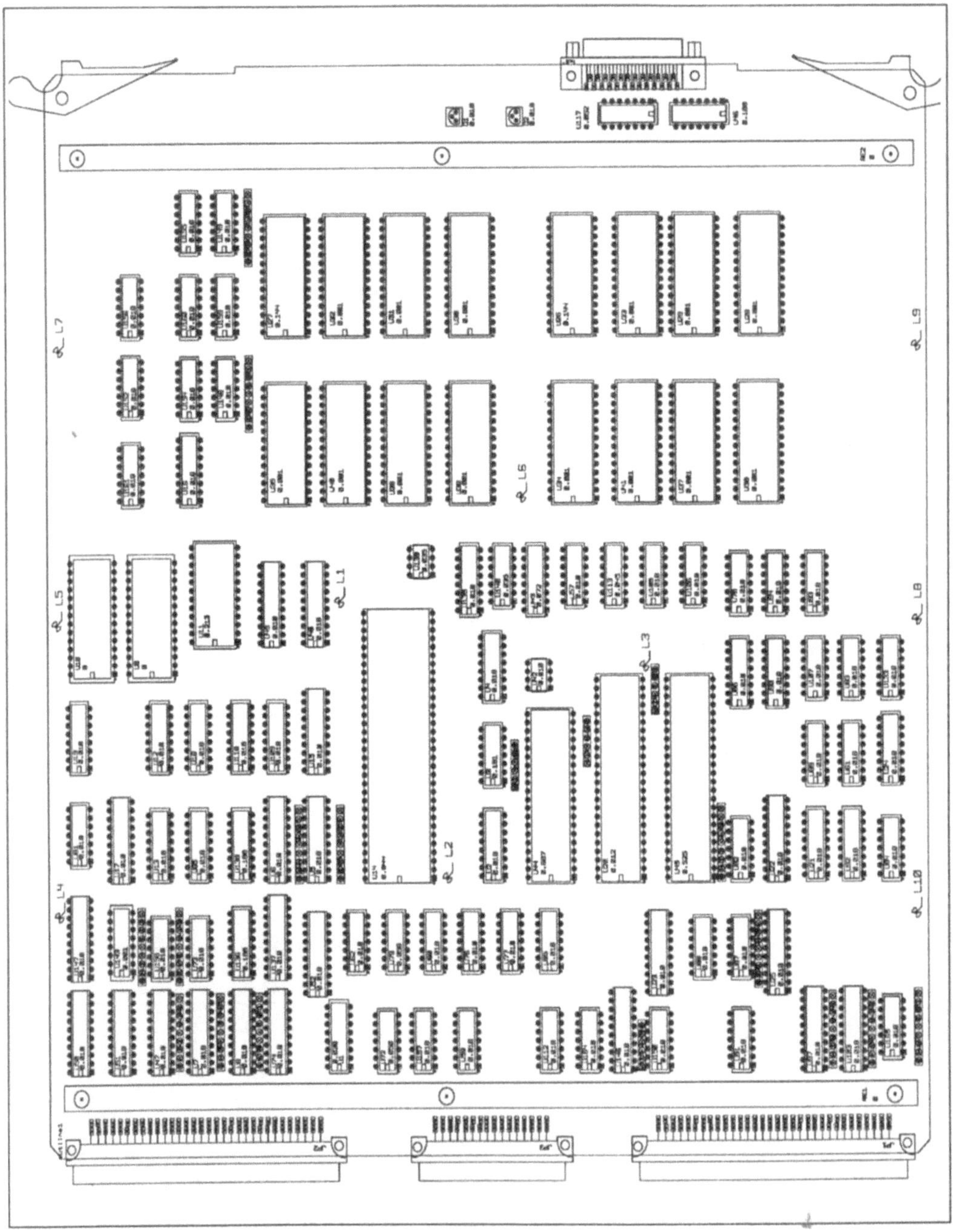

Fig. 2: PCB geometry

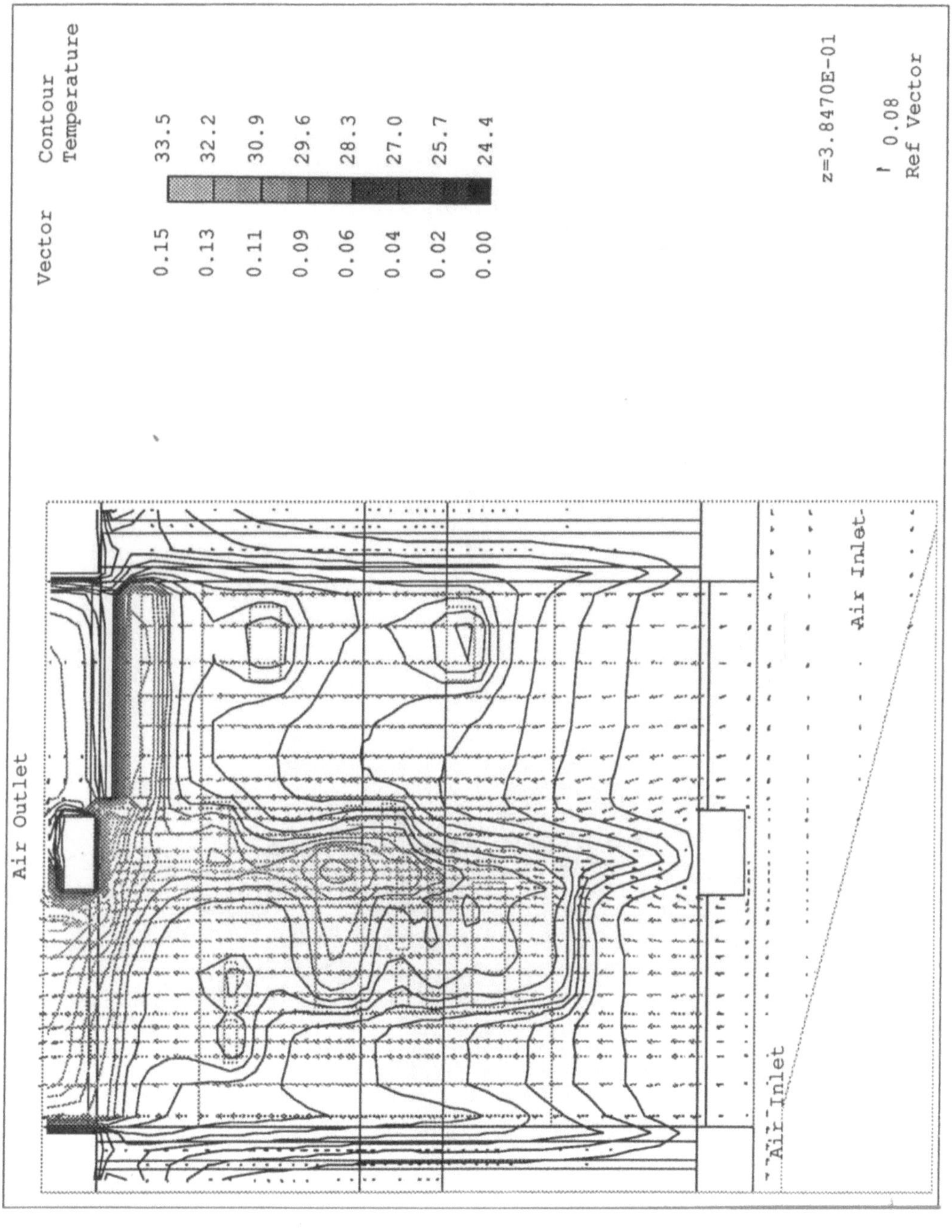

Fig. 3: Velocity map. Fluid Isotherms

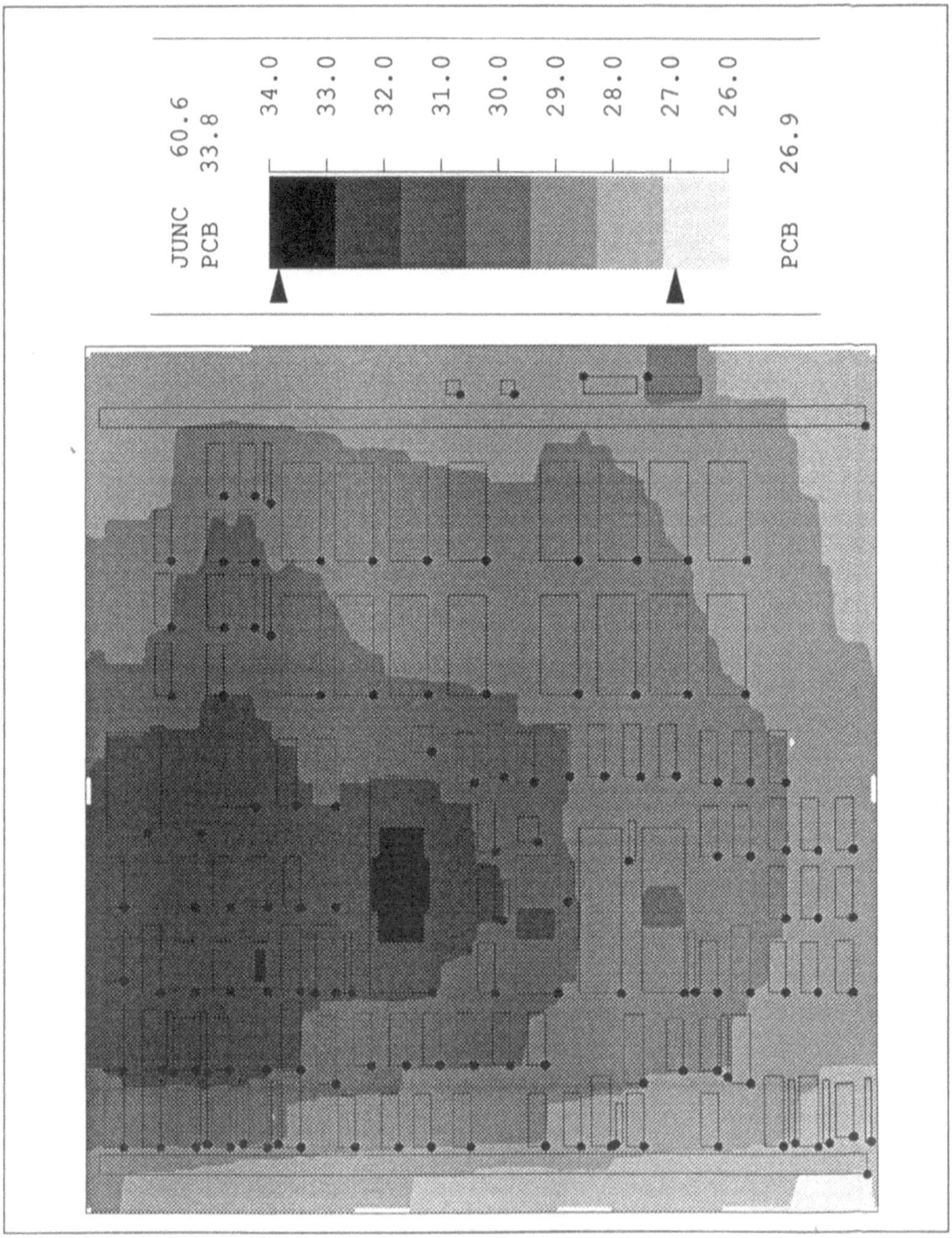

Fig. 4: PCB isotherms

Some considerations on practical formulas used for thermal design of natural ventilated enclosures

M. MISALE
Energy Engineering Department, University of Genoa, Via All'Opera Pia 15/a (I) 16145 Genova - Italy

Abstract - In this paper four practical formulas which are used for the thermal design of natural ventilated enclosures will be presented. For each correlation the main parameters involved such as the inner air temperature distribution, the coefficient of fluid resistance in the vents, the chimney height, and so on will be discussed. The theoretical volumetric flow rate evaluated with the four correlations have been compared with experimental data available in literature.

1. Introduction

Natural air convection is commonly applied as a cooling technique for electronic equipment of moderate power density. This heat transfer mode is interesting because of its intrinsic reliability, its simplicity, and its nearly zero operational cost.

However, due to the relatively low efficiency of this cooling technique, the thermal design of the electronic equipment must be optimized, i.e. the distribution of the heat sources and the ventilated enclosure must be selected such as to generate an air flow rate which minimizes the average and the maximum temperature rise inside the enclosure itself.

In fact, as well known the thermal behavior of a natural ventilated box depends on the balance between the net buoyancy force and the friction losses along the whole air path. In particular different correlations are reported in literature which can be used for thermal design of electronic equipment.

In this paper four relationships will be considered : General Electric [1], Ellison [2], Ishizuka et al. [3], and Guglielmini et al. [4]. Each correlation will be described evidencing the influence of the main parameters involved such as coefficient of friction losses, chimney height, inner air temperature distribution, and so on.

C. J. Hoogendoorn et al. (eds.), Thermal Management of Electronic Systems, 85–94.

2. Practical formulas for the thermal design of natural ventilated enclosures

In this part of the paper the four correlation considered will be described.

2.1 General Electric correlation [1]

The physical model of the electronic equipment adopted in [1] is schematically shown in Fig. 1. The cabinet consists of an enclosure provided of an inlet and outlet vent (hole shape) and uniform heat sources which simulate the electronic components.

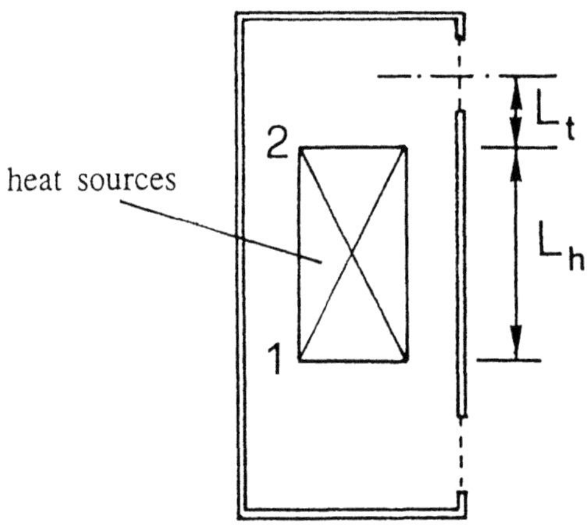

Fig. 1 - General Electric vent model

The correlation proposed in [1] is based on the concept of "critical free area". It is defined as the area just large enough to result in an out flow through of the holes in the upper part of the cabinet, that is just equal to the total boundary layer flow from the heat dissipating surfaces.

In [1] two correlations are reported, both valid in the case of atmospheric pressure, ambient air temperature T_0 not far from 21 (°C), and an air temperature rise less than 28 (°C).

The first correlation is applicable when the inlet vent area is the same as the outlet vent area:

$$A_0 = \frac{\sqrt{2\ C_D}\ \ \dot{V}}{127.2\ \sqrt{\Delta p_B^*}} \tag{1}$$

while the second correlation (2) is valid when the inlet vent area is much larger than the outlet vent area:

$$A_o = \frac{\sqrt{C_D}\ \dot{V}}{127.2\ \sqrt{\Delta p_B^*}} \tag{2}$$

In the expressions (1) and (2) the terms involved are: A_o the outlet vent area (m^2), C_D the flow resistance coefficient of the vent (dimensionless), $\dot{V}$ the volumetric air flow rate (m^3/s), and Δp_B^* the buoyancy pressure difference caused by the warm air (m H_2O). Δp_B^* can be obtained from the sum of $(\Delta p_B^*)_{L_t}$ and $(\Delta p_B^*)_{L_h}$. The expressions for Δp_B^* in the two regions in which has been divided the enclosures (see Fig. 1) are:

$$(\Delta p_B^*)_{L_t} = \frac{0.355\ \Delta T_a\ L_t}{T_o\ (T_o + \Delta T_a)} \tag{3}$$

$$(\Delta p_B^*)_{L_h} = \frac{0.178\ \Delta T_{1,2}\ L_h}{T_o\ (T_o + \Delta T_{1,2}/2)} \tag{4}$$

where L_t (m) is the height of the region containing air at uniform elevated temperature $\bar{T}_a$ in respect to the ambient temperature T_o ($\Delta T_a = \bar{T}_a - T_o$ (K)), L_h (m) is the height of the region with linear air temperature, and $\Delta T_{1,2}$ (K) is the rise in air temperature from plane 1 to plane 2 (see Fig. 1).

2.2 Ellison correlation [2]

In Fig. 2 the drawing of the simple cabinet considered by Ellison is depicted. The electronic equipment consists of an enclosure provided of two equal vents and some electronic cards.

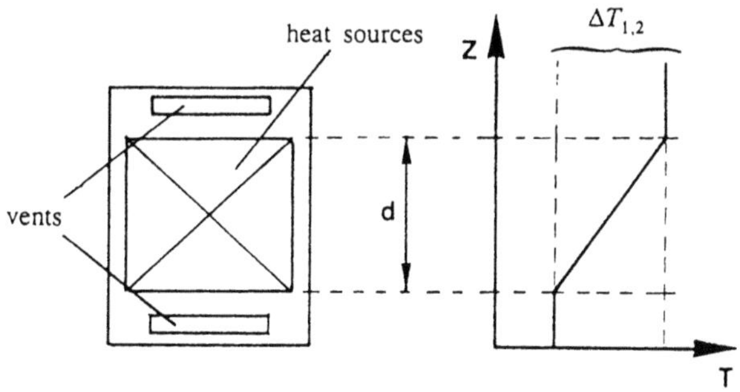

Fig. 2 - Ellison vent model and inner air temperature distribution.

The correlation is developed assuming a linear air temperature rise in the region where are placed the heat sources. The expression for the relationship is:

$$T_0 R\dot{V}^3 + 1.415\ 10^{-6}\ P'\bar{T}_B R\dot{V}^2 - 1.698\ 10^{-9} P' d\bar{T}_B = 0 \tag{5}$$

where T_0 is the inlet air temperature (K), R is the total resistance of the air flow (m $H_2O/(m^3/s)^2$). In particular Ellison considered only the concentrated pressure losses in the two vents, neglecting the distributed pressure losses along the electronic card channel. The R expression, in the case of two vents having the same area , is:

$$R = 2\,\frac{1.388\ 10^{-4}}{A^2} \tag{6}$$

where A is the free inlet or outlet vent area (m^2). The other terms involved in the (5) represent: $\dot{V}$ the volumetric air flow rate (m^3/s), P' the heat removed by air (W), $\bar{T}_B$ the well mixed inner air temperature (K), and finally d the chimney height which coincides with the dissipation height card (m). Equation (5) can be solved in only a few iterations by Newton's method.

2.3 Ishizuka et al. correlation [3]

The study conducted in [3] permitted to develop a correlation which takes into account either the conduction heat flux transferred to the ambient by the cabinet walls or the heat flux removed by air. In particular the part of the correlation that involves the air draft is reported here. The casing model is represented in Fig. 3.

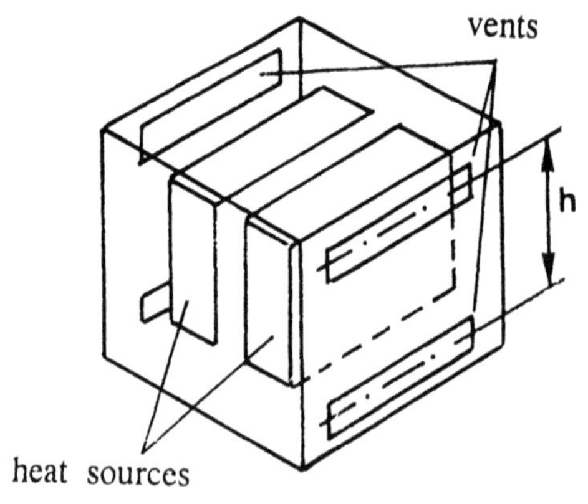

Fig. 3 - Ishizuka et al. vent model.

The relationship developed assuming the "vent model" is:

$$\dot{V}= A_0 \sqrt{\frac{2\ gh\ \Delta T_0}{T_0\ K}} \tag{7}$$

where $\dot{V}$ is the volumetric air flow rate (m^3/s), A_0 is the outlet free area (m^2), g is the gravitational acceleration (m/s^2), h is the chimney height, coincident with the distance between the center lines of the inlet and outlet vents (m), ΔT_0 is the difference between the outlet air temperature and the inlet one (K), T_0 is the inlet air temperature (K), and K is the coefficient of fluid resistance (dimensionless). Since the "vent model" assumes that the inlet vent area is larger than the outlet vent area, then K takes into account only the pressure losses in this last opening. To evaluate its value it is possible to use this expression:

$$K=2.5\ \frac{1-\beta}{\beta^2} \tag{8}$$

where β (dimensionless) is the throat area ratio (porosity).

2.4 Guglielmini et al. correlation [4]

The ventilated enclosure analyzed is reported in Fig. 4. It consists of a cabinet provided with two vents and containing six uniformly heated circuit board.

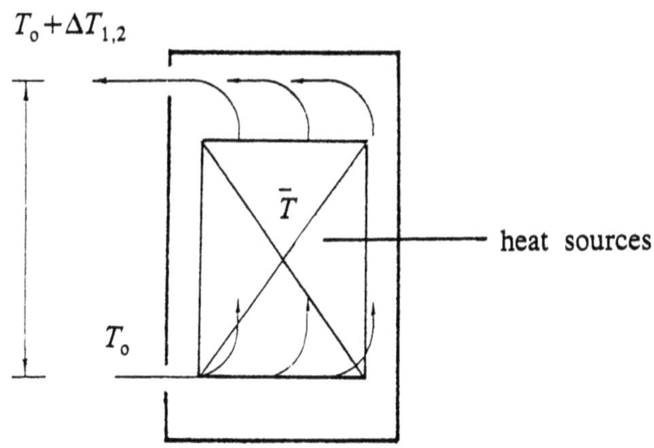

Fig. 4 - Guglielmini et al. vent model.

The correlation developed is:

$$\dot{V}^3 = 5.51\ 10^{-5}\ \frac{A_o^2\ h\ P'}{K_e\ C}\ \frac{\bar{T}}{T_o} \tag{9}$$

in which the symbols represent: $\dot{V}$ is the volumetric flow rate (m^3/s), A_o the outlet vent area (m^2), h the chimney height (m) (i.e. the distance between the centre lines of the vents), $\bar{T}$ the integral average air temperature over the whole volume of the box (K), C an experimental constant (dimensionless) defined as the ratio between the inlet-outlet difference $\Delta T_{1,2}$ and the mean inner air excess $\overline{\Delta T}$, and K_e the equivalent fluid resistance coefficient (dimensionless). K_e takes into account the distributed pressure losses along the electronic cards (K_{ch}) and the concentrated pressure losses for the two openings including the bending in the air path close to the vents (K_{con}); hence in the case of two equal vents $K_e = K_{ch} + 2K_{con}$. Since in literature correlations able to evaluate the K_{ch} term are not available then it is possible to utilize the relationship reported in [5] valid for forced laminar convection to calculate the K_{ch} values. To evaluate the K_{con} values it is possible to use some correlations, deducted for different vent shapes (circular holes, horizontal slots, and vertical slots) and two values of porosity β [6]. These expressions are valid in the range of Reynolds number of $30 \leq Re_{dh} \leq 120$ (Re_{dh} is based on hydraulic diameter of the single opening). In the same paper an expression is reported which permits to evaluate the C value in function only of the power removed by air P'. The formulas of $K_{con} = K_{con}(Re_{dh})$ and $C = C(P')$ are reported below:

Circular holes $\qquad K_{con} = 22.4\ Re_{dh}^{-0.58}$ (10)

Horizontal slots

$\beta = 0.2 \quad ; \quad K_{con} = 19.5\ Re_{dh}^{-0.54}$ (11)

$\beta = 0.33 \quad ; \quad K_{con} = 8.6\ Re_{dh}^{-0.38}$ (12)

Vertical slots

$\beta = 0.2 \quad ; \quad K_{con} = 14.4\ Re_{dh}^{-0.48}$ (13)

$\beta = 0.33 \quad ; \quad K_{con} = 8.2\ Re_{dh}^{-0.37}$ (14)

$C = 1.22 + 0.08\ \text{Log}\ P'$ (15)

The correlation (15) is valid for all vent geometry investigated in [6].

3. Discussion

As reported in the introduction the thermal behavior of a ventilated case depends on many parameters. However, as it can be noted from the correlations reported before, the authors suggest simple formulas which allow to perform the thermal design of electronic equipment. Since the internal air flow along the electronic path is very complex only by adopting very simple models it is possible to develop these practical formulas.

Analyzing the four correlations considered it turns out that all the expressions contain parameters such as the inner air temperature distribution, volumetric flow rate, chimney height, vent area, and friction losses.

More attention must be paid to the evaluation of the friction parameter. In fact this term allows to take into account all the pressure losses along the air path: the distributed pressure losses in the hot channel formed by electronic cards, the concentrated friction losses in the inlet-outlet vents and in the bending of the air path close to the vents. In many cases the card channel produces a negligible effect [2] while the vent loss is the principal parameter that must be considered in the correlations used for the thermal design of electronic cabinets. Nevertheless if the opening area is relatively large compared to the cross section area of the channels (in the case of vertical cards), the distributed pressure losses are generally not negligible with respect to those concentrated near the openings, but its contribution is comparable to the concentrated friction losses [4].

Concerning the concentrated friction losses due to the change of air direction (after the inlet vent and before the outlet vent) none of the correlations gives any quantitative information about its evaluation. For the concentrated friction losses in the vents all relations indicate how it is calculated. These indications are:

- General Electric - (Vent shape: circular holes in sheet metal wall). $2.2 \le C_D \le 2.6$ for Reynolds number of about 200; however in [1] is recommended, as a conservative value: $C_D = 2.8$;
- Ellison - (perforated or slotted plate). The total resistance of coefficient R utilized by Ellison, in the case of two equal hole vents, is just reported before. Nevertheless, transforming this expression to the classic fluid kinematic energy equation ($K_{EL} \rho u^2/2$) it is found that the value of the coefficient of vent resistance K_{EL} is equal to 2.26. This value is comparable with the C_D value utilized in the General Electric correlation.
- Ishizuka et al. - (vent shape: circular holes). An important parameter that is not present in the previous expressions is the vent porousness. In fact, this last quantity determines significant changes in the K value. The relation (8) allows to calculate the K values in the range between $\beta = 0.3$ up to 0.5 that are the ones usually adopted in the electronic applications. Thus the extreme β values correspond to K values of 19.4 and 5.0 respectively. It is important to note that in [7] another K relationship is reported which involves the porousness β and the Reynolds number. However, utilizing the relation (8) or the formula reported in [7], both expressions give values for the vent resistance coefficient larger than the ones employed in [1], [2], [4], and [6].
- Guglielmini et al. - (vent shapes: circular holes, horizontal slots, and

vertical slots). After preliminary information on the coefficient of fluid resistance [4], the vent geometry such as vent area, porosity, and hydraulic diameter has been studied in [6]. Assuming some simplifications, the coefficient of fluid resistance in the vents has been deduced by relating K_{con} to relation with the Reynolds number *Re* based on the hydraulic diameter and on the air velocity in the opening. The expressions obtained are valid in the range of Reynolds number between 30 up to 120 . The coefficient of fluid resistance, in the case of circular holes and in range of Reynolds number of $30 \leq Re \leq 120$ are of 3.12 and 1.39 respectively.

The coefficient of fluid resistance is quite similar for the correlations (1), (6), and (10-14) while it is sensibly lower than calculated by (8). This remarkable difference will give larger discrepancies for the values of the volumetric flow rate, calculated by the different correlations.

Another important parameter that can be involved in the practical formulas for thermal design of electronic equipment is the average inner air temperature. In [1] and [2] a linear air temperature rise, in the region where the electronic components are placed, has been assumed; in [3] the difference between the outlet air temperature and the ambient temperature is involved without specifying the inner air temperature distribution; in [4] a conventional average air temperature has been defined, which takes into account the air temperature distribution inside the cabinet and which utilizes an experimental constant *C*. Its definition is reported before (§ *2.4*).

The *C* value depends on the vent area [2] or can be a function only of the power removed by air [6]. However in [2] it has been recommended to assume a constant value of 1.3 for every ventilated box investigated; even if in [6] it has been reported a law of the kind $C=C(P')$ the value of *C* pointed out in [2] corresponds to the average *C* value introduced in [6]. Utilizing the *C* definition illustrated in § *2.4*, it is interesting to note that with the temperature distribution assumed in [1] and [2] *C* is equal to 2. The assumption of a *C* value equal to 2, i.e. linear slope of temperature, does not agree with the measurement reported in [4], [8], and [9]. In these three papers some air temperature and electronic surface temperatures are reported that are almost never linear.

Another parameter involved in the correlations is the chimney height. For this term there are not great differences among these practical formulas in which this term is well defined. Only a comment can be made on the Ellison correlation; in fact in this correlation the chimney height is coincident with the region where the electronic components are placed. No information is given for the region above these components. There is also no information about how that region can be considered for the thermal design of the ventilated enclosures. Obviously in the case of inlet and outlet vents placed at the extreme limit of the region where the electronic components are located there is no difference in the chimney height.

Finally it is noted that the theoretical results obtainable by these practical formulas would be compared to experimental data; thus their applicability can be extended to natural ventilated enclosures having different aspect ratio (ratio between height, width, and depth).

Having discussed the influence of some parameters involved in the practical formulas utilized for the thermal design of natural ventilated enclosures, it is interesting to compare these relationships with experimental data. The measurements were obtained for a natural ventilated cabinet of the following dimensions: 0.254 (m), height, 0.1 (m), width, and 0.152 (m) depth. The box was provided with two equal vents and it contained six uniformly heated cards [6].

The comparison can be conducted in a homogeneous way only in the case of vents realized with circular holes. In fact this vent geometry has been assumed in the different models proposed by the authors.

As described previously, the correlation (7) has been deducted for a box with an inlet vent area that is larger than outlet vent area. Working in this way only the outlet concentrated pressure losses have been involved in the formula. Since the experimental data utilized during the comparison have been obtained on a box with two equal vent areas, the coefficient of vent resistance K utilized in (7) has been multiplied by 2.

The result of this comparison is reported in Fig. 5. The X-axis gives the experimental flow rate $\dot{V}_{exp}$ while the Y-axis gives the volumetric flow rate $\dot{V}_{th}$ calculated with the practical formulas.

The best agreement between the experimental data and the theoretical results is obtained utilizing the correlation (9) while the relationships (1), (5), and (7) always give a theoretical air flow rate lower than the experimental data. The good behavior of formula (9) can be explained from the fact that the coefficient of fluid resistance in the vents has been deducted starting from the experimental data utilized for the comparison. The sensible discrepancies observed for the formula (7) are determined by the higher values of the coefficient of fluid resistance in the vents. Besides the small difference between the data obtained in eq.s (1) and (5) is determined by the values of the coefficient of fluid resistance introduced in the equations: eq. (1) --> C_D=2.8, eq. (3) --> K_{EL}=2.26.

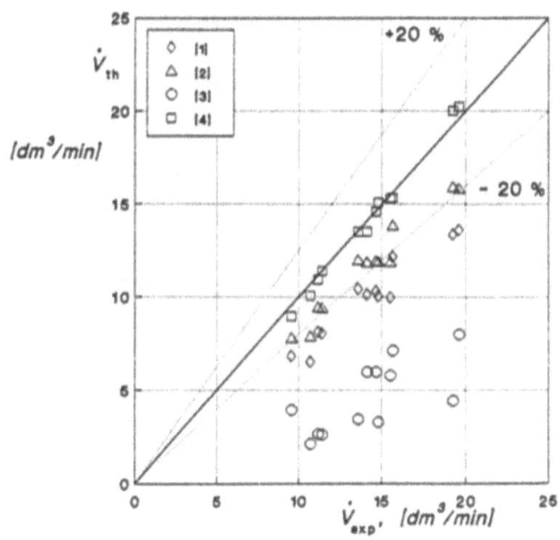

Fig. 5 - Comparison between experimental volumetric flow rate $\dot{V}_{exp}$ and the theoretical ones $\dot{V}_{th}$.

4. Conclusions

The use of four practical formulas for thermal design of natural ventilated enclosure containing electronic components has been illustrated. In particular the paper has shown the evaluation of some parameters such as: coefficient of fluid resistance in the vents, chimney height, and average air inner temperature. Further research should give information on the applicability of the correlations for different aspect ratios of the cavity and, for the distributed pressure along the electronic cards, and for the concentrated pressure losses in the change air path after the inlet vent and before outlet one.

Acknowledgements
This work has been supported by M.U.R.S.T. 60%.

REFERENCES

1. General Electric, *Heat Transfer and Fluid Flow-Data Books*, Section 504.4, 1-4, April, 1974.
2. Ellison, G.N., *Thermal Computations for Electronic Equipment*, Van Nostrand Reinold Company, New York 1984. Re-print edition by Robert Z. Kreises Publishing Company, Malabar, Florida, 1989.
3. Ishizuka, M., Miyazaki, Y. and Sasaki, T., On the Cooling of Natural-Air-Cooled Electronic Equipment Casing, *Bulletin of JSME,* Vol. 29, N. 247, pp. 119-123, 1986.
4. Guglielmini, G., Milano, G., and Misale, M., Some Factors Influencing the Optimum Free Air Cooling of Electronic Cabinets, in *Heat Transfer in Electronic and Microelectronic Equipment*, A.E. Bergles, Eds., Hemisphere Publications Corporation, pp. 311-325, New York, 1990.
5. Brater, M.E., and Patankar, S.V., Analysis of Laminar Mixed Convection in Shrouded Arrays of Heated Rectangular Blocks, *Int. J. Heat Transfer*, Vol. 28, N. 9, pp. 1699-1709, 1985.
6. Misale, M., Electronic Cabinet Cooling by Natural Convection: Influence of Vent Geometry, the paper will be presented at *3rd World Conference on Experimental Heat Transfer, Fluid Mechanics and Thermodynamics*, Honolulu, Hawaii, USA, Oct. 31-Nov. 5, 1993.
7. Ishizuka, M., Miyazaki, Y. and Sasaki, T., Air Resistance Coefficients for Perforated Plates in Free Convection, *Transaction of the ASME,* Vol. 109, pp. 540-543, May 1987.
8. Birnbreier, H, Experimental Investigation on Temperature Rise of Printed Circuit Boards in Open Cabinets with Natural Ventilation, *Heat Transfer in Electronic Equipment*, HTD. Vol. 20,pp.19-23, 1981.
9. Kato, K., Takarada, T., Miyazaki, H., Sato, H., and Nakagawa, N., Heat Transfer in a Channel Between Vertical Electronic Circuit Boards Cooled by Natural Air Convection, *J. of Chemical Engineering of Japan*, Vol. 24, N. 5, pp. 568-574, 1991.

3. NUMERICAL AND EXPERIMENTAL ANALYSIS OF CHANNELS

Natural Convection in a rectangular Enclosure with an Array of Chips Mounted on a Vertical Wall

C. DI PERNA, A. EVANGELISTI, M. PARONCINI, R. RICCI.
Dipartimento di Energetica - Università di Ancona , I- 60100
ANCONA - tel +39-71- 894780 - fax +39-71- 894677

Abstract Experimental and numerical studies were performed on heat transfer associated with a laminar free convective flow in a rectangular enclosure non-uniformly heated by an in line array of chips mounted on a vertical surface. A finite difference scheme was employed for computation: the QUICK method was used to solve the convective terms in energy and momenton equations, while time discretization was performed by means of explicit method. Holographic interferometry techniques were used to measure the temperature distribution in the air layer. Heat transfer coefficients were obtained for two different values of enclosure aspect-ratio. Experimental and numerical results were compared.

1. Introduction

The miniaturization of Integrated Circuits and the resulting increase in power density have raised the importance of the role played by heat transfer in electronics. Microelectronic circuits designers face the problem of maintainig the junction temperature below a predetermined limit, because the failure rate of microelectronic chips increases exponentially with the junction temperature increase. In applications involving low power and packaging densities, natural convection is a convenient and inexpensive way to cool a system. In these applications, the free convective heat transfer in confined spaces, as ducts, enclosures and cells, with distributed or concentrated heat sources, is of particular interest. In the last years increasing attention has been devoted to the issue of cooling multiple heaters mounted on the same vertical wall. Heat transfer by natural and forced convection in channels from an array of small heaters has been studied by many authors. Chu

C. J. Hoogendoorn et al. (eds.), Thermal Management of Electronic Systems, 97–106.

and Churchill [1] studied theoretically and experimentally the effect of both size and location of an isothermal horizontal strip on an insulated vertical surface of a rectangular channel. Ortega and Moffat [2] experimentally studied heat transfer form an array of simulated electronics components Recently Shaw et al. [3] numerically studied the laminar mixed convective heat transfer in a three dimensional channel with a thermal source. Magnusson et al. [4] experimentally studied, for low Reynolds numbers, the mixed natural and forced convective heat transfer in a channel with Integrated Circuits as heating sources. The investigation of heat transfer in a rectangular enclosure from an array of small heaters has received only limited attention. Poulikakos [5] reported a numerical simulation of free convection flow and heat transfer in an enclosure heated and cooled along a single vertical wall. Kuhn and Oosthuizen [6] presented a study concerned with three-dimensional unsteady natural convective flow and heat-transfer in a square enclosure wich has one vertical wall partially heated and the opposite wall kept at a lower uniform temperature. Cesini et al. [7,8] studied experimentally and numerically the heat transfer associated with a laminar free convective flow in a square or rectangular enclosure non-uniformly heated by isothermal strips mounted on a vertical surface. This paper reports experimental and numerical investigation on laminar natural convective flow in a rectangular enclosure with an array of chips mounted on one vertical wall, when the opposite wall is uniformly cooled; both top and bottom surfaces are adiabatic. The temperature distribution in the air layer is measured using holographic interferometry thecniques. Heat transfer coefficients are computed for two different values of enclosure aspect-ratio. The numerical study is carried out by solving the two-dimensional governing equation for laminar natural convection by finite difference methods. QUICK methods are used to solve the convective terms in energy and momentum equations, while time discretization is obtained by means of explict method. Numerical correlation of heat transfer coefficients are evaluated Numerical and experimental results are compared.

2. Experimental apparatus

The experimental apparatus is represented in fig.1. It consists of a test cell and a holographic interferometer. The test cell is a rectangular enclosure filled with air. The right vertical wall is a test plate. A schematic drawing of the test plate is showed in fig.2. The opposite vertical wall is an aluminium cold plate cooled by water circulation through a jacket attached to the back surface. The top and bottom surfaces of the enclosure are made of plexglass to model an adiabatic wall, while the remaining vertical walls are glass windows to accomodate holographic interferometry. The test plate is a printed circuit board upon which an array of chips is mounted. The array consists of 3 horizontal rows of 17 chips each. Beside the P.C.B., on the side without components, is a thermally insulating layer to reduce thermal loss through the P.C.B. Each chip is a 16 pin Plastic Dual Inline Package commercial resistor network. Each row is heated applying voltage from a DC power supply unit through a couple of wires running under the chips on the board. The test plate is purposely designed to produce a nearly bidimensional thermal field.Thermal analysis show that the maximum temperature difference between chips on the same row, supplied with tre same power, is less than 4% of the difference between the mean row and ambient temperatures. The dimension of the test cell in the direction parallel to the laser beam is large enough that end effects can be neglected. Then the air layer may be considered as a two-dimensional phase object. The temperature distribution in the air-layer is measured by a holographic interferometer. The ligth source is an Argon-ion laser with power rating of 5 Watts with etalon for the 514.5 nm wavelength. Both object and reference beams have a maximum diameter of 0.15 m. The optical set-up allows to use both double-exposure and real time holographic interferometry thechniques. The first method is used for steady-state measurements, the second one to study the dynamic evolution of transfer processes. A copper constantane thermocouple is located in the core region of the air layer to provide a reference temperature for the evaluation of the interferograms. Temperature measurements are not corrected for radiation effects, because of the low temperature differencies involved. The interferograms are experimentally evaluated by using a travelling microscope to investigate the recorded light intensity distribution. Density and temperature distribution are obtained by usual methods of inversion.

3. Numerical model

To solve, numerically, the partial differential equations connected with the physical problem we have used a finite difference technique based on the solution of Vorticity-Streamfunction formulation. The differential equations solved numerically are:

Vorticity-Streamfunction Equation

$$\frac{\partial^2 \Psi'}{\partial x'^2} + \frac{\partial^2 \Psi'}{\partial y'^2} = - \varpi'$$

Momenton Equation

$$\frac{\partial \varpi'}{\partial t'} + \frac{\partial}{\partial x'}(u'\varpi') + \frac{\partial}{\partial y'}(v'\varpi') = \frac{\partial^2 \varpi'}{\partial x'^2} + \frac{\partial^2 \varpi'}{\partial y'^2} + Gr \cdot \frac{\partial \theta}{\partial x'}$$

Energy Equation

$$\frac{\partial \theta}{\partial t'} + \frac{\partial}{\partial x'}(u'\theta) + \frac{\partial}{\partial y'}(v'\theta) = \frac{1}{Pr} \cdot \left(\frac{\partial^2 \theta}{\partial x'^2} + \frac{\partial^2 \theta}{\partial y'^2} \right)$$

The momenton and energy equations consist of similar differential terms. Therefore it is convenient to introduce a general variable, ϕ, and apply the same numerical algorithm on it; the general form is:

$$\boxed{\frac{\partial \phi}{\partial t'}} + \underbrace{\boxed{\frac{\partial}{\partial x'}(u' \cdot \phi) + \frac{\partial}{\partial y'}(v' \cdot \phi)}}_{\text{Convective Term}} = \underbrace{\boxed{A \cdot \left(\frac{\partial^2 \phi}{\partial x'^2} + \frac{\partial^2 \phi}{\partial y'^2} \right)}}_{\text{Diffusive Term}} + \underbrace{\boxed{S_\phi}}_{\text{Source Term}}$$

where : $\phi = q$ if $A = 1/Pr$ ed $S\phi = 0$

$\phi = w'$ se $A = 1$ ed $S\phi = Gr \cdot \partial q / \partial x'$

Integrating the equation on a control volume ($\Delta x' \cdot \Delta y'$) we obtain:

$$\frac{\partial \bar{\phi}}{\partial t'} \Delta x' \Delta y' + [(u' \phi)_r - (u' \phi)_l] \Delta y' + [(v' \phi)_t - (v' \phi)_b] \Delta x' =$$

$$= A \left\{ \left[\left(\frac{\partial \phi}{\partial x'} \right)_r - \left(\frac{\partial \phi}{\partial x'} \right)_l \right] \Delta y' + \left[\left(\frac{\partial \phi}{\partial y'} \right)_t - \left(\frac{\partial \phi}{\partial y'} \right)_b \right] \Delta x' \right\} + \bar{S}_\phi \Delta x' \Delta y'$$

Diffusive and source terms were discretized by central difference scheme. We used a QUICK (Quadratic Upstream Interpolation for Convective Kinematics) method, [9,10], on a Staggered Grid 40x30. For the grid points near the wall, where the velocity is very low, we have used SUDS (Second Upwind Difference Scheme) method, [11]. The SUDS-QUICK coupling was very efficient to solve our problem. The transient terms were discretized by an explicit finite difference technique; the maximum time step has been evaluated by applying the "infinite norm less than one" criterion to the coefficients' matrix.

4. Results

A number of computations have been carried out to verify the ability of the numerical code to describe the convective heat process inside the enclosure. Numerical results have been compared with the experimental data for $A^*= 2$ and $A^*= 3.3$. The Rayleigh number ranges between $5.8 \cdot 10^6$ and $1 \cdot 10^8$ for the numerical study, and between $1 \cdot 10^7$ and $7 \cdot 10^7$ for the experimental study. Fig. 3 shows typical interferograms obtained for $Q^*= 143\ W \cdot m^{-2}$. The holograms were obtained with the interferometer positioned in infinite fringe field, in order to display isothermal contours. The experimental temperature distribution can be directly compared with that obtained numerically for the same Rayleigh number, fig. 4. Data have been correlated employing the least-square technique. Results are showed in the tables below. Each pair of tables refers to a value of the aspect-ratio. The first table reports the correlations obtained from both numerical and experimental data with the mean percent standard deviation between data and correlations. The second table reports the mean percent standard deviation between numerical and experimental correlations. Fig.6 reports numerical and experimental results for the upper array, when $A^* = 2$. A first consideration related to both experimental and numerical results concerns the thermal influence of a single row of heaters on the others: downstream rows are hotter than upstream ones, although they are supplied with the same power. This phenomenon is partly due to air stratification near the top wall of the test cell; however it also happens when the test plate is positioned vertically in free air and stratification is non-existent. The temperature of the air flowing over the upstream rows rises, and this effect is not compensated by the buoyancy induced air velocity, referring to a row to row spacing of 2.5 cm. Air stratification and heating, cause a lower Nusselt number for the downstream rows with respect to the upstream ones, as evicts from all measures.

A* = 2	$9.721 \cdot 10^6 < Ra < 1.664 \cdot 10^8$		$1.524 \cdot 10^7 < Ra < 1.244 \cdot 10^8$	
	Numerical Correlations	Δ % Numer	Experimental Correlations	Δ % Exper
UP	$Nu = 0.021 \cdot Ra^{0.293}$	1.794	$Nu = 0.047 \cdot Ra^{0.242}$	3.622
MID	$Nu = 1.047 \cdot Ra^{0.087}$	1.091	$Nu = 0.348 \cdot Ra^{0.141}$	1.800
DOWN	$Nu = 0.598 \cdot Ra^{0.136}$	0.923	$Nu = 1.223 \cdot Ra^{0.086}$	1.789

A*=3.3	$5.816 \cdot 10^6 < Ra < 9.953 \cdot 10^7$		$9.098 \cdot 10^6 < Ra < 7.473 \cdot 10^7$	
	Numerical Correlations	Δ % Numer	Experimental Correlations	Δ % Exper
UP	$Nu = 0.002 \cdot Ra^{0.393}$	5.040	$Nu = 0.003 Ra^{0.365}$	2.498
MID	$Nu = 0.019 \cdot Ra^{0.297}$	1.435	$Nu = 0.016 Ra^{0.299}$	5.436
DOWN	$Nu = 0.009 \cdot Ra^{0.356}$	1.161	$Nu = 0.003 \cdot Ra^{0.356}$	2.956

Tab. 1 - Correlations from numerical and experimental data

A*= 2		**A*= 3.3**	
	Δ % Between Numerical and Experimental Correlations		Δ % Between Numerical and Experimental Correlations
UP	8.171	UP	15.898
MID	13.361	MID	10.729
DOWN	15.2	DOWN	20.492

Tab. 2 - Mean percent standard deviation

Aspect ratio of the enclosure also influences heat transfer; both numerical and experimental study show that enclosures with lower aspect ratio present higher Nusselt numbers. For high aspect-ratios, the narrower and narrower enclosure geometry gives boundary layers on the opposite walls closer and closer. At a certain point this effect can obstruct mass tranfer and cooling, through a pre-heating on falling air. These considerations are confirmed by the comparison of the interferograms showing air flow field for A*=2 and A*=3.3, fig.3. Experimental results are systematically lower than the numerical correlation. This is probably due to the difficulties in interpreting the interferograms near to the chips, where fringes are mostly concentrated. Furthemore the effect of thermal conduction in the P.C.B. and in the insulating walls should be taken into account.

5. Conclusions

Natural convection in enclosures with a Printed Circuit Board as one of the vertical walls, has been numerically and experimentally studied for two different values of the enclosure aspect-ratio. A finite difference scheme was employed for computation. Temperature distribution was obtained by holographic interferometry techniques. Heat transfer coefficients were obtained and correlated. Experimental results agree with numerical ones within a mean percent standard deviation ranging from 8 to 20.

Nomenclature

A^* = Aspect-Ratio
L = Cavity width, m
H = Cavity height, m
g = gravitational acceleration, $m \cdot s^{-2}$
Q^* = input power, $W \cdot m^{-2}$
x',y' = Dimensionless coordinates (x'=x/H; y'=y/H)
u',v' = Dimensionless velocity components $(u'=(u \cdot H)/\nu$; $v'=(v \cdot H)/\nu)$
Gr = $g\, \beta\, Q^*\, L\, H^3 / \nu^2 \lambda$ = Grashof number
Nu_x = $\frac{L}{H} \frac{\delta T}{\delta x'} \frac{1}{T}$ = Local Nusselt number
Nu = Average Nusselt number
Pr = Prandtl number
Ra = Gr·Pr = Rayleigh number

Greek Symbols

θ = temperature, K
ν = kinematic viscosity, $m^2 \cdot s^{-1}$
λ = thermal conductivity, $W \cdot m^{-1} \cdot K^{-1}$
β = thermal expansion coefficient, K^{-1}

References

1. Chu, H.H.S., Churchill, S.W.: The effect of size, location, aspect- ratio and boundary condition on two dimensional laminar convection in rectangular channel. *ASME Journal of Heat Transfer* 98 (1976) 194-201.

2. Moffat, R.J., Ortega, A.: Buoyancy induced convection in a non-uniformly heated array of cubical elements on a vertical channel wall. *ASME HTD* 57 (1986) 123-134.

3. Shaw, H.J., Chen, W.L., Chen, C.K.: Study on the laminar mixed convective heat transfer in tree-dimensional channel with a thermal source. *Journal of electronic packaging* 113 (1991) 40-49.

4. Magnusson, R., Haji-Sheikh, A., Snezana Konecni, Xuqum Wang.: Convection cooling of microelectronic chips. *Proceedings Thermal phenomena in electronic system* (1992) 138-144.

5. Poulikakos, D.: Natural convection in a confined fluid filled space driven by a single vertical wall with warm and cold regions. *ASME HTD* 107 (1985) 867-876.

6. Kuhn, D., Oosthuizen, P.H.: Three dimensional transient natural convection flow in a rectangular enclosure with localized heating. *ASME HTD* 63 (1986) 55-62.

7. Cesini, G., Paroncini, M., Ricci, R.: Experimental and numerical investigation on natural convection in square enclosure with a nonuniformly heated vertical surface. *Proceedings ICHMT XXth International Symposium on heat transfer in electronic and microelectronic equipment* (1988).

8. Cesini, G., Di Perna, C.,Paroncini, M., Ricci, R.: Analysis of free convective heat transfer in an enclosure with localized heat source. *Numerical methods in thermal problems* 6 (1989) 650-656.

9. Leonard, B. P.: A stable and accurate convective modelling procedure based on quadratic upstream interpolation. *Comput. Methods Appl. Mech. Eng.* 19 (1979) 59.

10. Leonard, B. P.: Adjusted quadratic upstream algorithms for transient incompressible convection. *AIAA Journal* (1979) 226-233.

11. Lilly, D. K.: On the computational stability of numerical solutions of time-dependent non-linear geophysical fluid dynamics problems. *U.S. Weather Bureau, Monthly Weather Review* 93, n.1, (1965) 11-26.

Fig. 1 Experimental apparatus.

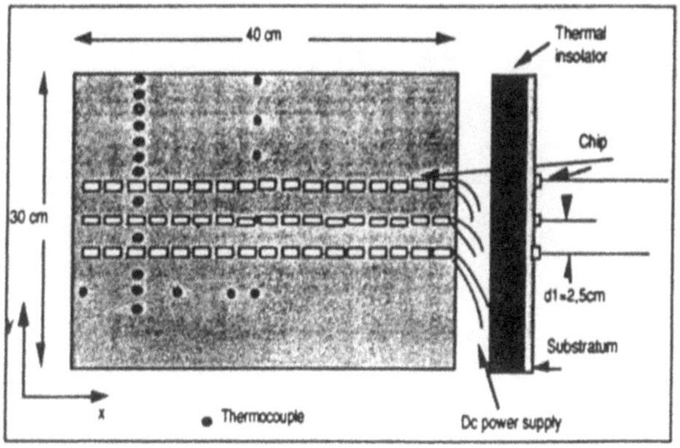

Fig.2 Schematic drawing of the test plate

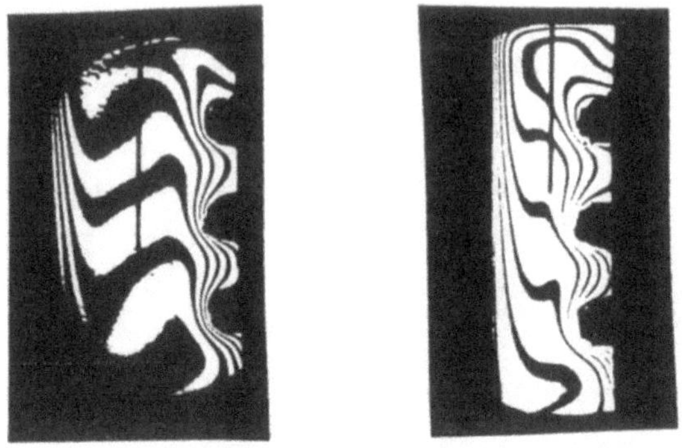

a) b)

Fig.3 Typical interferograms (a) $A^{*}=2$ and $Ra= 3.966 \cdot 10^{7}$
(b) $A^{*}= 3.3$ and $Ra= 2.372 \cdot 10^{7}$.

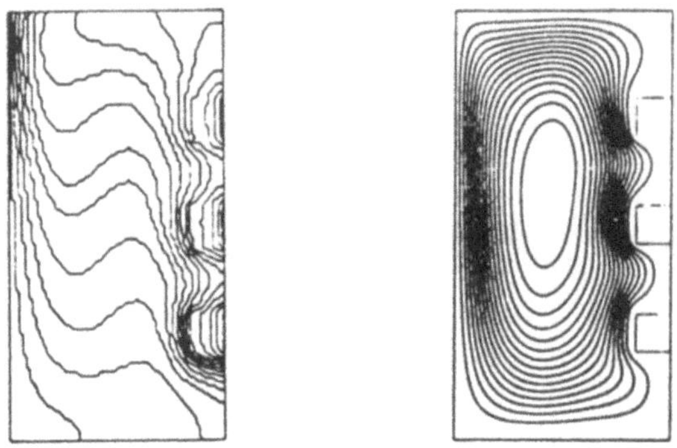

a) b)

Fig.4 (a) Isotherms (b) Streamlines for $A^{*}=2$ and $Ra= 3.966 \cdot 10^{7}$

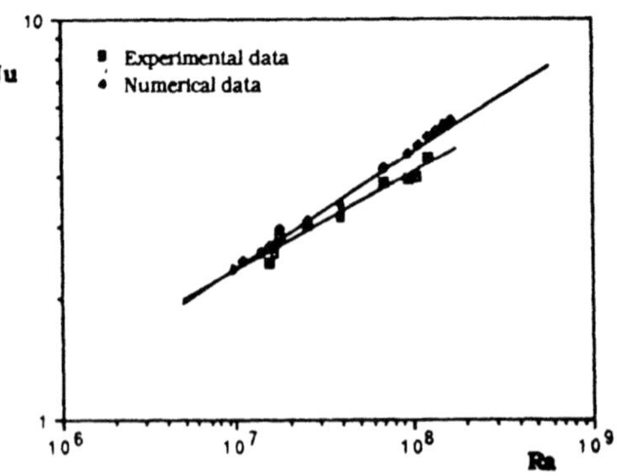

Fig.5 Comparison of numerical and experimental data for $A^{*}=2$.

Differences between 2D and 3D models predictions in a channel with a dissipating obstacle

J. B. SAULNIER, H. Y. WANG & B. FOURKA
Laboratoire d'Etudes Thermiques, URA 1403, C.N.R.S.
E.N.S.M.A, POITIERS, FRANCE

Abstract. In this paper, a numerical study is performed by an efficient multigrid technique, for laminar forced convection heat transfer in a two or three-dimensional channel with one heating element simulating a protruding integrated circuit. The predictions by the 2D and 3D models show a significant difference on the velocity, temperature and Nusselt number for the same physical problem.

I- Introduction

Conjugate conduction/forced convection heat transfer is of great interest in a variety of industrial applications, such as for instance in cooling of electronic equipment. The current trend for miniaturization of electronic solid state devices has introduced problems of heat dissipation from the devices themselves. Although the total energy generated by a single electronic Integrated Circuit (IC) may be small (about 1 W), the chip surface area of the devices is so small that it induces a very large surface density of heat flux of the order of 10^4 W/m^2 or even more. Sparrow et al. [1,2], Ortega and Moffat [3], and Garimella and Eibeck [4] have investigated experimentally the effects of air cooling on an array of square heating elements protruding from the Printed Circuit Board (PCB) in a three-dimensional, partially blocked channel. However, the difficulty lays in performing the experiment for the badly conductive heterogeneous ICs (for instance, the epoxy type ICs), and it consumes quite a lot of time particularly for parametric studies. So as complementary means, heat transfer characteristics of discretely heated protrusions in natural or forced channel flow have been studied extensively by numerical two or three-dimensional models (Habchi and Acharya [5], Afrid and Zebib [6,7], Wang [8]). Although, the influence of the presence of the ICs in a three-dimensional channel on the flow and temperature fields has been demonstrated by Shaw et al. [9], the thermal boundary condition was a simple one, with a given temperature imposed around the obstacle. Hence the objective of this paper is, on the one hand to show the conjugate effect of conduction/convection heat transfer in a 3D complex configuration, on the other hand to compare the results of the 3D model with the ones provided by 2D schemes that we used in previous studies [8]. The problem we present here simulates in a rather realistic way the cooling of electronic equipment (cf. Fig. 1) in forced convection: we consider the IC as an obstacle, with the presence of the rather small size heated chip (black layer in Figure 2) within the IC, and we take account of the different thermal conductivities of the electronic device materials.

Figure 1 shows the physical model to be investigated. Because both geometric properties and physical conditions are symmetric relatively to the surface (z=0), we consider half of the physical domain as our computional domain. The velocity and temperature of the cooling fluid at entrance are Ue = 1 m/s and Te = 20 °C, respectively. A heating element

C. J. Hoogendoorn et al. (eds.), Thermal Management of Electronic Systems, 107–116.

with length L_L, width L_T, height B and dissipating power Q = 1 W is located at a distance L_E from the channel entrance. Let us observe that a periodic boundary condition is applied to the surface at $z = L_W/2$ in order to simulate a series of components located side by side in the z-direction.

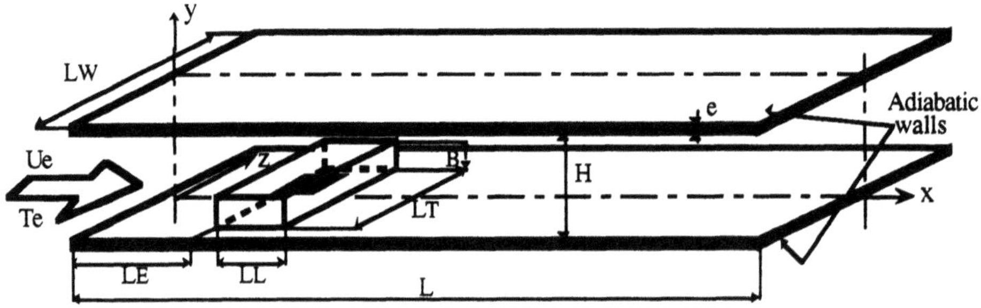

Fig. 1 Physical model and coordinate system

II- Description of physical model

In this study, two configurations of the IC are considered in the 3D channel flow. In the first one the larger obstacle lateral surface is perpendicular to the core flow (case I). Whereas in the second one it is the lateral smaller surface that is perpendicular to the core flow (cf. Fig. 2).

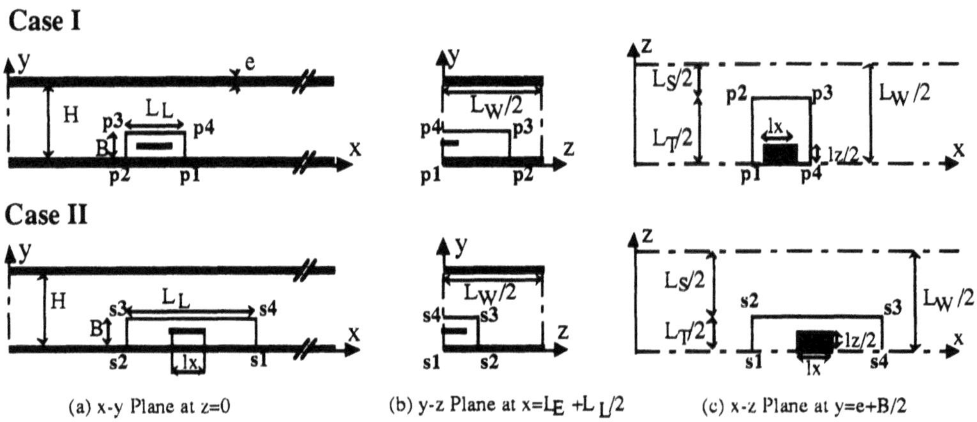

Fig. 2 Scheme of the electronic system for cases I and II

We shall now comment the way we have schematized the real 3D problem (cf Fig. 1) and stated a simplified 2D representation (cf Fig. 2a), used in the previous study (Wang [8]).

The main problem, when shifting from 3D to 2D, concerned essentially the estimation of the heat dissipation. The spanwise dimension of the chip was chosen here as L_z (Fig. 3), the heat flux is calculated assuming a uniform heat distribution in the transverse z-direction (cf.Fig.3). We transformed then the actual power Q (W) dissipated on a chip which is assumed a very thin sheet almost without thickness, into a surface density q (W/m^2), given by :

$$q = \frac{Q}{L_z \, lx} \tag{1}$$

This heat is dissipated uniformly at a constant rate inside the junction region represented by the black layer in Fig.3. Fig. 2a illustrates the 2D schemes (in the x-y plane) of the physical situation simulated. The width in the 2D model L_Z is chosen as lz, L_T and L_W respectively, in order to show the influence of the flux density definition on the chip temperature in comparison with the one obtained by the 3D model. Table 1 summarizes the heat flux density for all the 2D cases.

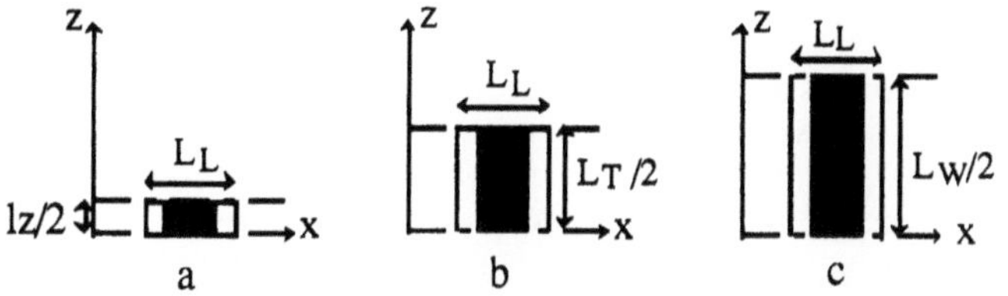

Fig.3 Two-dimensional schemes for the chip heat flux density estimation

q (W/m^2)	a (L_Z = lz)	b ($L_Z = L_T$)	c ($L_Z = L_W$)
Case I-2D	40000	10000	6666.67
Case II-2D	40000	22222.22	6666.67

Table 1: Heat flux density in the chip (2D models)

Table 2 summarizes the geometrical dimensions shown in Figs. 1 and 2, and the physical properties.

Symbol	Value	Symbol	Value
L (mm)	200	B (mm)	4
H (mm)	11	L_L (mm)	Case I: 9
L_E (mm)	20		Case II: 20
$L_W=L_T+L_S$ (mm)	30	L_T (mm)	Case I: 20
e (mm)	1.6		Case II: 9
lx (mm)	5	λs (W/m.K)	IC: 0.7
lz (mm)	5		PCB: 2
Q (W)	1	λf (W/m.K)	2.5x10-2

Table 2 The geometrical dimensions and physical properties

III- Mathematical model and numerical features

The problem studied is elliptic in nature and all the fluid properties are supposed uniform. The flow is steady, laminar, and the effects of heat radiation and viscous dissipation are neglected.

A unique form, used to express the governing equations, is as following:

$$u\frac{\partial\phi}{\partial x} + v\frac{\partial\phi}{\partial y} + w\frac{\partial\phi}{\partial z} = \Gamma_\phi\left(\frac{\partial^2\phi}{\partial x^2} + \frac{\partial^2\phi}{\partial y^2} + \frac{\partial^2\phi}{\partial z^2}\right) + S_\phi \qquad (2)$$

where ϕ is a general variable, Γ_ϕ is a diffusion coefficient, and S_ϕ is a source term.

The boundary conditions in the 3D model are:

inlet: u=1m/s, v=w=0, T=20°C

solid: u=v=w=0

symmetric surfaces: $\frac{\partial u}{\partial z}=\frac{\partial v}{\partial z}=\frac{\partial T}{\partial z}=0,\ w=0$

the chip: flux surface density $q=\frac{Q}{lz\ lx}$ (W/m2)

exit: $\frac{\partial u}{\partial x}=\frac{\partial v}{\partial x}=\frac{\partial w}{\partial x}=\frac{\partial T}{\partial x}=0$

The finite volume forms of the above differential equations are derived by integrating them over discrete control volumes (Patankar [10]) in the physical domain. A staggered mesh system is used in which the discrete velocities are located on the faces of the finite volume cells and the discrete pressures and temperatures are situated at the cells centres. The finite volume equations, expressed per unit volume can be written in the general form :

$$a_P\phi_P - a_N\phi_N - a_S\phi_S - a_E\phi_E - a_W\phi_W - a_T\phi_T - a_B\phi_B = S_\phi\, \Delta x\, \Delta y \Delta z \qquad (3)$$

The coefficients a_P, a_S... and a_B represent the combined effects of convection and diffusion, and are obtained by using the Hybrid scheme. The 2D model equations are included also in the eq. (2) by eliminating the terms concerning the z-direction. For laminar flows, conjugate conduction/convection heat transfer can be analyzed easily by using the harmonic averaging thermal conductivity at interface between solid and fluid.

The procedure, we have retained in our multigrid technique (Brandt [11]) to solve the equation (2) on each level of grids, is the SIMPLEC algorithm (Doormaal and Rathby [12]). The restriction to the coarse grid velocities is defined as the arithmetic mean of four variables. The prolongation operator is derived in all cases by using bilinear interpolation.The full multigrid procedure FAS-FMG (Vanka [13]), which starts at the coarsest grid and progresses in an adaptive manner to the finest grid, is appropriate.

IV- Results and discussion

All our analyses were performed using our computer program on a DEC 3100 (15 Mips), running under ULTRIX. The case I-3D (cf. Fig. 2) was solved with the grids 77x29x21 (case A), 125x33x21 (case B), 125x41x21 (case C) and 125x41x25 (case D) to determine the impact of the grid size on the chip temperature and the recirculating zone near the heating element surfaces. It was found that the grid 125x41x21 practically reproduces the same chip temperature (cf. Fig. 4a) and velocity field (for example, just behind the obstacle, shown in Fig. 4b) as the ones of the grid 125x41x25. The maximum CPU time is about two hours for each tested case. So we have chosen a 125x41x21 grid for our solution with a uniformly distributed grid over most of the domain and finer grids near the solid. Under-relaxation was necessary in the solution of the momentum equations. After having a converged dynamic solution, the flow field data is then served as an input to the solution of the energy equation. In addition, overall conservation of momentum and energy is satisfied to within 1% with this 125x41x21 grid.

Now we shall discuss some characteristic results concerning velocity profiles and vectors, temperature and average and local Nusselt number distributions.

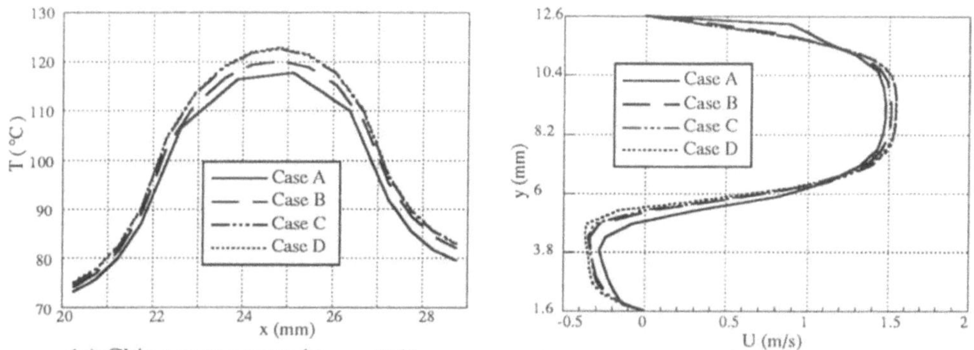

(a) Chip temperature (at y=e+B/2) (b) u-velocity profile at x= L_E+L_L+5 mm

Fig. 4 Distribution of the chip temperature and u-velocity profile in the plane at z=0 for different grids

Distribution of the velocity field:

The u-velocity profile across the channel at z=0 for different cases is plotted in Fig. 5. We can observe that in the blockage region (x= L_E+L_L/2 and L_E+L_L+5 mm), the velocity levels obtained by the 2D model for the two implementations of the component in the channel are practically identical. However, in this same region the influence of the position of the component in the 3D flow channel on the velocity profiles is clearly shown. The acceleration of the dynamic field is stronger in the case I-3D in comparison with the one in the case II-3D. It can be explained because the transverse channel flow section is narrower in case I-3D. In all cases, the velocity value provided by a 2D model is always greater than the one obtained by a 3D model, and the overestimated velocity value attached to the 2D model could then induce a false "better heat exchange" prediction. Finally, a global view of the velocity field is shown thanks to Figure 8. It can be seen that the flow is strongly perturbated by the presence of the IC, both in the longitudinal plane and in the transverse plane. A strong recirculating region is always present just behind the IC but we can observe that

* for case I, near the lower electronic board the extension of this zone is less in 3D than in 2D (see Fig. 8a and b); but, on the contrary is on a longitudinal line at the chip elevation, the recirculation becomes less extended in 2D (see Fig. 6): the air flowing over the IC flattens this recirculation zone in 2D.

* for case II its extension is always greater in 2D (Fig. 6).

* besides for 3D models the recirculation is more extended for case I than for case II (Fig. 6).

Distribution of the chip temperature:

Distributions of the chip temperature provided by the 2D and 3D models are shown in Fig. 7. Among the three definitions of the chip flux density in 2D model, the chip temperature is irrealistic (about 300 °C and not represented on Fig. 7) if we chose the chip flux density as identical to that of the 3D model (case a of Fig. 3). We observe also that the evolution of the temperature obtained by 2D and 3D models in the case I does not follow the same law as in the case II. In the case I-3D the maximum temperature value is always higher than in the 2D schemes (about 55 °C in the case I-2D-c and 30 °C in the case I-2D-b). But in the case II-3D it is higher (about 60 °C) than the one in case II-2D-c, and lower (about 40

°C) than the one in the case II-2D-b. This shows how difficult it is to estimate the surface flux density dissipated by the chip if we use the 2D schemes.

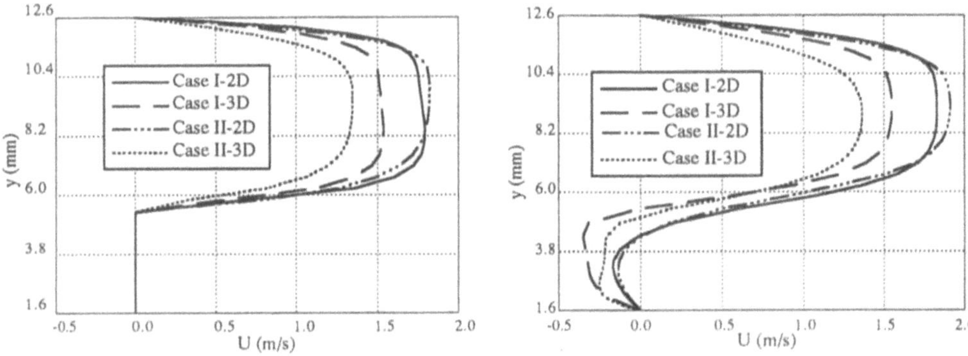

(a) at $x=L_E+L_L/2$ (b) at $x=L_E+L_L+5$ mm

Fig. 5 A u-velocity profiles across the channel in the plane at z=0

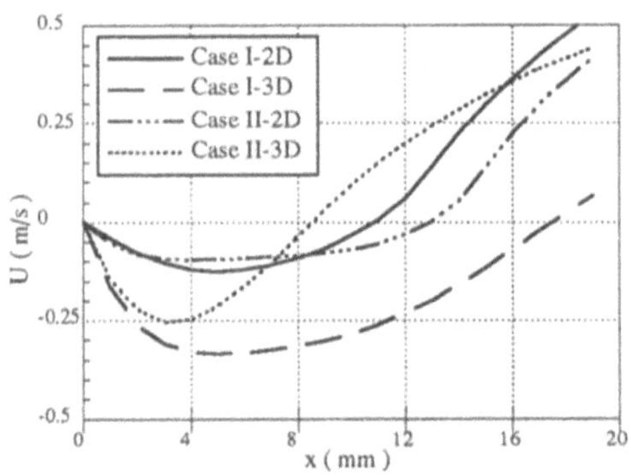

Fig. 6 A u-velocity profile behind the IC in the plane at z=0 at y=e+B/2

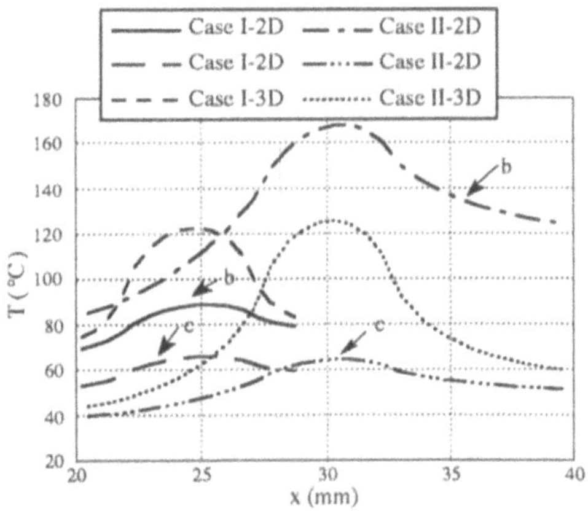

Fig. 7 A chip temperature in the plane at z=0 for different cases

Distribution of the Nusselt number:

The local Nusselt number at the points located on the surface normal to the x-axis is defined by:

$$Nu_x = \frac{h_x\ H}{\lambda_f} \qquad \text{where } h_x = \frac{-\lambda_f(\frac{\partial T}{\partial x})_p}{T_p - T_e}$$

and the average Nusselt number is: $\overline{Nu_x} = \frac{\overline{h_x}\ H}{\lambda_f}$ where $\overline{h_x} = \frac{1}{S}\int_S h_x\, dS$

S represents the area of surfaces of the IC. Nu_y and Nu_z are defined in a similar way.

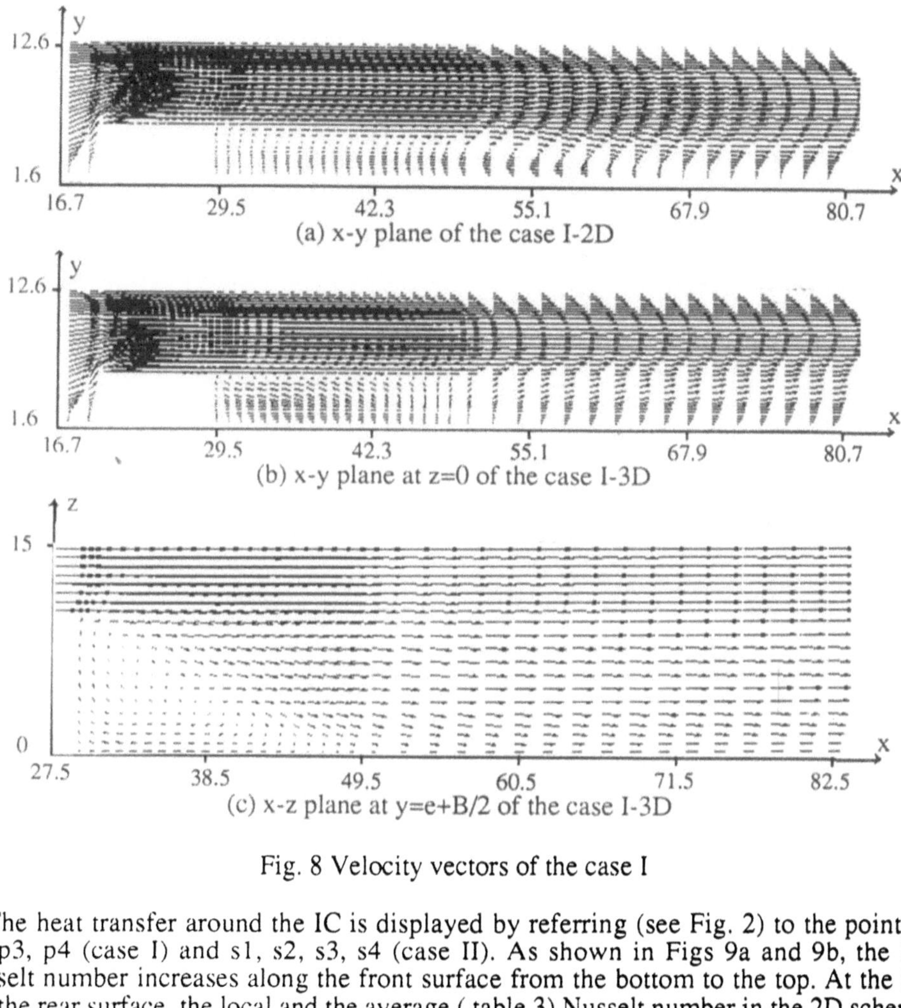

(a) x-y plane of the case I-2D

(b) x-y plane at z=0 of the case I-3D

(c) x-z plane at y=e+B/2 of the case I-3D

Fig. 8 Velocity vectors of the case I

The heat transfer around the IC is displayed by referring (see Fig. 2) to the points p1, p2, p3, p4 (case I) and s1, s2, s3, s4 (case II). As shown in Figs 9a and 9b, the local Nusselt number increases along the front surface from the bottom to the top. At the front and the rear surface, the local and the average (table 3) Nusselt number in the 2D scheme is smaller than in the 3D scheme. The local Nusselt number decreases rapidly along the upper surface, but the difference between case I-2D and case I-3D appears only near the rear surface. But, on the same surface, the local and average (table 3) Nusselt number in the case II-3D is always smaller than in the case II-2D. Let us observe that the average Nusselt number in the upper surface has about the same values in the 2D and 3D schemes (because of a small increase near the edge: see the branch p4p3 on figure 9d), and it is higher in case I than in case II.

Figure 9c shows the local Nusselt number around the IC but in an horizontal plane. The local Nusselt number of the lateral surface seems to be larger than the one of the front surface in the case I-3D. But in the case II-3D, the same remark is also observed in the first part of the lateral surface. Whereas in the second part the conclusion is inverted. We can also observe in this figure 9c that in the rear surface, the Nusselt number is higher in the symmetry plane (z=0, point p4) than on the edge (point p3). This can be due to the higher reverse circulating flow towards the IC, which in this symmetry plane acts as an impinging jet (see Fig. 8c and 10).

Figure 9 shows the local dimensionless heat flux (local convected heat flux density divided by the heat flux density dissipated at the chip level) which follows the same variations as the Nusselt numbers. Let us now comment the table 4 which summarizes the different percentages of heat flux convected around the IC. We first observe that the main difference between 2D and 3D, is that a 2D model is clearly unable to evaluate the flux convected by the lateral surface, which can be as high as 8 to 18 % of the total dissipated power. Besides, the flux convected by the upper surface is highly overestimated by 2D models, due to a strong flow evaluation between IC and PCB (see Fig. 5a). But even with the presence of the lateral surfaces, the convection in 3D models remains at the IC level less efficient than in 2D ones. As a consequence, conduction from the IC to the PCB has then a greater importance. We may suppose that the PCB acts as a 2D fin and that it finally enhances the conduction from the PCB itself, all around the IC.

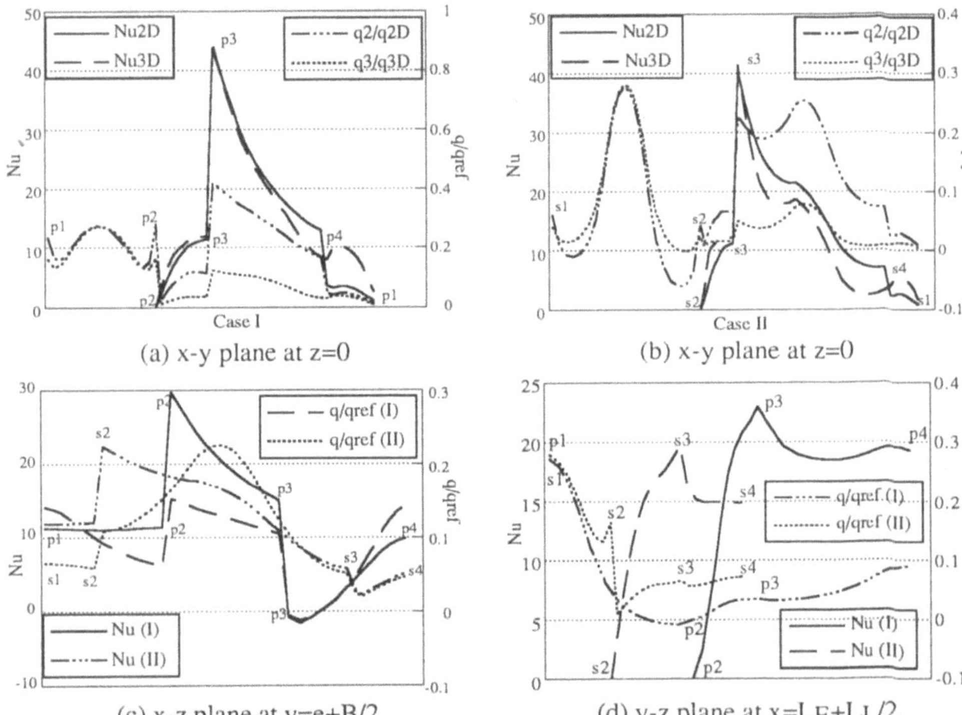

Fig. 9 The local Nusselt number and the local heat flux along the IC surfaces

Cases	front surface	upper surface	rear surface	lateral surface
I-2D	8.47	23.93	2.77	///
I-3D	9.58	22.87	2.96	16.98
II-2D	8.22	18.17	2.07	///
II-3D	10.82	15.48	3.19	12.37

Table 3: Average Nusselt numbers of the IC

Cases	front surface	upper surface	rear surface	lateral surface	interface (conduction)
I-2D	7.36	51.27	2.93	///	38.43
I-3D	6.44	39	3	7.37	44.18
II-2D	4.22	68.3	1.75	///	25.73
II-3D	2	26.98	1.03	18.53	51.44

Table 4 Percentage of the heat flux repartition around the IC

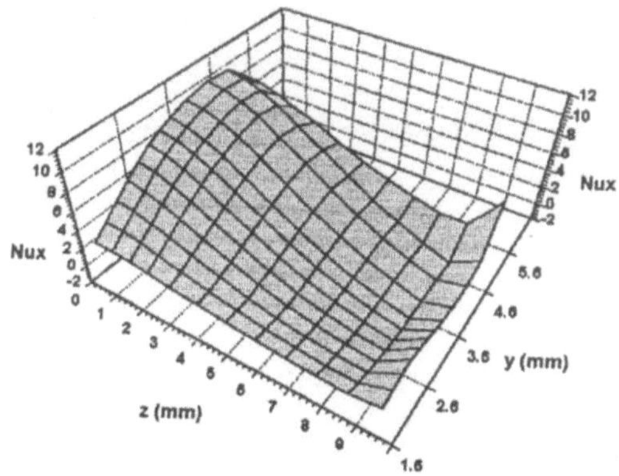

Fig. 10 The local Nusselt number of the case I-3D in the rear surface

7- Conclusion

The cooling by forced convection, of integrated circuits mounted in a channel has been discussed here by using a numerical multigrid technique, and the main trends can be summarized as follows. The influence of the use of 2D and 3D models for a given physical problem, on the velocity and temperature field is significant. Concerning the sensitivity to the definition of the flux density due to the dissipating chip in the 2D model, we observed that the chip temperature provided by 2D models is either overestimated or underestimated, as compared to that the one of 3D model, whatever may be the flux density definition. Thanks to 3D model the study shows that the presence of obstacle in a channel induces a perturbation over all around the IC, resulting in an increase of convected flux along the PCB. A stronger velocity field is observed between the IC and the PCB in the 2D model, as compared to that of the 3D, inducing a false conclusion leading to a better heat exchange.

Finally, all these results are given just for one heating element, and the way is now opened to an interesting complementary study with several heating elements.

Nomenclature

B : height of the IC
H : channel height
L_W : channel width
L_E : distance between the entrance and the IC
lx : chip length
Q : heat dissipation of the IC
T_p : temperature of surfaces
U_e : inlet velocity (m/s)
x, y, z: coordinates

$\overline{Nu_x}$, $\overline{Nu_z}$, $\overline{Nu_z}$: average Nusselt number in the x, y and z directions

e : plate thickness
L : channel length
L_T : IC width
L_S : transversal distance betwen the ICs
lz : chip width
q : flux surface density (W/m^2)
T : temperature (°C)
T_e : inlet temperature (C°)
u, v, w : velocity components

Nu_x, Nu_y, Nu_z : local Nusselt number in the x, y and z directions

Greek symbols

λ_f : thermal conductivity of the fluid (W/m.K)
ϕ : general variable

λ_s : thermal conductivity of solid (W/m.K)

References

[1] Sparrow E. M., Niethammer J. E. and Chaboki A., 1982, "Heat transfer and pressure drop characteristics of array of rectangular modules encoutered in electronic equipment", *Int. J. Heat Mass Transfer*, Vol. 25, No.7, pp.961-973.
[2] Sparrow E. M., Vemuri S. B. and Kadle D. S., 1983, "Enhanced and local heat transfer, pressure drop, and flow visualization for arrays of block-like electronic components", *Int. J. Heat Transfer*, Vol. 26, No.5, pp.689-699.
[3] Ortega A. and Moffat R. J., 1986, "Buoyancy induced convection in a non-uniformly heated array of cubical elements on a vertical channel wall", *Heat Transfer In Electronic Equipment*, HTD-Vol. 57, pp.123-134.
[4] Garimella S. V. and Eibeck P. A., 1990, "Heat transfer characteristics of an array of protruding elements in single phase forced convection", *Int. J. Heat Mass Transfer*, Vol. 33, No.12, pp.2659-2669.
[5] Habchi S. and Acharya S., 1986, "Laminar mixed convection in a partially blocked, vertical channel", *Int. J. Heat Mass Transfer*, Vol. 29, No.11, pp.1711-1722.
[6] Afrid M. and Zebib A.,1989, "Natural convection air cooling of heated components mounted on a vertical wall", Numerical Heat Transfer, part A, Vol. 15, pp.243-259.
[7] Afrid M. and Zebib A.,1989, "Three-dimensional laminar and turbulent natural convection cooling of heated blocks", Numerical Heat Transfer, part A, Vol. 19, pp.405-424.
[8] Wang H. Y., 1991, "Simulation Numérique par Multigrilles des Transferts Conjugués dans les Cartes de Composants Electroniques", *Thèse N° d'ordre : 405*, Université de Poitiers, 53-56.
[9] Shaw H. J., Chen W. L. and Chen C. K., 1991, "Study on the laminar mixed convective heat transfer in three-dimensional channel with a thermal source", *J. of Electronic Packaging*, Vol. 113, pp.40-49.
[10] Patankar S.V., 1980, Numerical Heat Transfer and Fluid Flow, Mc Graw Hill Book Company.
[11] Brandt A., 1977, "Multi-Level Adaptive Solution to Boundary- Value Problems", *Mathematics of computation*, Vol. 31. Num. 138, pp.333-390.
[12] Doormaal J. P. V. and Rathby G. D., 1984, "Enhancements of the SIMPLE method for predicting incompressible fluid flow", *Numerical Heat Transfer*, Vol.7, pp.147-163.
[13] Vanka S. P., 1986, "Block-Implict Multigrid Solution of Navier-Stokes Equations in Primitive Variables", *J. of Computational Physics*, 65, pp.138-158.

Effect of radiation on natural convection in tilted channels

O. MANCA, S. NARDINI & V. NASO

DETEC - Università degli studi Federico II - Piazzale Tecchio - 80125 - Napoli - Italy

Abstract. Laminar air natural convection between two inclined parallel plates with uniform, both symmetric and fully asymmetric, heat flux has been experimentally investigated. Tests have been run at various channel spacings and heat fluxes and at two emissivity values. The effect of emissivity on wall temperature and on air flow was high at all angles in the asymmetric heating, in particular when it was the top wall to be heated. A marked effect of radiation transfer on the average Nusselt number was noticed at high emissivity.

Notation

A,B,C,D	coefficients in equation (9)
b	channel spacing, m
g	acceleration of gravity, m/s^2
Gr	channel Grashof number, equation (5)
k	thermal conductivity, W/mK
L	channel length, m
Nu	average Nusselt numbers, equations (6)
Pr	Prandtl number
q	heat flux, W/m^2
r	regression coefficient
Ra	channel Rayleigh number
T	temperature, K
v	velocity of the air, m/s
W	channel width, m
x	coordinate along the length, m
y	coordinate along the spacing, m
z	coordinate along the width, m

Greek symbols

β	volumetric coefficient of expansion, 1/K
ε	wall emissivity
θ	angle of inclination from the vertical, °
ν	kinematic viscosity, m^2/s

Superscripts

*	convective and radiative
+	dimensionless

Subscripts

b	refers to bottom wall
c	convective
k	conductive
max	maximum
o	ambient air
r	radiative
t	refers to top wall
w	wall
Ω	Ohmic dissipation

A bar over a symbol indicates a mean value on the walls

1. Introduction

Air natural convection in inclined open ended channels has been scarcely investigated. Tilted channels are widely used for heat removal from electronic equipments; nearly-horizontal channels are often met in telephones, modems and desk top applications. The heat transfer from the walls of the channel is affected by their radiative exchange with the surroundings;

C. J. Hoogendoorn et al. (eds.), Thermal Management of Electronic Systems, 117–126.

the radiant contribution can be remarkable at the rather high emissivity values of the typical PCBs surfaces.

Maximum wall temperatures and Nusselt numbers in laminar natural convection in an open ended vertical channel with uniform wall heat flux were correlated to the channel Rayleigh number numerically by Aung et al. [1], for both symmetric and asymmetric heating, and experimentally by Wirtz and Stutzman [2] and Webb and Hill [3].

The effect of inclination from the horizontal on the thermal performance of a channel asymmetrically heated with air natural convection was experimentally investigated by Kennedy and Kanhel [4]. Temperature profiles of the uniformly heated bottom plate were presented, which show that the higher the inclination angle the lower the temperature, up to a 70° angle. For angles greater than 20° the flow regime changed from a two-dimensional behaviour into three-dimensional longitudinal vortex rolls. Azevedo and Sparrow [5] performed heat transfer and flow visualization experiments in water to study the effect of inclination on natural convection in an isothermal, parallel walled channel. Various modes of heating, aspect ratios and Rayleigh numbers were investigated. Nusselt numbers were evaluated for all experimental conditions in the range 0°-45° of the inclination angle from the vertical and they were correlated to within ±10 percent. A two-dimensional recirculating flow was observed near the unheated wall of one sided top heated channels. Cheng and Kimura [6] presented a study on convective instability phenomena by flow visualization in a tilted channel with 10° of inclination from the horizontal and isothermal heating from below. Local and average air natural convection heat transfer characteristics for a uniformly heated upward-facing horizontal plate, shrouded by a parallel adiabatic surface situated above the plate, were determined experimentally by Sparrow and Carlson [7]. Lately the effect of the inclination angle from the vertical on laminar air natural convection between two tilted parallel plates with uniform, both symmetric and fully asymmetric, heat flux, has been experimentally investigated by Manca et al. [8, 9]. Experiments showed a sharp increase in maximum wall temperature for inclination angles above 60°. Nusselt numbers were evaluated and correlated to the channel Rayleigh number and to the inclination angle, in the range 0°-60° in the former paper and in the range 0°-90° in the latter one; all heating modes were taken into account .

The interaction of radiation with developing laminar free convective heat transfer on vertical parallel plate channels with isoflux asymmetric heating was investigated numerically by Carpenter et al. [10]. Radiation to the ends of the channel and to the cooler wall was found to alter the non radiation results by reducing the maximum wall temperature by as much as 50 percent. Numerical predictions were in good agreement with some experimental data. The same configuration was studied analytically and experimentally by Yamada [11]. Results were in a fair agreement, showing a major effect of the wall emissivity on the combined heat transfer. Sparrow et al. [12] determined numerically the heat transfer rates for pure natural convection and for interacting natural convection and radiation in a vertical parallel plate channel of finite height. Radiation effects were explored for the case of a channel having one isothermal wall. The direct radiation from the channel walls to the surroundings via the inlet and exit cross section was about 10 percent of the radiation-enhanced convection. Manca and Naso [13] reported measurements of wall temperatures and air temperatures and velocities of air natural convection between two vertical plates with uniform, both symmetric and fully asymmetric, heat flux. Surface emissivity of the plates was varied. Local and average Nusselt numbers, both including the effects of radiation at various plate emissivity values or not, were evaluated and correlated to the channel Rayleigh number.

Turbulent convective heat transfer to air from two vertical parallel plates, one adiabatic and the other with a uniform heat flux, was investigated experimentally by Miyamoto et al. [14]. Temperature and velocity of the air were measured by thermocouples and laser doppler velocimetry, respectively; thermal radiation effects were taken into account. Measured local

Nusselt numbers were compared with those either numerically calculated or measured by other authors. A relation is given for the Nusselt number near the channel outlet.

The combined effect of the channel inclination angle and plate emissivity seems to be not yet studied. In this paper laminar air natural convection between two inclined parallel plates, with uniform either symmetric or fully asymmetric heat flux, has been experimentally investigated. Measured wall temperatures along the channel length are presented for different values of spacing, angle of inclination, wall emissivity and channel Rayleigh number. Heat transfer coefficents have been evaluated and average Nusselt numbers are correlated to the inclination angle and to the channel Rayleigh number, either taking into account the radiative contribution or not. Midplane velocity and temperature profiles of the air along the spacing are also presented for different values of wall emissivity and tilting angle.

2. Experimental apparatus

The experimental apparatus is shown schematically in fig.1. The channel was made from two principal parallel walls with uniform heat flux; its side walls were unheated. Each principal wall consisted of two sandwiched phenolic fiberboard plates, 400 mm long and 530 mm wide. The plate facing the channel was 3.2 mm thick and its surface adjacent the internal air was coated with a 35 μm thick copper layer. Two different wall emissivities were investigated; one was that of the bare phenolic fiberboard plate (ε=0.8), the other one was obtained by nickel plating it (ε=0.05). The backside plate was 1.6 mm thick. Its back surface was coated with a copper layer, which was the heater. In order to reduce heat losses, a 150 mm polystyrene block was affixed to the rear face of each plate. The side walls were made of plexiglass rectangular rods. The channel was 400 mm long, 475 mm wide and the spacing varied between 20.00 mm and 40.00 mm; it was open to the ambient along its top and bottom edges, fig.1b. It was secured to a tilting support frame. Inclination angles of the channel in the range 0°-90° were obtained by brackets and pinned joints, with an accuracy of ±0.5°. A finer adjustment (±0.05°) was obtained in the range 85°-90° from the vertical by a micrometric screw system. The entire apparatus was located within an enclosed room, sealed to eliminate extraneous air currents.

The plates were heated by passing a direct electrical current through the heaters. The dissipated heat flux per board was evaluated with an accuracy of ±2 percent by measuring the voltage drop across the heaters and the current passing through them. Wall temperatures

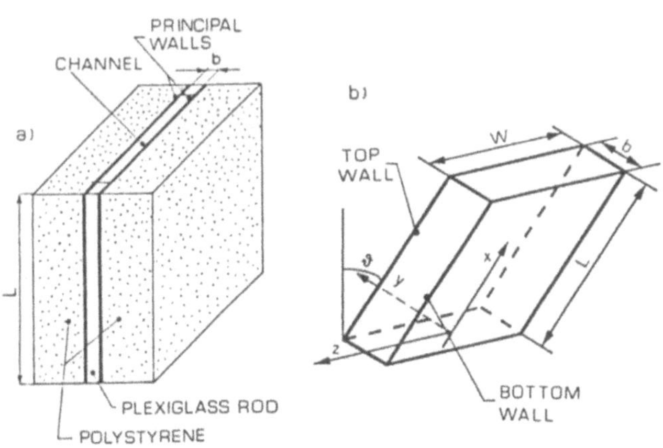

Fig.1. Experimental apparatus: a) test section; b) geometry of the channel.

were measured by twelve 0.50 mm OD ungrounded iron-constantan thermocouples embedded in the fiberboard plates and in contact with the outer layer. They were located in the vertical center-line of each plate at different distances above the bottom edge. In preliminary tests, carried out at all inclination angles, additional thermocouples placed horizontally outward the xy plane indicated that the wall temperature distribution was symmetric with reference to z=0 within ±0.2°C. Moreover, wall temperatures turned out to be independent of z within ±100 mm, their maximum deviation from the vertical centerline temperature being ±2 percent. Within the range of inclination angle 30°-70°, for which Kennedy and Kanhel [4] and Azevedo and Sparrow [5] observed three-dimensional longitudinal vortex rolls, no variation of the wall temperature along the z axis higher than the aforementioned ±2 percent was measured. Therefore, in the following reference has been made to the centerline wall temperature. Fifteen thermocouples were affixed to the rear surface of the plates and embedded in the styrene to enable evaluation of conductive heat losses. Their maximum value was 28 percent of the Ohmic heat rate dissipated in each plate, in the range of inclination angle 0°-90°. Ambient air temperature was measured by shielded thermocouples placed near the leading edge of the channel. Thermocouples voltages were recorded to 1 μV. Calibration of the temperature measuring system showed an estimated precision of the thermocouple-readout system of ±0.1°C. The velocity and the temperature of the air in the channel were measured by a hot wire system. In every case the air velocity could be determined with a maximum uncertainty of ±10 percent. The air temperature was measured with a maximum uncertainty of ±0.1°C. A more detailed description of the experimental apparatus is reported in Manca et al. [8].

3. Data reduction

The dimensionless coordinates are

$$x^+ = x/L \quad ; \quad y^+ = y/b \tag{1}$$

The dimensionless wall temperature is

$$T_w^+ = (T_w - T_o)k/\bar{q}_c b \tag{2}$$

where $\bar{q}_c$ is the mean value of the spatially-averaged convective heat flux on each heated wall

$$\bar{q}_c = \frac{1}{2L}\left[\int_0^L q_{c,t}(x)\,dx + \int_0^L q_{c,b}(x)\,dx\right] \tag{3}$$

The dimensionless temperature and velocity of the air are

$$T^+ = \frac{T - T_o}{T_{w,max} - T_o} \qquad \text{and} \qquad v^+ = \frac{v}{v_{max}} \tag{4}$$

where $T_{w,max}$ denotes the maximum surface temperature on the two walls and v_{max} is the maximum air velocity value at a given x.

Heat transfer results are presented in terms of the so called channel Rayleigh number

$$Ra = Gr\,Pr = \frac{g\,\beta\,\bar{q}_c\,b^5}{\nu^2\,k\,L}\,Pr \tag{5}$$

Nusselt numbers will be based on the difference between wall and inlet fluid temperatures ather than on that between wall and bulk fluid temperatures, since the last one usually is not vailable in practice. Two average Nusselt numbers can be defined. The former accounts for ıe convective heat transfer from the walls and the latter takes into account the convective and adiative heat transfer. Thus we have

$$Nu = \frac{\bar{q}_c}{\bar{T}_w - T_o}\frac{b}{k} \quad \text{and} \quad Nu^* = \frac{\bar{q}_c + \bar{q}_r}{\bar{T}_w - T_o}\frac{b}{k} \tag{6}$$

where $\bar{q}_r$ is defined in the same way as $\bar{q}_c$ and

$$\bar{T}_w = \frac{1}{2L}\left[\int_0^L T_{w,t}(x)\,dx + \int_0^L T_{w,b}(x)\,dx\right] \tag{7}$$

The properties of the air are evaluated at the reference temperature $(T_w + T_o)/2$.

Local actual convective heat flux, $q_c(x)$, is not uniform because of radiation and onduction. Experimental data were reduced by first introducing, in the above presented quations, the local convective heat flux

$$q_c(x) = q_\Omega(x) - q_k(x) - q_r(x) \tag{8}$$

where $q_\Omega(x)$ is the local heat flux due to the Ohmic dissipation, which is assumed to be niform, $q_k(x)$ are the local conduction heat losses from the plate and $q_r(x)$ is the local adiative heat flux from the plate. For each run, the terms $q_k(x)$ were calculated by a umerical procedure, a three-dimensional distribution of the temperature being assumed in ıe styrene. The predicted temperatures in significant configurations of the system had been reviously compared with those measured by thermocouples embedded in the polystyrene nsulation and the agreement was very good. The $q_r(x)$ terms were calculated for each emperature distribution of the walls, ambient temperature, wall emissivity and channel pacing, according to the procedure described by Webb and Hill [3]. With both walls heated, ıe minimum value of the ratio $q_k(x)/q_\Omega(x)$ was 0.060, at x=125 mm with ε=0.05, b=32.25 nm, $q_\Omega(x)$=121 W/m^2 whereas its maximum value was 0.35 at x=200 mm with ε=0.8, b=20.00 mm and $q_\Omega(x)$=60 W/m^2.

According to the procedure suggested in Kline and McClintock [15], the uncertainties in convective Nusselt number, convective and radiative Nusselt number and channel Rayleigh umber are estimated to be 12, 5 and 15 percent, respectively.

. Results and discussion

Experiments were performed at different angles of inclination in the range 0°-90° and various pacings (20.00, 32.25 and 40.00 mm). Ohmic heat fluxes were 60 W/m^2 and 121 W/m^2. The cooling medium was air, whose inlet temperature into the channel was nearly 22.1°C in ıll tests; its fluctuations were within ±0.5°C during each test. Three heating modes were tudied for each configuration. According to what proposed by Azevedo and Sparrow [5], ıey will be designated as heating modes I, II and III and are: I) both walls heated; II) top wall heated with bottom wall unheated; III) top wall unheated with bottom wall heated. The op and bottom wall designations are referred to the relative position of the walls shown in Fig.1b. When the channel is vertical, modes II and III represent the same heating mode: IV) one wall heated with the other wall unheated.

Wall temperature rise above ambient temperature as a function of the length and the angle of inclination, for different spacings, ohmic heat fluxes and heating modes and for a 0.8 emissivity value of the walls, is presented in fig.2. For mode I (symmetric heating) at the lowest spacing the figure shows that the higher the angle the higher the temperature of both the walls, due to the reducing chimney effect. One can notice that up to 75° the wall temperature profiles are very similar except near the exit; it is worthwile noting that the chimney effect is still present in the 88° inclined channel. The dip in the temperature profile of the bottom plate near the exit at high tilting angles is due to the air inflow at the outlet edge of the wall. At the same spacing small differences between the temperatures of the two walls are exhibited, but near the channel outlet; wall temperatures in the central region of the horizontal channel are nearly equal. Still for mode I the middle and upper parts of the figure show that in the 0° and 75° tilted channel wall temperatures are practically independent of the spacing at the same heat flux while in the highly inclined channel the temperatures of both the

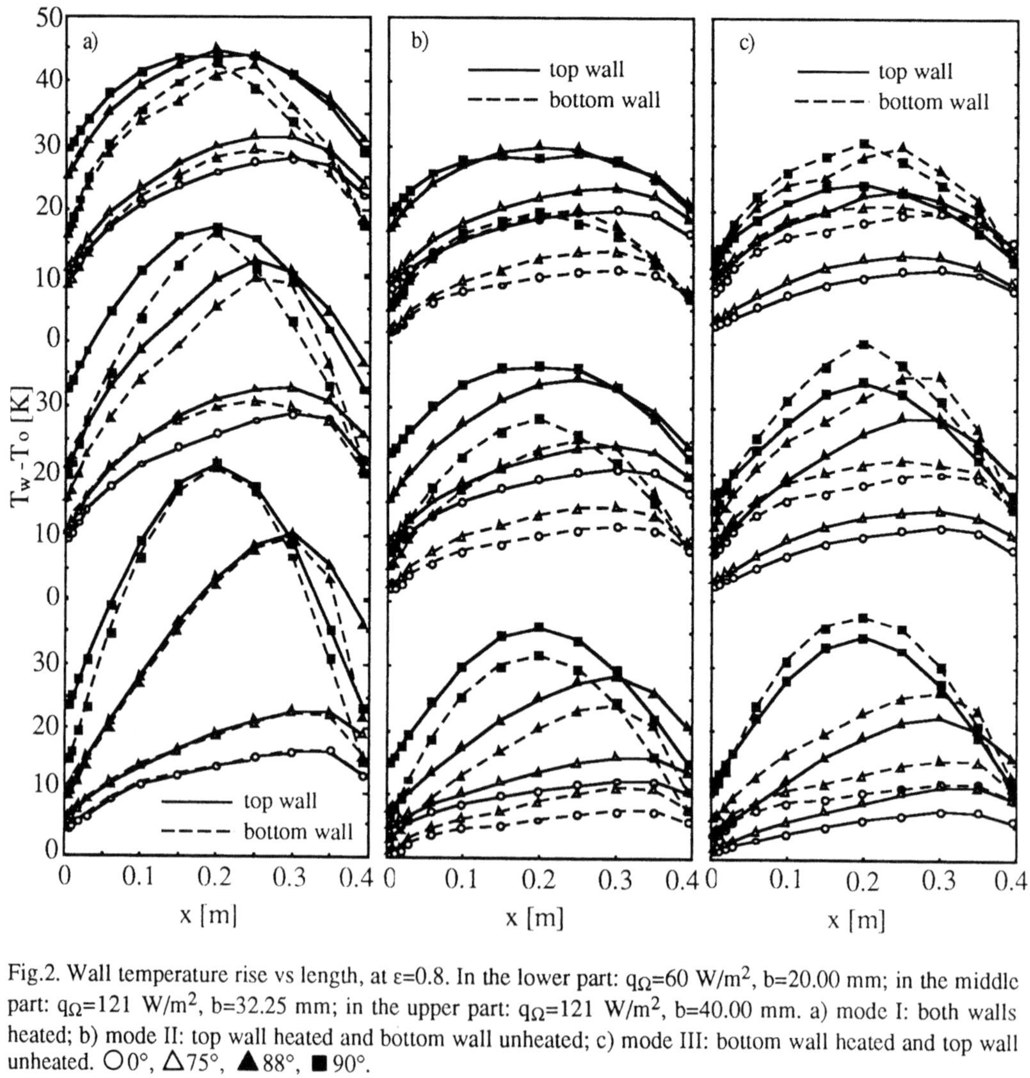

Fig.2. Wall temperature rise vs length, at ε=0.8. In the lower part: q_Ω=60 W/m^2, b=20.00 mm; in the middle part: q_Ω=121 W/m^2, b=32.25 mm; in the upper part: q_Ω=121 W/m^2, b=40.00 mm. a) mode I: both walls heated; b) mode II: top wall heated and bottom wall unheated; c) mode III: bottom wall heated and top wall unheated. ○ 0°, △ 75°, ▲ 88°, ■ 90°.

upper and the lower walls become nearly equal when the inclination angle turns from 88° into 90°. This is likely due to the greater direct radiation from the channel walls to the surroundings via the edge sections as well as to the easier convective heat removal from the top surface at the higher spacings. As far as the asymmetric heating is concerned (modes II, III and IV) one can see that temperature profiles are not dissimilar up to the 75° inclination angle. At higher spacing and heat flux values the raising of the unheated wall temperature due to its irradiation from the heated wall enhances the chimney effect in the channel. We can notice that, in the same way as in the symmetric heating, the higher the spacing the lower the differences between the 88° and the 90° temperature values of each wall. The figure shows that at the highest gap the temperature profiles of the heated wall at 88° and 90° are almost coincident when the upper wall is heated (mode II); in this case the chimney effect at 88° is very poor and, as a consequence, the maximum wall temperature is attained not far from the mid-length line. On the contrary, at the same 88° inclination angle, the chimney effect is still present in mode III (lower wall heated). For heating mode II at the highest gap the maximum wall temperatures at 88° and 90° are very similar. As a matter of fact, the temperature of the unheated bottom wall is raised by the irradiation from the top wall; this causes an air inflow from both the ends of the 90° tilted channel towards its middle part, which contributes to drop the temperature of the upper wall. Furthermore, in the 88° tilted channel there is an air inflow only at the upper edge, which throttles down the channel flow rate, thus lowering the axial coordinate at which the maximum temperature is attained. Finally, the figure points out that in mode III at high inclination angles the bottom wall temperature profiles exhibit a change of their concavity.

Wall temperature rise above ambient temperature, as a function of the length, the angle of inclination and the heating mode, is presented in fig.3, for a 32.25 mm channel spacing, a 121 W/m^2 Ohmic heat flux and for two different values of the walls emissivity (ε=0.05 and 0.8). The comparison between the wall temperatures in a vertical channel symmetrically heated shows that they are higher at the higher emissivity. Thus, in this case the net radiation heat transfer from the channel walls to the ambient is higher because of the lower radiative thermal resistance and the higher driving force. On the contrary, the convective heat removal from the walls is lower, the Ohmic heat flux being the same and the conductive heat losses being nearly equal in the two cases. In the 75° inclined channel the variation in the maximum

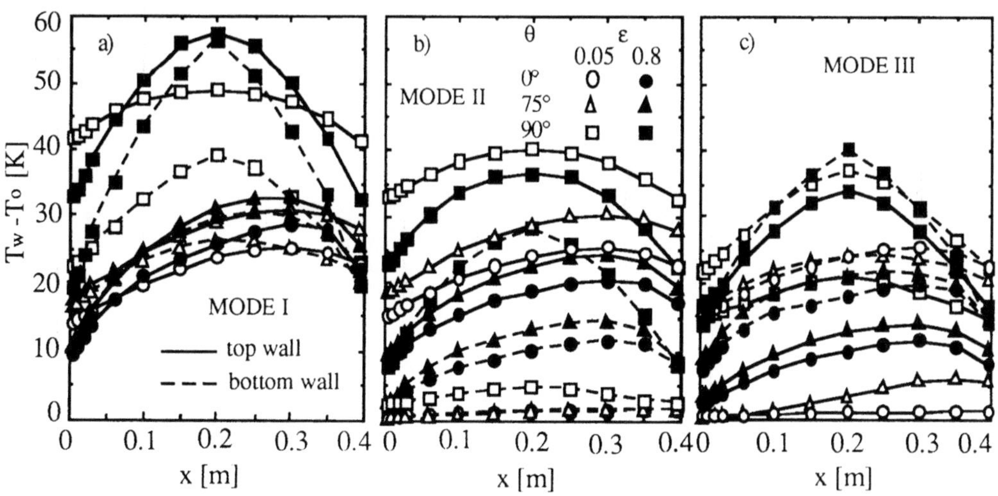

Fig.3. Wall temperature rise vs length, at ε=0.05 and 0.8, q_Ω=121 W/m^2, b=32.25 mm. a) both walls heated; b) top wall heated and bottom wall unheated; c) bottom wall heated and top wall unheated.

wall temperatures passing from ε=0.05 to ε=0.8 is lower on the top wall while the difference between the temperatures of the top and the bottom wall is, obviously, lower at the greater emissivity. The 0° and 75° asymmetrically heated channel exhibits higher temperatures of the unheated wall at the higher emissivity due to the fact that the unheated wall, irradiated from the heated one, can transfer heat to the air by convection, thus enhancing the chimney effect. In mode II, up to 75°, the shape of the wall temperature profiles practically is not affected by the emissivity whereas the unheated wall temperature increases at increasing angles only at the higher emissivity. As far as the horizontal channel is concerned, one can notice temperatures in the middle region of the ε=0.8 walls higher than those of the ε=0.05 walls; the contrary occurs in the entrance and exit regions for modes I and III. In mode II the temperatures of the heated upper wall are everywhere lower at the greater emissivity because of the convection from the lower, unheated, wall which helps cool the top plate.

The midplane dimensionless temperature and velocity of the air as a function of the dimensionless coordinate along the spacing for the symmetric heating at two values of the coordinate along the length (x^+=0.875 and x^+=0.9875) and at Ra=$1.4\cdot10^3$ (q_Ω=60 W/m^2 and b=20.00 mm), for different emissivity and inclination angle values, are presented, respectively, in figs. 4 and 5. Figure 4 shows that, at all angles, temperature profiles are practically independent of emissivity at x^+=0.875 whereas near the exit (x^+=0.9875) they are considerably different since radiative end effects are higher for ε=0.8. One can also notice that near the top edge the temperature of the air is almost unaffected by the angle and by the emissivity in the middle of the channel, while the temperature distribution is highly

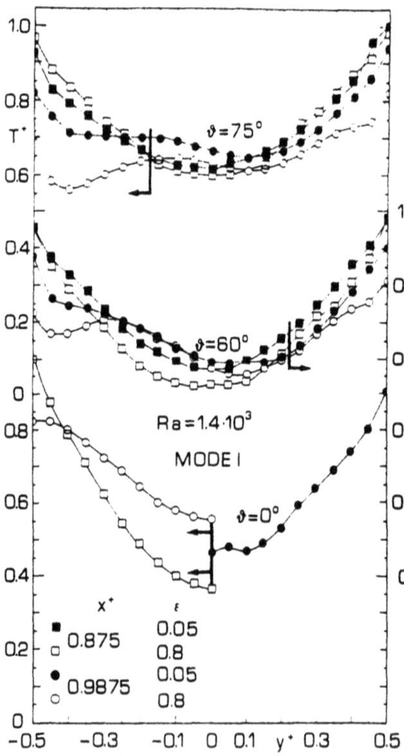

Fig.4. Midplane temperature of the air vs. coordinate along the spacing, at q_Ω=60 W/m^2, b=20.00 mm.

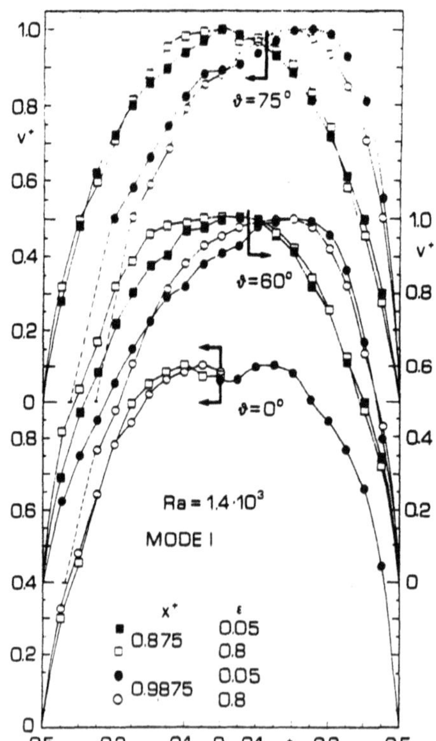

Fig.5. Midplane velocity of the air vs. coordinate along the spacing, at q_Ω=60 W/m^2, b=20.00 mm.

'able I. Coefficients in equation (9) for Nu at $\varepsilon=0.8$ with D=1.3°.

Mode	A	B	C	r^2
I	-0.991	0.668	-0.0628	0.980
I + IV	-0.886	0.533	-0.0455	0.977
I + IV	-0.838	0.550	-0.0502	0.983
All	-0.876	0.554	-0.0465	0.957

Table II.Coefficients in equation (9) for Nu* at $\varepsilon=0.8$ with D=1.7°.

Mode	A	B	C	r^2
I	-0.939	0.672	-0.0640	0.960
II + IV	-0.539	0.486	-0.0410	0.989
III + IV	-0.627	0.515	-0.0428	0.982
All	-0.638	0.524	-0.0453	0.960

ependent on the inclination angle in proximity of the wall. The same figure points out some istortion of the air temperature profiles near the upper end of the bottom wall: the higher the nclination angle the greater the deformation of the profile and the depth of the spacing oncerned. The afore described effect is caused by the external air inflow along the upper dge of the bottom plate, which is, of course, more pronounced at higher tilting angles. The ir temperature distribution is deformed, albeit less markedly, also at the lower emissivity alue, as it was confirmed by some flow visualization tests. Figure 5 exhibits no dependence of velocity profiles on the emissivity except that in the inflow region. In fact, for $\varepsilon=0.8$ a ignificant reverse flow could be already noticed at 60° tilting angle whereas for $\varepsilon=0.05$ it tarted only at 75°. The dashed lines indicate that in proximity of the bottom wall the neasured values of the air velocity were highly affected by the presence of the hot wire. Vear the channel exit ($x^+=0.9875$) the greater the angle the greater the skewing of the elocity profiles towards the top wall while the velocity distribution seems to be independent of the inclination angle at $x^+=0.875$.

The convective as well as the convective and radiative average Nusselt numbers were correlated to the channel Rayleigh number and the inclination angle, for any heating mode nd for $\varepsilon=0.05$ and 0.8 in the range 0°-90° of the tilting angle and in the range $6.0\cdot10^2$-$.0\cdot10^4$ of the channel Rayleigh number, by the equation

$$\log Nu = A + B \log[Ra \cos(\theta - D)] + C \{\log[Ra \cos(\theta - D)]\}^2 \qquad (9)$$

he coefficients in equation (9) and the regression coefficients are reported in tables I, II and II. The same equation correlates data for all heating modes by the coefficients in the last line of the tables. A better fit of experimental data was obtained by correlations in terms of $Ra\cdot\cos\theta$ rather than in terms of Ra. Diminishing the tilting angle by a constant value, D, llows to correlate also experimental data at $\theta=90°$. The values introduced in equation (9) are hose which give the highest regression coefficient when all data are taken into account. Either convective and convective-radiative Nusselt numbers as a function of the channel Rayleigh number and the cosine of inclination angle are plotted in fig.6.

'able III. Coefficients in equation (9) for Nu at $\varepsilon=0.05$ with D=2.7°.

Mode	A	B	C	r^2
I	-0.509	0.401	-0.0254	0.992
I + IV	-0.435	0.373	-0.0217	0.991
II + IV	-0.456	0.372	-0.0192	0.989
All	-0.463	0.382	-0.0225	0.998

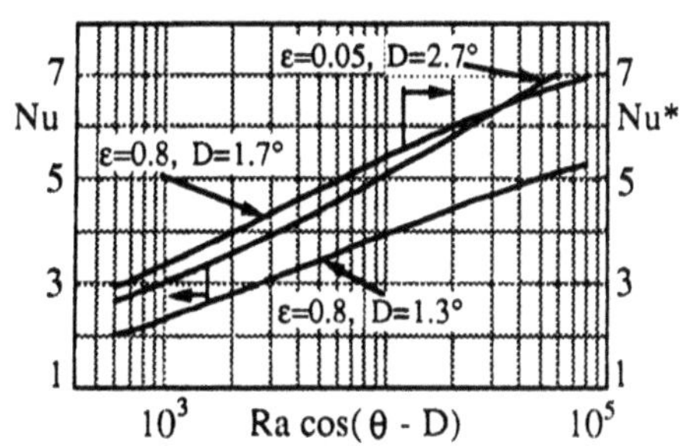

Fig.6. Nusselt numbers vs. Rayleigh numbers.

5. Conclusions

Wall temperature profiles up to 75° tilting angles are very similar at all heating modes and at a given emissivity.

In the symmetric heating the chimney effect is slightly lower at the higher emissivity whereas in the asymmetric one the irradiation from the heated wall allows the unheated one to participate to the convective heat transfer, thus lowering the temperature of the heated wall. In the asymmetric heating the maximum temperature of the heated wall decreases at increasing emissivity at a rate which depends on the heating mode. For b=32.25 mm and q_{Ω}=121 W/m^2, passing from ε=0.05 to ε=0.8 the maximum heated wall temperature drop is 18% in a vertical channel and 20% in a 75° tilted channel in mode II; its decrease is 21% in a vertical channel and 11% in a 75° inclined channel in mode III.

A high emissivity plays a considerable role on the air flow since a reverse flow near the exit of the channel starts at an inclination angle ($\simeq$60°) lower than that ($\simeq$75°) which determines the reverse flow at the lower emissivity.

Acknowledgement. The C. N. R. funded this work with the Grant N. 90.01625.CT07.

References

1. Aung, W., Fletcher, L. S., and Sernas, V.: Developing laminar free convection between vertical flat plates with asymmetric heating. *International Journal of Heat and Mass Transfer* 15 (1972) 2293-2308.
2. Wirtz, R. A., and Stutzman, R. J.: Experiments on free convection between vertical plates with symmetric heating. ASME *Journal of Heat Transfer* 104 (1982) 501-507.
3. Webb, B. W., and Hill, D. P.: High Rayleigh Number Laminar natural convection in an asymmetrical heated vertical channel. ASME *Journal of Heat Transfer* 111 (1989) 649-656.
4. Kennedy, K., J., and Kanhel, J.: Free convection in tilted enclosures. ASME *HTD* 28 (1983) 43-47.
5. Azevedo, L. F. A., and Sparrow, E., M.: Natural convection in open-ended inclined channels. ASME *Journal of Heat Transfer* 107 (1985) 893-901.
6. Cheng, K. C. and Kimura, T.: Observations of convective instability phenomena in slightly inclined air layers heated from below: effect of air layer thickness. ASME *HTD* 178 (1991) 55-63.
7. Sparrow, E. M., and Carlson, C., K.: Local and average natural convection Nusselt numbers for a uniformly heated, shrouded or unshrouded horizontal plate. *International Journal of Heat and Mass Transfer* 29 (1986) 369-379.
8. Manca, O., Nardini, S., and Naso, V.: Experiments on natural convection in inclined channels. ASME *HTD* 212 (1992) 41-46.
9. Manca, O. and Naso, V.: Experimental analysis of natural convection and thermal radiation in vertical channels. ASME *HTD* 145 (1990) 13-21.
10. Carpenter, J. R., Briggs, D. G., and Sernas, V.: Combined radiation and developing laminar free convection between vertical flat plates with asymmetric heating. ASME *Journal of Heat Transfer* 98 (1976) 95-100.
11. Yamada, Y.: Combined radiaton and free convection heat tranfer in a vertical channel with arbritary wall emissivities. *International Journal of Heat and Mass Transfer* 31 (1988) 429-440.
12. Sparrow, E. M., Shah, S. and Prakash, C.: Natural convection in a vertical channel: I. Interacting convection and radiation. II. The vertical plate with and without shrounding. *Numerical Heat Transfer* 3 (1980) 297-314.
13. Manca, O., Nardini, S. and Naso, V.: Experimental analysis of natural convection in a tilted channel. *Proceedings, 11th Australasian Fluid Mechanics Conference*, Hobart 1 (1992) 131-134.
14. Miyamoto, M., Katoh, Y., Kurima, J. and Sasaki, H.: Turbulent free convection heat transfer from vertical parallel plates. *Proceedings, 8th International Heat Transfer Conference*, C. L. Tien et al., ed., Hemisphere Publishing Corp., Washington, D.C. 4 (1986) 1593-1598.
15. Kline, S. J., and McClintock, A. F.: Describing Uncertainties in Single Sample Experiments. *Mech. Engin.* Jan. (1953) 13-21.

4. NUMERICAL AND EXPERIMENTAL ANALYSIS OF ELECTRONIC PARTS

THERMAL STUDY OF A LASER DIODE USING A FINITE ELEMENT METHOD ASSOCIATED WITH A MESHING SUPERIMPOSITION METHOD

N. OTTAVY, M. BOURHRARA, Ecole Nationale Supérieure de Mécanique et d'Aérotechnique, Laboratoire d'Informatique, 86034 POITIERS Cedex, France

J-P. LE JANNOU, P. PARIS, Centre National d'Etudes des Télécommunications, 22301 LANNION Cedex, France

Abstract

Temperature is an important parameter as regards the functioning of laser diodes. However, since the power is dissipated in a very small volume in comparison with the whole laser, digital simulations require a meshing which is complex and long to realize. For these reasons, a meshing superimposition method (MSM) was used to calculate the temperature of the laser diodes. This method is a variant of the finite-element method with make it possible to superimpose a coarse mesh (on the whole field), and a fine mesh in areas with drastic temperature profiles. The connection between the two meshes is made analytically which avoids connection problems regarding meshes of different sizes.

1 - INTRODUCTION

The temperature is an important parameter in the operation of laser diodes. The power output is a particularly sensitive factor. It is thus necessary to thermally optimize the circuit in which the diode is located. When the geometry of the circuit is fairly simple, analytic methods are frequently used to solve the temperature problem. But it becomes very difficult to use analytic solutions for complicated geometries, and digital simulations are then more convenient.

However, the disadvantage of using digital simulations is the necessity of preparing a mesh, and this is not easy since power is dissipated within very small dimensions relative to surrounding materials. A very fine mesh has to be used in this area, gradually leading to a coarser mesh over the entire domain to give a reasonable number of elements and consequently a reasonable CPU time. In order to solve this problem in the case of the laser diode, we used a finite element method in which it is possible to make a window in the heating area, within which a fine mesh can be created. This fine mesh is then connected to the coarse mesh using the Meshing Superimposition Method (MSM-Réf [1]). This method has the advantage that it is not needed to solve connection problems as for classical finite elements. This also leads to a minimum number of nodes and consequently calculation time savings. This method was compared with an analytical solution obtained using the integral transformation method as described by Ozisik [2].

C. J. Hoogendoorn et al. (eds.), Thermal Management of Electronic Systems, 129–138.

2 - THE PROBLEM STUDIED

The laser diode studied is shown in Figure 1.

This is an indium phosphide (InP) parallelepiped, with the main dimensions shown on Figure 1. The active part of the diode that emits the laser beam is located within a rectangular volume with length equal to U, centered on the x axis. The laser diode is placed on a base which removes the heat dissipated inside the active zone. Layers of different natures are placed between the active zone and the base.

Convection phenomena on the diode surface are insignificant compared with the removal of heat toward the base and may be neglected : the various

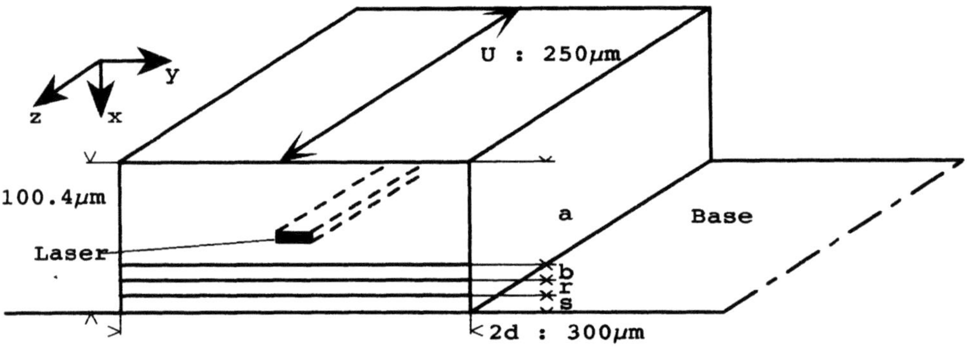

FIGURE 1

faces of the diode in contact with the ambient air are therefore considered to be adiabatic. Since it is assumed that the power dissipated in the laser is the same in the various x and y planes, the resulting problem becomes two-dimensional. Moreover, since the laser is symmetrical about the ox axis, the temperature distribution is also symmetrical about the x axis.The resulting system to be solved is shown in figure 2.

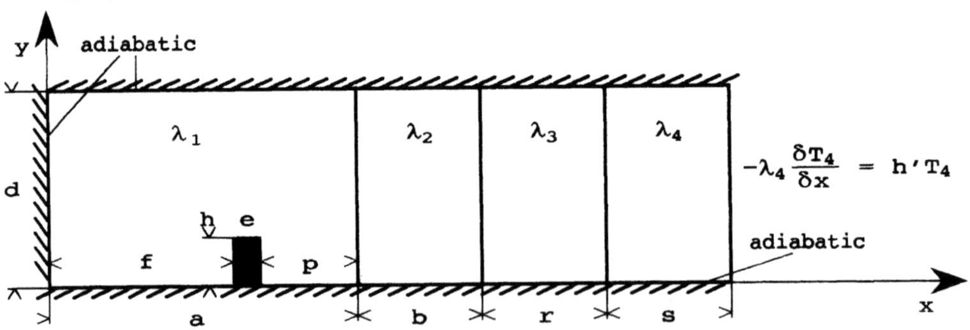

FIGURE 2

The difficulty in solving this problem is due to the small size of the laser forming the heat source : h is 1 µm while e is 0.2 µm (the figure is not drawn to scale), and the power density dissipated in it is very high (of the order of 1.3×10^{12} kW/m^3). Classical digital methods requiring the creation of meshes lead to very complicated meshes and a large number of nodes. In order to avoid these disadvantages, we used MSM method which can create a fine mesh within a normal sized mesh, and in which the connection of the function representing the temperature is made analytically and transparently to the user. In order to validate this method, it was compared with an analytical solution of the problem posed, obtained by using the integral transformation method [2]. The problem posed is relatively simple and consequently an analytical solution is possible. But these solutions become inextricably complex for more complicated geometries, whereas it is still possible to use a MSM type method.

3 - THE MESHING SUPERIMPOSITION METHOD (MSM)

Use of the Finite Elements Method to process the laser diode problem requires connecting a very fine mesh around the heat source area, to a much coarser mesh in the rest of the material.

The program uses bi-linear 4-node finite elements as an approximation method. These elements have the peculiarity of allowing a power injection over part of their surface area. Using the MSM method , solutions for two different sized meshes can be accurately connected. In the problem studied, a rectangular window is made around the heat source area by covering the area with the highest temperature gradient. If the mesh has to be modified, the modification needs only to be made inside the window. This method has already been described in detail [3]. We will mention its main features.

3.1 - The MSM method

The MSM method has two objectives :

- to facilitate coupling of distinct approximation methods defined on different subdomains which cover the main domain.
- to accelerate the sequential solution of a series of similar problems, each characterized by a specific placement of singularities characterizing the physical problem.

For the laser diode study, we only use the features resulting from the first objective. But the second objective is very useful when the problem to be studied includes singularities of the same type. For instance this is the case for each component on an electronic board, or for each transistor on an integrated circuit.

These objectives are achieved by making a structural decomposition of the P problem into subproblems P_i, each of which is as independant as possible (Figure 3). This decomposition is selected such that :

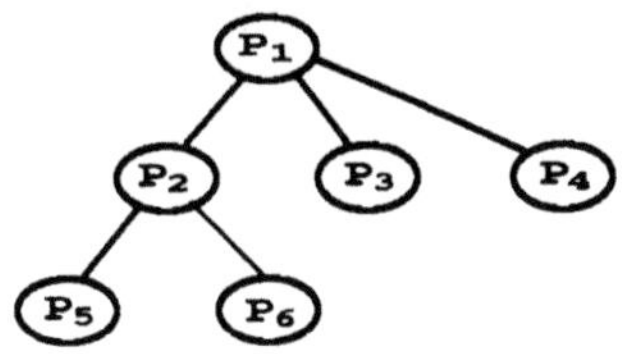

Structure of problem P

FIGURE 3

- a subproblem P_i is associated with each singularity liable to be moved.
- a subdomain and a suitable approximation method are attached to each subproblem P_i.

Since the problem P is global, it is obvious that subproblems P_i associated with a structural decomposition of P, cannot be completely independent. The MSM method then consists in separating the data for each subproblem into two distinct parts :

- the first is made up of data which are independent of the context within which subproblem P_i is located ; these data can then be the subject of a decentralized calculation for P_i.

- the second is made up of data which are dependent of the context
within which subproblem P_i is located, these data provide coordination between the various subproblems involved in the decomposition of P.

The advantages of this approach by decentralization and coordination are known [4, 5] and in particular make it possible to :

- execute in parallel a large proportion of the calculations necessary for the solution of the global problem. From this point of view, the MSM method therefore facilitates parallelization of finite element calculations.

- reuse partial results associated with unmodified subproblems, when the global problem has to be solved several times. From this point of view, the MSM method can quickly solve a series of similar problems and also facilitates the creation of data libraries associated with frequently used decentralized subproblems.

3.1.1 - MSM windowing

The use of the decentralization-coordination approach in the MSM method depends on the use of geometrically independent "windows" added into the support domain. In order to illustrate this approach, consider the example of a support made up of two components (for example two laser diodes), and attach a window to each component (Figure 4) ; we can then deduce a tree structure of the support domain.

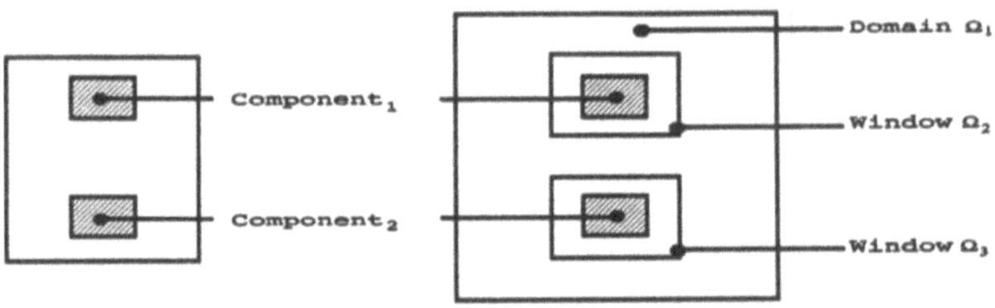

FIGURE 4

The problems P_i making up problem P are then built up as follows : P_1 is the problem deduced from P by eliminating the two components and P_2 is the problem P restricted to window Ω_2 (the same applies for P_3 and window Ω_3). Then assume that the windows are separate or included to deduce a tree structure for the decomposition of the Ω_1 domain (Figure 5).

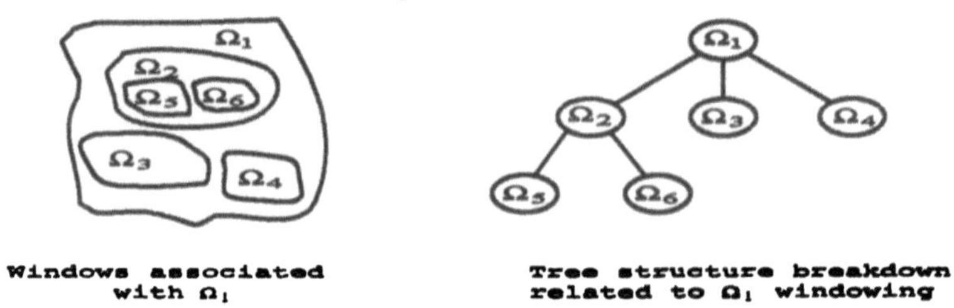

FIGURE 5

In general when P is a problem posed on the Ω_1 domain, a procedure similar to the above procedure (Figure 5) can be used to define a tree structure decomposition of the problem P. Note that there are many possible decompositions of a problem, all concerning the choice of windows and the definition of the problems P_i (particularly when portions of the boundary related to the description of boundary conditions for problem P are covered by a window). Note also that a modification to one of the parameters associated with a problem P_j (for example moving the Ω_j window) only affects the branch of the tree connecting P_j to the root. The result is that all decentralized calculations carried out to solve problems P_i are preserved, except possibly those belonging to the branch connecting P_j to the root of the tree (Figure 6). The result is that the MSM method can result in substantial time savings compared with standard methods for the solution of a series of similar problems.

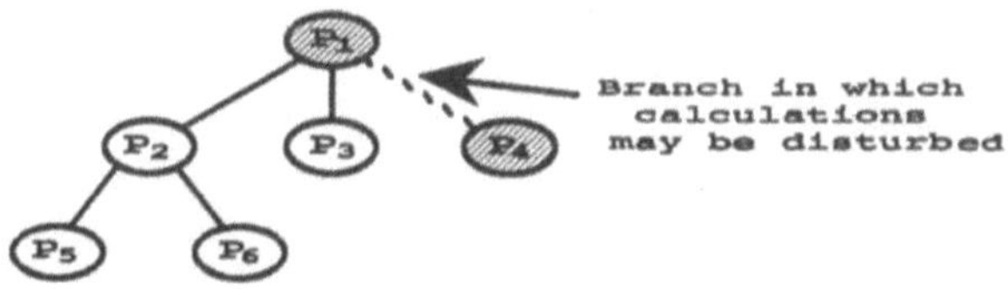

Case of displacing the window Ω_4 to which the problem P_4 is related

FIGURE 6

3.1.2 - <u>MSM approximation space</u>

In order to describe details of the mechanism for the construction and manipulation of the tree associated with processing the problem using the MSM method, we will now consider the problem of a support with a single component, and we will associate a window Ω_2 with the component. The windowing method can be used to build a tree structure decomposition of the domain Ω and the problem P.

In order to solve this problem using the MSM method, we then use a classical finite element method on each domain Ω_1 and Ω_2. This leads to defining a mesh specific to each domain (Figure 7) :

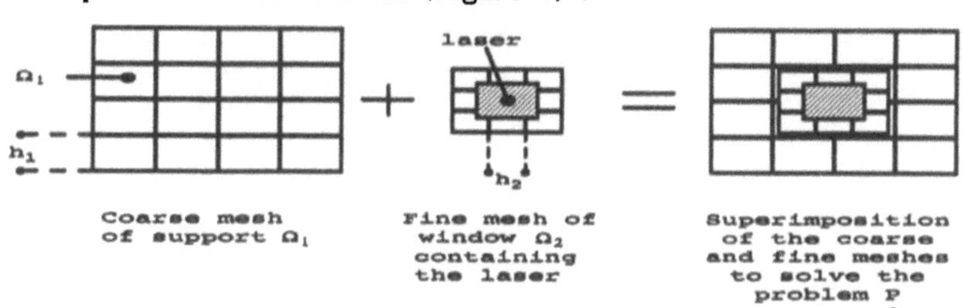

FIGURE 7

Taken separately, the meshes used are typically finite element meshes and are treated as such. The essential difference with classical meshes is due to the fact that these meshes superimpose and are completely independent both geometrically and functionally. The "fine mesh" is related to the component and covers the critical area in which the temperature field varies quickly, whereas the coarse mesh is related to the support and represents the temperature field in an area in which it is fairly uniform (Ω/Ω_2). The global approximation space V_h ($h = h_1, h_2$) is defined by a functional aggregation technique of approximation spaces V_{h1} and V_{h2}, usually associated with the coarse mesh and fine mesh by the finite elements method :

$$V_h = \lambda^- V_{h1} + \lambda^+ V_{h2} \tag{1}$$

Functions λ^- and λ^+ involved in the definition of V_h are designed to ensure uniformity of functions of V_h ; they partition the unit of the Ω domain ;

$$\lambda^-(x) + \lambda^+(x) = 1 \qquad ; x \in \Omega \tag{2}$$

such that the V_h space becomes coincident with V_{hi} on $\Omega\backslash\Omega_2$ (the same applied for V_{h2} on $\Omega_2\backslash\Lambda$).

In practice, the first step in defining λ^+ and λ^- is to obtain an application λ defined on Ω_2 (see Figure 8) ; the application λ^+ is then defined by extension through zero :

$$\lambda^+(x) = \lambda(x) \qquad ; x \in \Omega_2$$

$$\lambda^+(x) = o \qquad ; x \in \Omega \text{ et } x \notin \Omega_2 \tag{3}$$

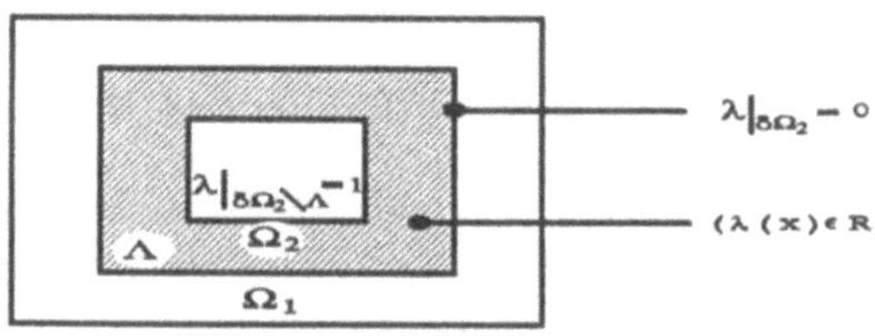

Conditions satisfied by λ

FIGURE 8

The application λ^- is deduced from λ^+ by formula (2).

Consequently, for each element $u_{hi} \in V_{hi}$ ($1 \le i \le 2$), the function u_h of V_h associated with them by the MSM method, satisfies the following relations :

$$u_h(x) = \lambda^-(x)u_{h1}(x) + \lambda^+(x)u_{h2}(x) \qquad ; x \in \Omega$$

$$u_h(x) = u_{h1}(x) \qquad ; x \in \Omega \text{ et } x \notin \Omega_2 \tag{4}$$

$$u_h(x) = u_{h2}(x) \qquad ; x \in \Omega_2 \text{ et } x \notin \Lambda$$

The overlap zone Λ of supports associated with the two functions $\lambda^- u_{h1}$ and $\lambda^+ u_{h2}$ is called the attachment strip. Note that this attachment strip may vary in size and that it will always be selected so as to reduce the calculation size. The concept of the attachment strip is the basis of the functional aggregation technique used in the MSM method. Although this functional aggregation approach is frequently used in theoretical work, there are very few examples of its application in practice. However work related to the definition of layers for approximating a set of given values on a domain forms another example of the application of this approach in practice, in a similar domain.

3.1.3 - Linear system

The linear system associated with the related problem defined on V_h, inherently possesses a block structure :

$$\begin{vmatrix} A_{11} & A_{12} \\ A_{21} & A_{22} \end{vmatrix} \begin{vmatrix} U_1 \\ U_2 \end{vmatrix} = \begin{vmatrix} L_1 \\ L_2 \end{vmatrix} \tag{5}$$

The blocks forming the linear system may be classified in three groups depending on whether they contain completely decentralized (A_{22}, L_2), partially decentralized data (A_{11}, L_1) or coordination data (A_{12}, A_{21}). In practice :

1) Since the definition of functions $\lambda^+$and λ^-depends essentially on the Ω_2 domain, it is obvious that quantities associated with blocks A_{22} and L_2 are composed of completely decentralized data which depend only on the problem P_2, the domain Ω_2 and the window related to Ω_2.

2) Similarly, quantities A_{11}, L_1 are partially decentralized blocks since they are composed of decentralized elements and coordination elements. Decentralized elements are uniquely related to problem P_1, whereas coordination elements depend both on problem P_1 and the attachment strip Λ. The decentralized element calculation is classical and therefore causes no problems. Similarly, the coordination elements calculation only involves problem P_1 and the attachment strip Λ, which has the effect of reducing the size of the calculation for this coordination phase. In practice, blocks A_{11}, L_1 can be quickly determined, since the decentralized elements have already been calculated when a window is added or moved.

3) Blocks A_{12} and A_{21} only are composed entirely of coordination data depending on P_1 and P_2, the window Ω_2 and the attachment strip Λ. However, calculations for these quantities involve only the attachment strip Lambda, thus reducing the size of the calculations in this final coordination phase.

This decomposition demonstrates the difference between completely decentralized elements (A_{22}, L_2), partially decentralized elements (A_{11}, L_1), which may be completely or partially calculated (or precalculated) for each subproblem, and coordination data A_{12} and A_{21} which must be recalculated whenever window Ω_2 is to be moved. For more accuracy, note that the linear system associated with the MSM method has a specific block structure that must be taken advantage of in solution methods. The crude use of an elimination type of method would make it necessary to consider the global matrix which would be unacceptable both in terms of space and calculation time. However, the separate inversion of each block associated with the submesh can reduce the size of the calculations and maintain independence of subproblems P_1.

This can result in a very important calculation time saving which has been estimated in some cases (1).

3.2 - Numerical example

3.2.1 - Reference example

The problem posed in § 2 (Laser n° 1) has been solved both by the analytical method and using the MSM method. Values of parameters are shown on Figure 9.

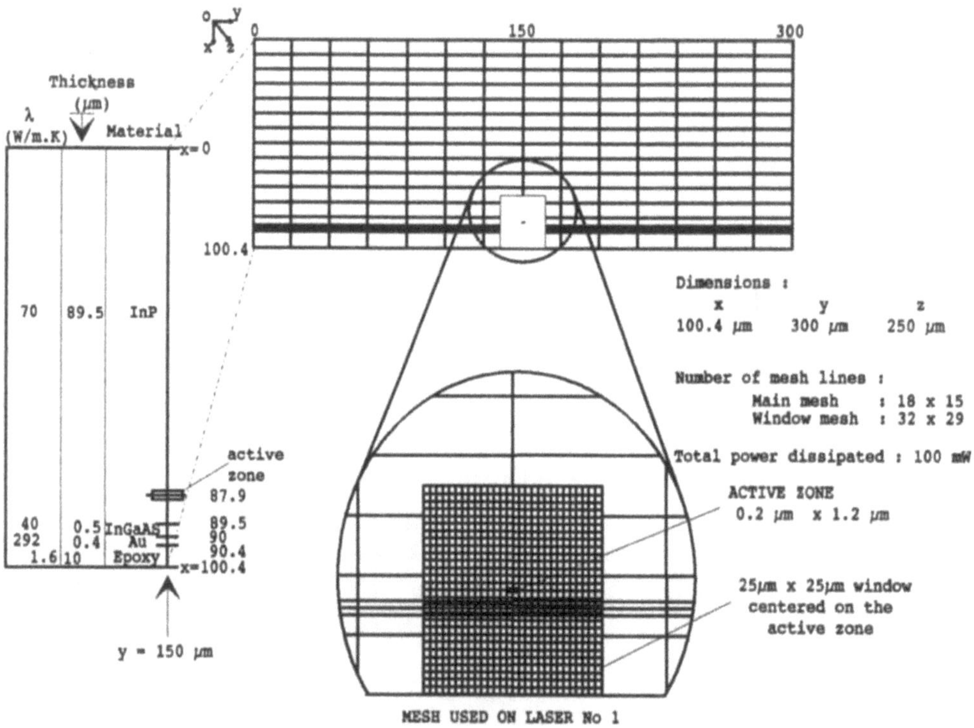

FIGURE 9

With the MSM method, the following meshes are used :

coarse mesh : 18x 15 meshes

fine mesh (see inserted figure 9) : 32 x 29 meshes in a 25 μm x 25 μm window.

Table 1 shows the temperatures calculated along the axis of symmetry. Note the excellent agreement between the two calculations : temperatures are identical to the third significant digit.

x µm	0	87,8	87,9	88,0	89,5	90	90,4	100,4
$T_{analytical}$	8,92	14,68	14,79	14,70	13,29	12,99	12,97	0
T_{MSM}	8,93	14,64	14,80	14,66	13,23	12,94	12,92	0

Table I

3.2.2 - Example of a laser diode with a more complicated structure :

In practice, the structure of laser diodes is frequently not as simple as that for laser N°1. Structures may be similar to those shown on Figure 10 (Laser N° 2).

An analytical solution can also be used to process this geometry, but it would be much more complicated. This problem can be handled easily by the MSM method. It was solved using the data shown in Figure 10.

The maximum temperature within the heating area (270 mW in a volume of 0.1 x 0.8 x 1000 µm) was calculated to be 4,2° using MSM. Zero reference temperature were imposed at the bottom surface of copper. The coarse mesh was of 15 x 35 and in the window of 100 x 100 µm, the mesh was of 25 x 25.

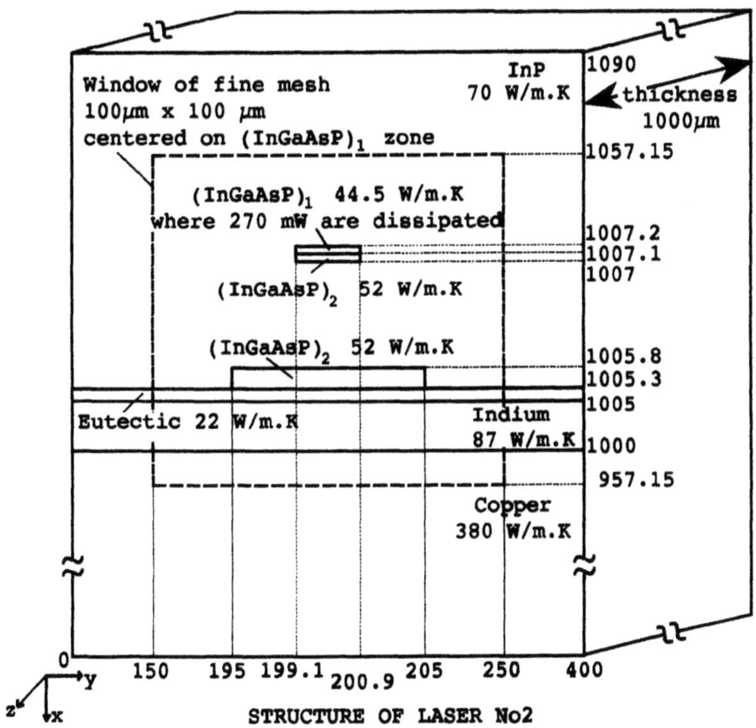

STRUCTURE OF LASER No2
(THE DRAWING IS OUT OF SCALE, DIMENSIONS ARE GIVEN IN µm)

FIGURE 10

Values of some conductivity materials are not well known. This is the case for the conductivity of $(In\ Ga\ As\ P)_1$ and for $(In\ Ga\ As\ P)_2$. If we use calculation method from [6] the values are the ones taken in the preceeding calculation because :

$$(\text{In Ga As P})_1 \text{ exactly : } In_{0,60}\ Ga_{0,40}\ P_{0,15}\ As_{0,85}$$

$$\text{and } (\text{In Ga As P})_2 : \quad In_{0,73}\ Ga_{0,27}\ P_{0,40}\ As_{0,60}$$

If we use [7], we find for $(\text{In Ga As P})_1$: 4,5 W/m°-C

$(\text{In Ga As P})_2$: 4 W/m°-C

With these values, the maximum temperature becomes 5,88°C, an important increase of 40 %.

4 - CONCLUSION

A new type of method based on finite elements has been used to study the thermal behavior of laser diodes. This is the meshing superimposition method (MSM) for connecting different size meshes without being concerned with how they are connected ; the continuity of the function representing the phenomenon is done analytically, and transparently to the user. This feature is particularly suitable for determining the temperature in a laser diode in which the power dissipation area is very small compared with the total area ; a rectangular window with a fine mesh is placed around the heating area, in the area with the highest temperature gradient. A coarser mesh is used elsewhere.

This method was used to solve a laser diode problem with a fairly complicated geometry (multilayer structure), but which can also be analytically solved. Very good agreement is obtained between the results for the two methods. The MSM method was then applied to the case of a typical industrially manufactured laser for which analytical solutions are complicated to obtain.

Finally, it is noted that the MSM method can be used for many applications other than those considered in this study :

. it can be used to associate different size meshes, and also different approximation methods and analytical solution methods.

. moving a window does not make it necessary to repeat all calculations, such that a response can be obtained quickly when modifications are made, making it very suitable for CAD applications.

REFERENCES

1 - M. Ottavy, G. Pierra, Proc Heat Transfer Electronic Equip, 953 (1988)

2 - M.N. Ozisik, Heat Conduction (J. Wiley and sons, 1980)

3 - M. Ottavy, "Contribution à l'algorithme numérique en environnement CAO : Méthode de Superposition de Maillage en éléments finis", Ph. D. Thesis, University of Poitiers (1989)

4 - M.D. Mesarovic, D. Macko, Y. Takahora, Theory of hierarchical multilevel Systems (Academic Press, New-York, 1970)

5 - J.L. Lions, G.I. Marchouck, "Sur les méthodes numériques en sciences physiques et économiques" (Dunod, Paris, 1974).

6 - H. Burkhard, H.W. Dinges, E. Kuphal, J. Appl. Phys. 53, 655 (1982).

7 - Properties of indium phosphide - Inspec - EMIS Data reviews n° 6-1991.

Computer-Aided Optimization of Electrothermal CMOS-Compatible Infrared Sensors

G. WACHUTKA, J. FUNK, AND H. BALTES
Physical Electronics Laboratory, ETH Zurich, 8093 Zurich, Switzerland

Abstract: We present a detailed study of the operation of miniaturized electrothermal infrared detectors. On the basis of a three-dimensional numerical thermal analysis, the sensor performance is optimized with respect to several structural parameters. Moreover, we validate an analytical model of integrated thermopiles from which we deduce general scaling laws for the practical design. We find that an unconventional topography, namely a meander geometry, may significantly enhance the sensor output signal.

1. Introduction

Using industrial CMOS technology for the fabrication of miniaturized thermoelectric infrared sensors [1] is attractive as it allows the much desired cointegration of sensor element and signal processing circuitry on the same chip. With this approach, the inexpensive batch fabrication of microsystems with high sensitivity, low power consumption and small overall size becomes feasible [2,3]. The basic operational principle consists in converting the absorbed infrared radiation heat into an output voltage by means of integrated thermopiles. In consequence of the small feature sizes, temperature variations in the range of a few mK can be detected.

However, because of the narrow technological base provided by CMOS compatible materials, an optimized design of the thermoelectric component turns out to be indispensible for a satisfactory sensor performance. The optimization is complicated by surface heat transfer effects such as thermal conduction and convection by the ambient gas or radiative losses which have a crucial and, unfortunately, not easily controllable influence on the sensor signal.

Using three-dimensional numerical thermal analysis as well as an analytical equivalent network model, we studied the sensor operation in detail. The calculations allow us to discuss the optimum sensor performance with respect to several structural parameters. In particular, we deduce general scaling laws for the practical design, suggesting a new topography with an enhanced output signal.

2. Sensor Structure and Operational Principle

Fig. 1 shows the typical structure of the infrared radiation sensors investigated in this work. It consists of a $3\mu m$ thick cantilever beam composed of several layers of silicon dioxide, nitride, polysilicon and aluminum, as it can be manufactured by means of a standard CMOS process with subsequent anisotropic etching. The cavity underneath the free-standing beam structure effects thermal isolation between sen-

C. J. Hoogendoorn et al. (eds.), Thermal Management of Electronic Systems, 139–148.

sor element and supporting wafer.

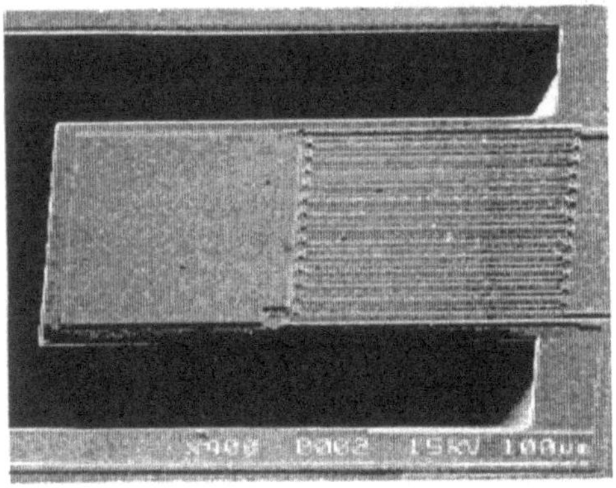

Fig. 1: SEM micrograph of a thermoelectric infrared radiation detector.

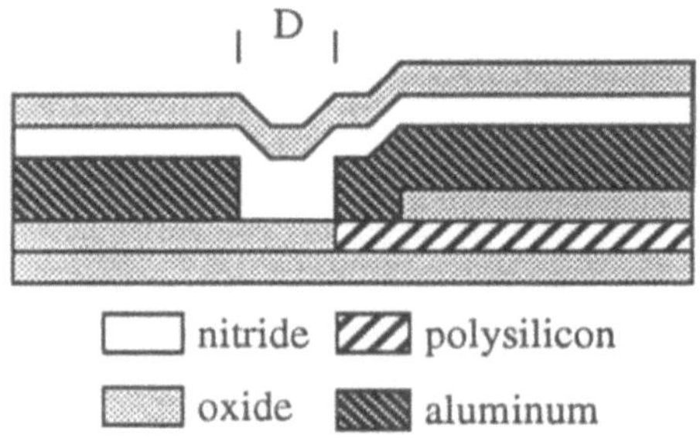

Fig. 2: Cross-section of the structure on the left showing IR-absorber (left) and thermopile (right). D denotes the width of the oxide/nitride barrier in between.

The infrared radiation absorber is located on the foremost half of the cantilever beam. It comprises a buried aluminum layer acting as radiation reflecting mirror, and a nitride/oxide sandwich structure on top, where the incoming and the reflected radiation is dissipated into heat. The topside of the other half of the beam is occupied by a thermopile which detects the temperature difference between absorber and bulk silicon (chip). The thermopile is composed of a sequence of aluminum/polysilicon thermocouples arrayed in parallel and connected in series, with the "hot" contacts situated near the absorber and the "cold" contacts on the bulk silicon. For electrical isolation, the radiation reflector and the thermopile are separated by a thin oxide/nitride wall of width D (Fig. 2).

The sensor operation is based on the effect that the incoming absorbed radiation power induces a temperature rise on the foremost part of the cantilever, which again is converted into an thermovoltage due to the Seebeck effect in the thermopile. In the steady-state, the temperature gradient along the thermopile is balanced by a corresponding heat flow from the absorber region to the bulk. Since this heat current is controlled by the impinging radiation flux, the temperature drop along the cantilever beam per absorbed radiation power is the higher, the lower we make the thermal conductance of the cantilever sandwich. Unfortunately, this quantity is predominantly determined by the contribution coming from the thermoelement materials, which are excellent thermal conductors. Thus, the optimization of the sensor response is subject to stringent constraints resulting from the properties of the employed materials and the CMOS process design rules.

3. Three-Dimensional Numerical Analysis

3.1 Model equations

Finding the optimum sensor structure is, in essence, equivalent to the problem of maximizing the temperature difference ΔT_{thp} between "hot" and "cold" contacts of the thermopile as a function of the structural parameters. For an accurate treatment of this problem, we calculated heat flow and temperature distribution in the interior of the sensor structure using the finite element package SESES [4]. It allows to solve the stationary heat flow equation

$$div\,(\kappa\,(x,y,z)\,\nabla\,T) \;=\; -P\,(x,y,z) \tag{1}$$

for complex structured three-dimensional domains. Here, κ, T and P denote the thermal conductivity, temperature distribution, and heat generation, respectively.

Assuming uniform absorption of the incoming radiation throughout the absorber layer, this heat current J_{in} is incorporated by the boundary condition

$$\kappa \cdot \frac{\partial T}{\partial n} \;=\; J_{in} \tag{2}$$

effective along the boundary of the absorber. As a value typical of the application envisaged, $J_{in} = 5\,W/m^2$ was used in our simulations. In view of this very small incoming radiation flux, it is essential to allow for heat losses by conduction and convection through the ambient air and, in linearized form, also for radiation losses by the boundary condition

$$-\,\kappa \cdot \frac{\partial T}{\partial n} \;=\; J_{out} \;=\; (h + 4\,\epsilon\,\sigma\,T^3_{amb})\,(T - T_{amb}) \tag{3}$$

Here, h denotes the effective surface heat transfer coefficient between sensor surface and ambient gas, including heat conduction to the walls of the surrounding cavity through the air in between (Fig. 3). T_{amb} is the ambient temperature, σ the Stefan-Boltzmann constant, and ϵ the emissivity. At the "cold" end of the thermopile, located on bulk silicon, an ideal heat sink is assumed ($T = T_{bulk} = 300\,K$).

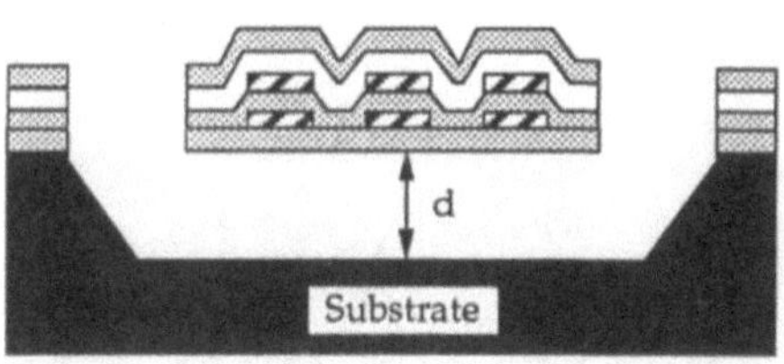

Fig. 3: Lateral cross-section of the structure shown in Fig. 1.

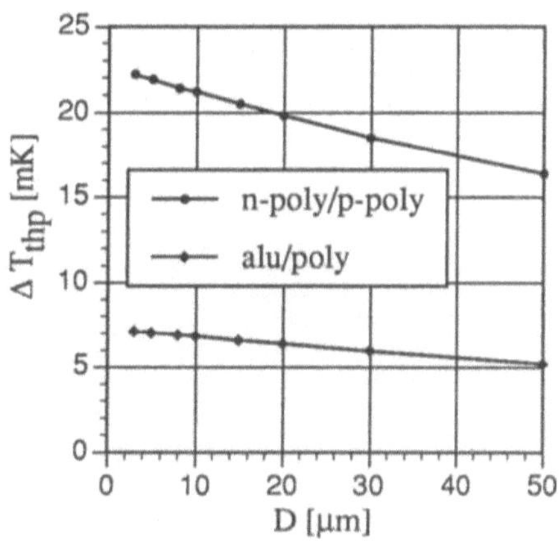

Fig. 4: Temperature drop ΔT_{thp} along the thermopile vs. width D of the oxide/nitride barrier for different thermocouples ($h = 50\,W/m^2K$, $J_{in} = 5\,W/m^2$).

Since the midplane perpendicular to the cantilever bisects the sensor structure into two symmetric halves, it is sufficient to restrict the simulation to one half. Along the mirror plane, reflecting boundary conditions are used. In this way, CPU-time and memory requirements can be economized.

The boundary value problem (1)-(3) is discretized using a hybrid finite element method [4]. Typical numbers of nodes for the entire 3D structure amount to about 50,000. The grid generation is performed partly by user-defined macroelements, partly by adaptive regridding algorithms.

3.2 Effective One-Dimensional Temperature Distribution

Fig. 5 shows a cut of two temperature profiles along the principal axis of the cantilever beam (i.e., the straight line from the cantilever tip to the bulk silicon, located in the midplane of the cantilever). We recognize a flat part with slightly convex curvature in the left half of the curve, where the absorber is situated, followed by a steep gradient over the oxide/nitride barrier and a moderately inclined slope along the thermopile. This is easily understood by considering the effective heat flow in the direction of the thermal gradient from the tip of the cantilever to the bulk. To this end, we average eq. (1) over cross-sections perpendicular to the principal axis (= x-axis). By virtue of the boundary conditions (2)-(3) we arrive at the effective one-dimensional heat flow equation

$$\frac{\partial}{\partial x}\left(\kappa_{av}(x)\frac{\partial\Theta}{\partial x}\right) = (h + 4\,\epsilon\,\sigma\,T_{amb}^3)\,\frac{U}{A}\,\Theta \;-\; \frac{U_{abs}}{A}\,J_{in} \;=:\; -\,P(\Theta,x) \qquad (4)$$

where $\kappa_{av}(x)$ denotes the average thermal conductivity and $\Theta(x) := T_{av}(x) - T_{amb}$ the average temperature elevation above T_{amb} at cross-section position x. U and A are the total circumference and area of the cross-section, and U_{abs} is the circumference of the absorber.

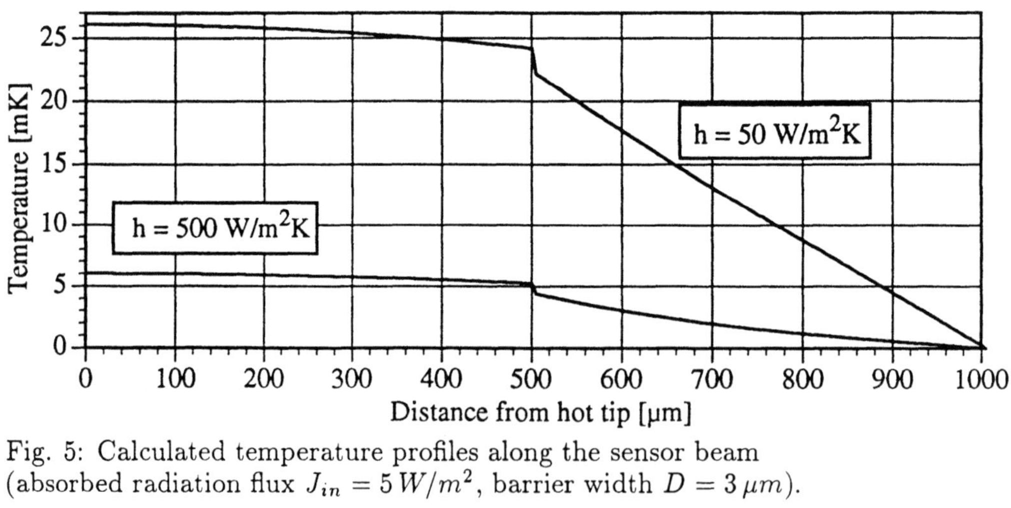

Fig. 5: Calculated temperature profiles along the sensor beam (absorbed radiation flux $J_{in} = 5\,W/m^2$, barrier width $D = 3\,\mu m$).

In the absorber region the average heat generation $P(\Theta, x)$ is positive, and so we obtain a downwards bended (convex) temperature profile, whereas along the thermopile P is negative resulting in an upwards bended (concave) curve. The step-like temperature drop in between is caused by the poor thermal conductivity of the oxide/nitride barrier. For moderate values of the heat transfer coefficient (i.e.,

$h \leq 50\,W/m^2K$), the temperature profile along the thermopile shows a nearly linear behavior, whereas for larger values of h a concave curvature appears, indicating considerable cooling by heat conduction and convection through the ambient gas.

Hence, we conclude that neither the temperature drop along the full beam structure, ΔT_{abs}, nor that along the thermopile alone, ΔT_{thp}, obeys a simple thermal resistance model relating it to $I_{in} = A_{abs}\,J_{in}$, the total IR radiation power received by the absorber area A_{abs}, by R_{th}^{thp}, the thermal resistance of the thermopile structure, or R_{th}^{tot}, the total thermal resistance of the oxide/nitride barrier and the thermopile in series. For linear heat flow, neglegible thermal losses through the surface would be a necessary condition, which is obviously not the case in the structure considered. A more accurate treatment including heat leakage through the surface is therefore necessary (see next section).

For small values of h, however, surface losses mainly occur above and underneath the absorber structure and the oxide/nitride barrier, because $T - T_{amb}$ is high there (cf. Fig. 5). So, for an approximate analysis, we may relate the temperature drop along the thermopile, ΔT_{thp}, by the pseudo-linear relation

$$\Delta T_{thp} = R_{th}^{thp} \cdot I_{in}^{*}\,(\,\Delta T_{abs}) \tag{5}$$

to the effective heat current I_{in}^{*} entering the thermopile on the hot side. I_{in}^{*} is given by I_{in} diminished by the heat losses between cantilever tip and thermopile which, of course, depend on ΔT_{abs}.

This approach is confirmed by three-dimensional numerical simulations for different thermopile materials. For n-type/p-type polysilicon thermopiles, we have a thermal resistance which is about 3.6 times larger than that of an alu/poly structure. Correspondingly, we obtain a value of ΔT_{thp} for the first structure which exceeds that of the latter by nearly the same factor, namely 3.2 (see Fig. 4). The remaining difference between the two factors results from the fact that the temperature level along the thermopile is larger for the n-poly/p-poly structure. This leads to slightly higher heat leakage through the surface and, consequently, to a smaller ΔT_{thp}. Since, in addition, the relative Seebeck coefficient of the n-poly/p-poly thermocouples is three times larger than that of the alu/poly junctions ($\alpha_{poly/poly} = 200\,\mu V/K$, $\alpha_{alu/poly} = 65\,\mu V/K$), we expect a total increase of the output voltage by a factor of about 10, when substituting n-poly/p-poly for aluminum/polysilicon thermocouples.

Fig. 4 and Table 1 elucidate that the effective heat flux I_{in}^{*} is indeed lowered by heat losses through the surfaces between cantilever tip and thermopile. With larger width of the oxide/nitride barrier the thermal resistance increases and, consequently, so does the difference temperature ΔT_{abs}. This gives rise to higher heat leakage through the surface (as expressed in eq. (3)), which in turn lowers the heat current I_{in}^{*} through the thermopile and, hence, the temperature gradient ΔT_{thp}.

The influence of the heat transfer coefficient h on the balance of heat flow and, thereby, on ΔT_{thp} is illustrated in Fig. 6. Again, we see that ΔT_{thp} degrades with increasing heat leakage through the sensor surface. So the sensor signal is reduced by a factor of 2 for values of $h \geq 100\,W/m^2K$. Unfortunately, reliable experimental values of h for micromachined structures are not available to us. For the sensor type considered, a lower bound for h may be estimated from $h \approx \kappa_{amb}/d$, where

Width [μm]	J_{in} [W/m²]	J_{out} [W/m²]	J^*_{in} [W/m²]	J_{bulk} [W/m²]	ΔT_{thp} [mK]	ΔT_{cant} [mK]
3	1.23E-7	4.30E-8	9.91E-8	7.98E-8	22.2	29.7
10	1.23E-7	4.68E-8	9.35E-8	7.60E-8	21.2	31.8
30	1.23E-7	5.65E-8	8.14E-8	6.64E-8	18.5	37.3

Table 1: Heat flux distribution for three different widths of the thermally insulating barrier between absorber and thermopile. I_{bulk} is that portion of the incoming heat flux I_{in} (= absorbed radiation power) that reaches the bulk silicon, while I^*_{in} is the heat flux passing the hot end of the thermopile. I_{out} is the heat flux leaving the sensor through the exterior surface ($h = 50\,W/m^2K$ for all cases considered).

κ_{amb} is the thermal conductivity of the ambient air and d is the distance between sensor beam and bottom of the surrounding cavity (cf. Fig. 3). For a typical value of $d = 100\,\mu m$, we find $h \approx 260\,W/m^2K$ indicating that this distance is still too small for sufficient thermal isolation. As countermeasure we suggest a sensor structure which is etched from the back of the wafer. This would largely reduce the conductive contribution to the h-value. Generally it seems that packaging is the crucial issue for this type of sensor. Placing the sensor into an evacuated or special gas-filled housing might be necessary for stable operation.

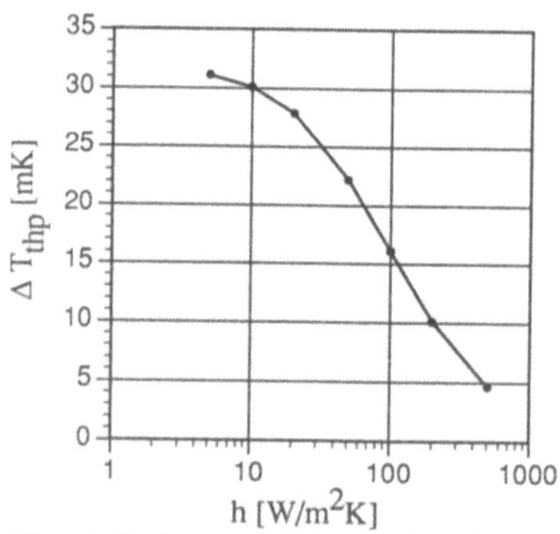

Fig. 6: Influence of the surface heat transfer coefficient on the temperature drop ΔT_{thp} along the thermopile ($D = 3\,\mu m$, $J_{in} = 5\,W/m^2$).

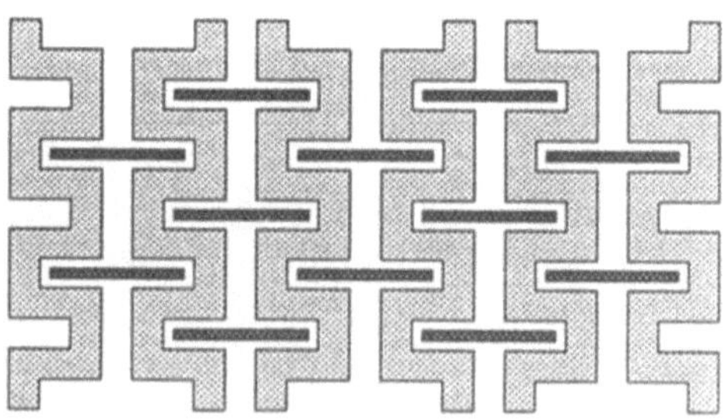

Fig. 7: Meander-shaped topography of thermopiles. Etched slots (black) impede the detrimental by-pass of heat.

4. Influence of the Topography of the Thermoelement

The conventional design of thermopiles relies on the intuitive assumption that straight-lined thermocouples such as shown in Fig. 1 yield the optimum thermovoltage V_{thp}. When N_0 thermocouples are connected in series and the heat current I^*_{in} enters the thermopile on the hot side, the generated voltage V_{thp} is, according to eq. (5), approximately given by

$$V_{thp} = \alpha\, N_0\, \Delta T_{thp} = \alpha\, N_0\, R^{thp}_{th}\, I^*_{in} \tag{6}$$

Here, α is the relative Seebeck coefficient between the thermocouple materials. Eq. (6) holds under the premise that, along the thermopile, surface heat losses are neglegible. The latter are accounted for by a multiplicative correction function F as it will be discussed below (cf. eq. (7)).

For a given technological process, the value of the Seebeck coefficient α and the maximum number N_0 of thermocouples per given area are determined by the properties of the employed materials and the process design rules. Consequently, an enhancement of the output voltage can be achieved primarily by maximizing the overall thermal resistance R^{thp}_{th} between the hot and the cold end of the thermopile. This requires a judicious design, where unconventional topographies have also to be considered.

A three-dimensional numerical study shows that R^{thp}_{th} significantly grows by introducing meander-shaped thermocouples as shown in Fig. 7. The underlying idea is to increase the effective length of the thermocouples, thereby enlarging their thermal resistance. However, holes have to be etched through the supporting structure in order to force the heat current to follow the course of the meanders and to inhibit a detrimental by-pass between neighboring bends. Of course, with a meander topography, less thermocouples can be placed on a given area. If we proceed from straight-lined to meander-shaped thermopiles, the arc length L of the bended thermocouples increases by a factor of n, and so does the width W of a single thermoelement (Fig. 8). Consequently, the maximum number of thermoelements per area is N_0/n, and so we expect that the output voltage V_{thp} will increase if the thermal resistance per unit width grows by more than a factor of n.

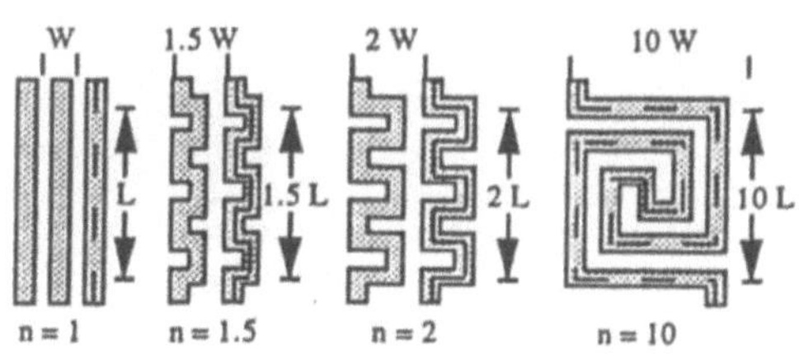

Fig. 8: Arc length L of thermoelements for several meander-shaped structures characterized by the scaling factor n. L/W is constant for all structures.

Fig. 9: Simulation domain of a meander-shaped thermocouple with $n = 3$. White: polysilicon meanders; grey: etched air-slots; black: oxide/nitride.

Due to the complicated meander topography, only short structures (total length $\leq 70\,\mu m$) could be modeled in view of both sufficient numerical accuracy and acceptable computational expense. By symmetry arguments the simulation domain may be restricted to one single thermocouple only (Fig. 9). Despite of the large number of finite elements (30,000 – 80,000), the discretization error bound might still be about 10% for higher values of the structure parameter n.

The temperature distribution along a thermocouple with $n = 3$ (cf. Fig. 9) exhibits a step-like shape (Fig. 10) which is typical for all meander structures. It indicates that the heat current, $-\kappa \nabla T$, indeed follows the course of the meanders; this is confirmed by the calculated heat flow distribution. The steep temperature gradient across the air gaps illustrates that they really constitute an effective barrier against heat flow.

Fig. 11 shows the temperature drop ΔT_{thp} along the thermopile, normalized to a given area, for different values of the surface heat transfer coefficient h and the structure parameter n. We recognize that smaller values of h lead to a significant shift of the maximum sensor response towards higher values of n. Hence, for short thermopile length ($\leq 70\,\mu m$ in the case considered), a meander topography turns out to be superior to the traditional straight-lined layout. Broader meanders (up to $n = 4 \ldots 8$) are the more preferable, the less heat is lost by leakage through the sensor surface.

5. Scaling Laws for Thermopiles

To gain a more general understanding of the simulation results above, we set up an analytical thermal model that allows an interpretation of the thermopile sensitivity in terms of simple scaling laws. In particular, the basic relation (6) must be modified so that surface heat losses are properly included. To this end, we consider the effective one-dimensional heat flow equation (4) between the hot end ($x = x_{hot}$) and the cold end ($x = x_{cold}$) of the thermopile. Supplemented with the appropriate boundary conditions $\Theta(x_{cold}) = 0$ and $\partial\,\Theta/\partial\,x\,(x_{hot}) = I^*_{in}/(\kappa_{av}\,A)$ eq. (4) can be solved analytically yielding

$$\Delta T_{thp} \;=\; R_{th}^{thp}\, I^*_{in}\, F\left(\sqrt{2\, R_{th}^{thp} h_{tot} W_0 L_0}\right) \tag{7}$$

where $F(x) := tanh\,(x)\,/\,x$ is a monotonically decreasing function smaller than 1 correcting the law of linear heat flow, $\Delta T_{thp} = R_{th}^{thp}\, I^*_{in}$, for heat losses through the surface. To derive eq. (7) we set $R_{th}^{thp} = L_0/(\kappa_{av}\,A)$ and $U \simeq 2\,W_0$, where L_0 and W_0 denote the overall length and width of the thermopile, respectively. h_{tot} is defined as $h + 4\,\epsilon\,\sigma\,T^3_{amb}$.

The simulated temperature profile (Fig. 10) suggests an interpretation in terms of an equivalent thermal network that allows calculating the total thermal resistance $R_{th}^{thp}(n)$ of meander-shaped thermopiles for any value of the structure parameter n. The final result, as described elsewhere [5], is an analytical expression of the sensitivity of meander thermopiles:

$$S \;:=\; \frac{V_{thp}}{I^*_{in}} \;=\; \alpha\, N_0\, \frac{L_0}{2\, l_0}\, R_{elm}(n)\, F\left(L_0\,\sqrt{h_{tot}\, n\, R_{elm}(n)}\right) \tag{8}$$

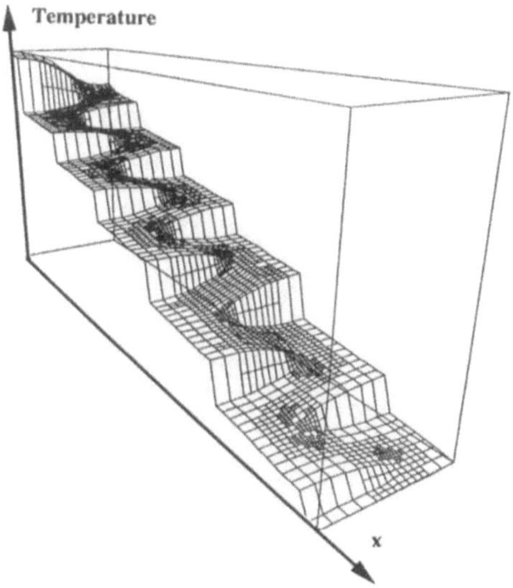

Fig. 10: Perspective spatial view of the temperature distribution along the thermocouple shown in Fig. 9.

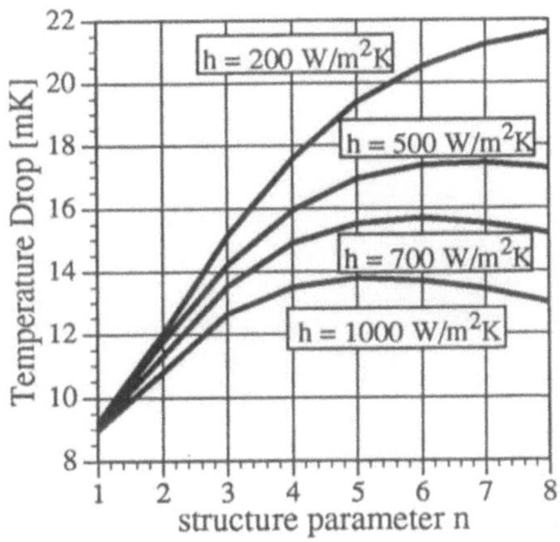

Fig. 11: Normalized temperature drop along meander-shaped thermopiles with geometric length $L_0 = 63\,\mu m$.

Here, L_0 denotes the geometric length of the thermopile, and l_0 is one half of the period length of a meander along the principal axis (e.g., the distance between subsequent air slots in Fig. 7). $R_{elm}(n)$ is the thermal resistance of a single elementary cell of the periodic meander structure; the explicit expression is a rather lenghty fractional polynomial of the variable n as derived in [5].

The typical behavior of the sensitivity S as function of the structural parameters L_0 and n is shown in Fig. 12. With respect to L_0, S is a monotonically increasing function proportional to $L_0 \cdot F(c\,L_0) \sim tanh(c\,L_0)$, which saturates within a length scale $L_{sat} = (h_{tot}\,n\,R_{elm}(n))^{-1/2}$. For straight thermopiles and moderate h-values, L_{sat} is about $400\,\mu m$; hence one should maximize L_0 up to the limit of mechanical stability. For broad meanders (e.g., $n = 8$), L_{sat} is smaller than $100\,\mu m$. Exceeding this length will not significantly improve the sensitivity. Instead one should find the optimum value of n.

The sensitivity S depends on n primarily through $R_{elm}(n)$. It shows [5] that, for structure dimensions encountered in practice, $R_{elm}(n)$ attains the maximum at a value of n much higher than 1. By the action of the correction factor F, the maximum of R_{elm} is shifted towards smaller n values if either the total length L_0 or the heat transfer coefficient h increase. This behavior agrees well with the above-dicusssed numerical simulations (Fig. 11). We conclude that, above a certain critical length of L_0, the maximum sensitivity is always attained at $n \simeq 1$ (i.e., straight thermopiles are preferable), while below this critical length a meander topography is superior. The critical length itself is a function of the heat transfer coefficient h. Therefore, the optimal topography depends on the ambient conditions and on the embedding of the thermopile in the entire sensor structure.

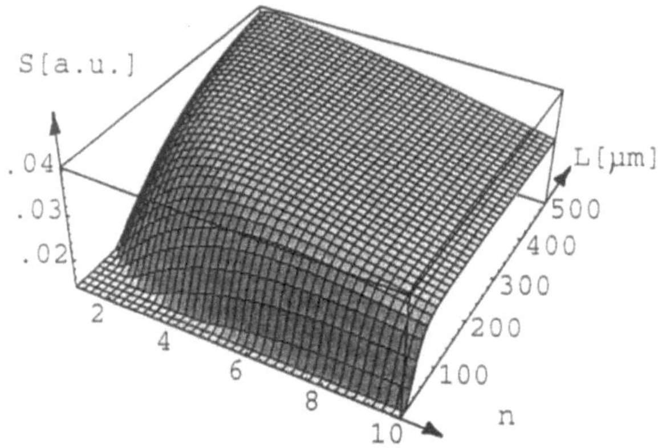

Fig. 12: Thermopile sensitivity as a function of the structure parameter n and the overall length L_0 ($h = 100\,W/m^2K$).

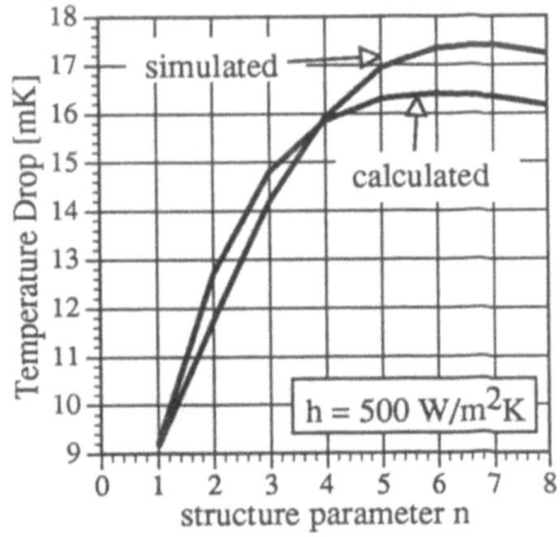

Fig. 13: Analytical calculation and 3D-simulation of the temperature drop along the thermopile vs. the structure factor n.

Conclusion

Our analysis has shown that a judicious design is indispensible for satisfactory performance of electrothermal IC-compatible IR sensors. Considering Seebeck coefficient and thermal conductivity, n-poly/p-poly thermocouples are in general preferable to alu/poly-couples. Heat leakage through the surface is very detrimental to the sensor response and, hence, must be impeded either by appropriate packaging (e.g., an evacuated or gas-filled housing) or by providing a sufficiently large separation of the sensor structure from the walls of the surrounding cavity, possibly by etching from the back of the wafer. A new design with pairwise arranged meander-shaped thermocouples promises an enhanced output signal.

References

[1] Lenggenhager, R., Baltes, H., and Elbel, T.: Thermoelectric infrared sensors in CMOS technology. *Sensors and Actuators* A 37–38 (1993) 216–220.

[2] Baltes, H.: Microtransducers by industrial IC technologies and micromachining. *Tech. Digest 10th Sensor Symposium*, IEE Japan, Tokyo (1992) 17–23.

[3] Baltes, H., *et al.*: Silicon microsensors and microstructures. *Proc. 1992 IEEE Int. Symp. on Circuits and Systems*, IEEE, Piscataway (1992) 1820–1823.

[4] Anderheggen, E. *et al.*: SESES - a two- and three-dimensional semiconductor sensor simulation program. *Proc. NASECODE VIII*, Vienna (1992) 76-78.

[5] Funk, J., Wachutka, G., and Baltes, H.: Scaling laws for optimized design of integrated thermopiles. *Digest of Technical Papers*, Transducers'93, Tokyo (1993) 742–745.

Mixed convection and heat transfer on vertical parallel printed circuit boards.

S. MERGUI[1], F. PENOT[1]
[1] *Laboratoire d'Etudes Thermiques, (URA CNRS 1403), Université de Poitiers, 40 Avenue du Recteur Pineau, 86022 Poitiers CEDEX, France*

Abstract. We examine the problem of the roughness represented by components mounted on vertical electronic circuit boards in airtight enclosures and which are cooled by pure natural convection. Analysis of the magnitude of the dynamic phenomena and visualization of airflow in the region of the components show that it can be characterized by natural, forced or mixed convection according to the position of the components on the circuit board and the position of the latter in the enclosure. However, temperature measurements reveal no presence of winglet, obstacle or cavity effects, be the components mounted in horizontal or vertical rows. Convective transfer is identical to that on smooth faces.

1. Introduction

Manufacturers are sometimes led to enclose circuit boards in airtight boxes to protect them from an aggressive environment. Mechanical ventilation of the circuit board accompanied by air extraction is often not possible under such conditions and natural cooling is used. The heat dissipated on the electronic components is evacuated by conduction, radiation and natural convection.

In such facilities, one seeks to optimise each mode of transfer. In particular, boundary conditions which generate significant convective flow can be found. Among the conceivable situations are those where negative vertical temperature gradients are generated, with cold zones above hot zones. Such is the type of configuration studied here. We examine the case of a cubic enclosure with a cooled ceiling and which contains boards equipped with heat dissipating components. The boards are mounted vertically with enough space above and below each board for air to circulate freely.

We examined the thermal and dynamic problems that arise in such a configuration both experimentally and numerically [1]. Numerical investigation demands good

C. J. Hoogendoorn et al. (eds.), Thermal Management of Electronic Systems, 149–158.

knowledge of local conditions to accurately determine convective transfer. Analysis is made particularly difficult when circuit boards have numerous protuberances and there are localised heat sources, [2, 3, 4, 5]. Nevertheless, we show that it is possible to extract a certain number of general criteria which can be used to improve analysis of the dynamics in the vicinity of the circuit boards. We also carried out an experimental study of the influence of a board's roughness on heat transfer.

2. Dynamic analysis

There are two ways of looking at the dynamics of components mounted on circuit boards [6] – (i) the component is considered to be an obstacle with a characteristic dimension d, placed in the flow in the immediate vicinity of the board, the flow having a velocity Ue; (ii) the space between the components is considered to be a valley, i.e. an open cavity in the more or less regular surface represented by the upper faces of the components. In either case, the obstacle or the cavity is considered to be the seat of heat dissipation causing local temperature variation between the component and the air, of magnitude ΔT. One observes the generation of buoyancy forces, which are characterized by a GRASHOF component number, Gr_d, defined by

$$Gr_d = \frac{g\beta\Delta T d^3}{\nu^2} \tag{1}$$

Given the component sizes commonly encountered (of the order of $5x10^{-3}$m) and the admissible temperature variations (of the order of 25°C - 50°C), GRASHOF numbers remain very low (of the order of 200 - 400) and (as expected) natural convection is highly localised and in the vicinity of the asperities. The asperities will no doubt never be of the boundary layer type as viscosity forces are still dominant everywhere and, in any case, these sub-layers remain laminar. Flow in the vicinity of the components also has a characteristic REYNOLDS number, defined by

$$Re_d = \frac{U_e d}{\nu} \tag{2}$$

The dynamics of the flow in the vicinity of the asperity (cavity or protuberance) is characterized by the local balance of inert and buoyant forces, i.e. by the parameter : Gr_d/Re^2. According to these values, we encounter pure natural convection ($Gr_d/Re^2 >> 1$), local mixed convection ($Gr_d/Re^2 \approx 1$) or pure forced convection ($Gr_d/Re^2 << 1$) [7], and consequently behavior proper to each flow type, i.e. flow following the lines of the steepest slope, the appearance of wakes, and dominance of flow along the streamlines of inviscid exterior flow respectively. In the case at hand, mixed convection is favorable. The boundary conditions described for the enclosure and the circuit boards generate ascendant flow in the vicinity of the boards in the same direction as the local natural convection on the component.

On the scale of the component, the regime encountered thus depends heavily on the magnitude of velocity U_e. We must consequently distinguish two cases : that of a board set in a medium considered infinite (the space between the boards or the space separating the board from the enclosure is great compared to the thickness of the boundary layers), or that of boards set close together, thus forming a vertical thermosyphon. In the first case, flow is considered to be solely produced by body forces which induces natural convection boundary layer flow. If we take a temperature variation between board and surrounding air of magnitude $\Delta T'$, the characteristic velocity in the boundary layer at height z (counting from the bottom of the board) is given by

$$U_e^2 \approx g\beta\Delta T' z \qquad (3)$$

At first approximation, the result is that one can take $\Delta T = \Delta T'$ for a practically isothermal wall, and consequently :

$$\frac{Gr_d}{Re_d^2} \approx \frac{d}{z} \qquad (4)$$

Thus the position of the component on the board significantly influences the rate of flow likely to be encountered in the locality. If the component is situated on the lower part of the board, z being small, we observe behavior characteristic of pure natural convection, and the air, having very low velocity, assumes the shape of the obstacle. As one moves higher on the board, convective flow intensifies and we move into mixed convection on the component with separation at the leading edge and highly localized recirculation appearing. If the component is placed high enough, the d/z ratio can become quite small and flow will be of the forced convection type with wake effects or significant recirculation in the cavities.

In the case of flow between the circuit boards, characteristic of a thermosyphon, characteristic velocity U_e is quite close to mean velocity U_q. This velocity is constant over the whole height L of the duct and depends on global dissipated heat flux on the whole board and on the geometry of the vertical channel. According to [8], a first approximation of the magnitude of velocity U_q can be obtained using

$$U_q^2 = g\beta\Delta T' L \qquad (5)$$

Consequently, we have

$$\frac{Gr_d}{Re_d^2} = \frac{\Delta T}{\Delta T'}\frac{L}{d} \qquad (6)$$

This ratio will generally be small (the temperature variations involved are of the same order of magnitude) and practically identical dynamics will be found whatever the position of the component on the board. In particular, separation is observed on components placed at the bottom of the board.
This analysis shows that one can encounter a wide diversity of regimes. We carried out visualization experiments in an enclosure (described below), equipped with real electronic circuit boards with components mounted at regular intervals in horizontal rows. Downstream from the components, we observed the formation of more and more active recirculatory eddies as we moved up the board, signalling the onset of mixed convection. However, uniformity of behavior could not be satisfactorily observed for components in a thermosyphon due to technical problems with laser sheet lighting of confined spaces, although visualization does globally vindicate the analysis.
Does this dynamic diversity have important consequences for convective transfer? The series of experiments described below were carried out to seek an answer to this question.

3. Convective transfer on rough vertical plates

3.1 Experimental apparatus

The apparatus used in this experiment comprises two main components : the box and the circuit boards.
The box (described elsewhere [9]) shown in *figure 1*, is made up of a 30cm cubic enclosure with an 8cm thick polystyrene isolating floor. The vertical walls are transparent and are rendered thermally passive with the aid of a double layer of air. The ceiling is a water exchanger cooled at uniform temperature.

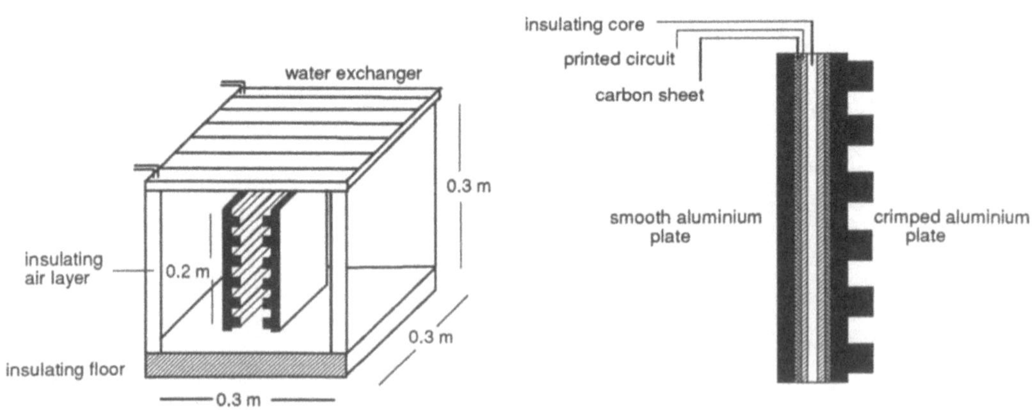

Figure 1. Schematic of experimental set-up and cross section of a circuit board.

False electronic circuit boards 20cm high x 30cm wide are set vertically in the enclosure. There is thus a 5cm space above and below each board. Two identical boards were used, and placed so as to create a duct (thermosyphon) of variable width. Each board is equipped so that a uniform density of heat can be dissipated on both sides and the integrated mean temperature measured on each of five horizontal strips.
One face is covered with smooth, polished aluminium foil 2mm thick, and the other with a 6mm thick aluminium plate into which horizontal furrows 4mm deep and 4mm wide are hollowed out. This simulates horizontal rows of electronic components 4mm high and 4mm wide in rows set 4mm apart. It is to be noted that this roughness doubles the exchanger surface of the corrugated face.
A cross section of the circuit boards (*figure 1*) shows them to be made up successively of a smooth aluminium sheet, a printed circuit specially designed for measuring the temperatures on given surfaces using the variation in temperature of the electrical resistance of the circuit and, finally, a sheet of carbon 20cm x 30cm, less than 1mm thick which is charged with a fixed continuous electric current to provide uniform heat flux. The electrical resistance of this material does not vary with temperature. The central layer of the board is a 5mm thick sheet of polystyrene, 20cm x 30cm, which thermally insulates each face of the circuit board from the other. Next, there is another sheet of carbon, a printed circuit and the hollowed out aluminium making up the other face.
Several series of experiments were carried out with dissipation of 3W, 7W and 10W, respectively on both faces of each board. To minimize heat loss through the floor and lateral walls, the temperature of the enclosure ceiling was adjusted (to. 10°C, 5°C and 2°C respectively) so that the temperature of the air outside the boundary layer regions remained close to ambient temperature. Each trial was carried out twice – once with the smooth surfaces facing each other, once with the rough surfaces facing each other. In both configurations, we left a space of 16mm between the boards for the thermosyphon. (In the case where the corrugated surfaces face each other, the width of the channel is taken to be between the surfaces of the components, i.e. the narrowest part of the duct.)

3.2 Temperature measurement results

Consider first the case where the rough surfaces face each other. The evolution in mean temperatures per horizontal strip (*figure 2*) in terms of z for the three levels of heat dissipation studied lead to the following observations :

(i) The vertical temperature profile is non-linear and this means that the thermal regime is never developed, otherwise the temperature profile would tend towards the linear since the free convection or thermosyphon boundary layers are generated by fixed heat dissipation. Although we do come close at the lowest levels of dissipation.

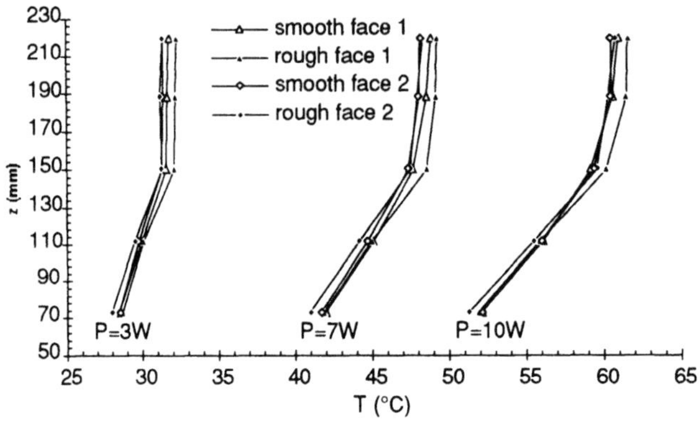

Figure 2. Temperature profiles on each face for three levels of heat dissipation. Components in horizontal rows and surfaces face to face.

(ii) What is more surprising is that temperatures evolve identically on both the rough and smooth faces for the three dissipation levels studied. The thermal power injected into each face being identical, and the exchange surfaces being doubled on the rough faces, this must mean that the convective heat transfer coefficient is not the same on each face of the circuit board. This cannot be explained by the fact that one of the faces is placed in a thermosyphon flow while the other is in pure natural convection since, when the smooth surfaces are placed facing each other, exactly the same phenomenon occurs (cf. fig. 3). The direct comparison of temperature profiles on the rough surfaces, be they facing each other or not (cf. figure 4), confirms that convective transfer is independent of both surface aspect and flow type encountered.

Two distinct conclusions must be drawn. (i) Looking at the global energy balance, it has to be admitted that the source is fixed by the power applied to the circuit boards. Mean temperature and velocity levels in the enclosure are then naturally adjusted so that exchange is balanced on the ceiling (the cold source).The significant pressure loss in the airstream inside the box is essentially a result of the singularities encountered and the sudden changes in direction. This would indicate that the level of general circulation is not significantly affected by local modification of the singularities, such as placing rough surfaces face to face or not. We thus find identical temperatures for all the configurations studied for a given level of power injected into the circuit boards.

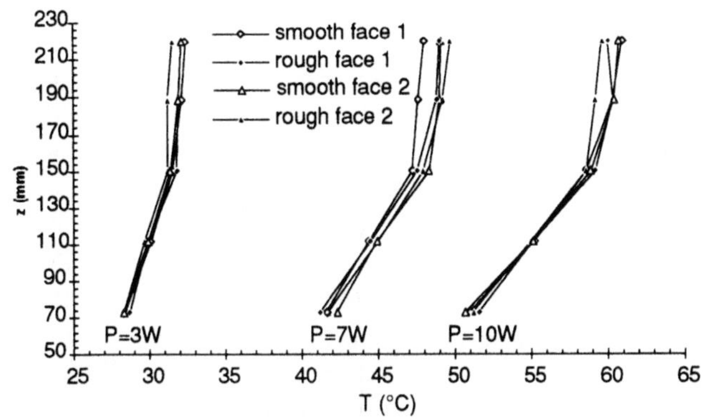

Figure 3. Temperature profiles on each face for three levels of heat dissipation. Components mounted in horizontal rows on surfaces back to back.

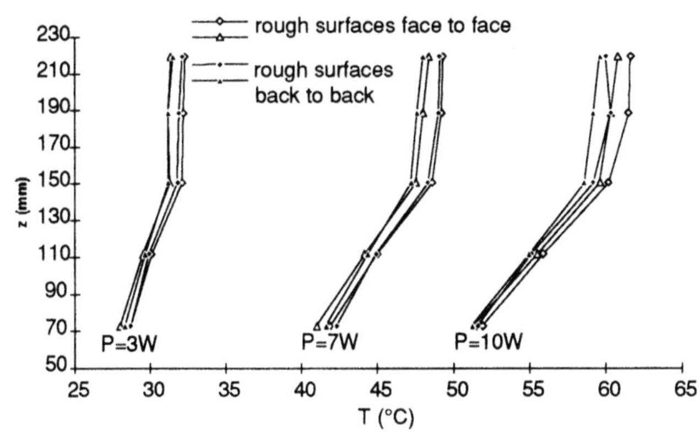

Figure 4. Comparison of face to face and back to back temperature profiles.

(ii) The second conclusion concerns the fact that temperatures are identical, which is in contradiction with the doubling in exchanger surface. Mounting the components in rows perpendicular to the main flow does not generate the winglet effect expected. In the small cavities downstream from the components, the creation of practically stagnant zones modifies local transfer in such a way that we find the same temperature, with or without roughness. This thermal behavior of cavities has already been observed for pure natural convection [10]. According to this set of trials, the same would seem to be the case with mixed convection in small volumes (laminar flow dominated by viscosity effects). Moreover these measurements are in perfect agreement with results presented in [11] concerning one or two asperities on a vertical wall.

3.3 Extra measurements

Given the context discussed above, it seemed worthwhile to carry out trials for other configurations. In particular, we examined the case of circuit boards mounted with components forming vertical ducts, i.e. veritable vertical winglets which double the exchanger surface of the plate. The same trials were carried out, adapting the enclosure and the circuit boards to the new configuration – the height of the box was raised by 10cm to accommodate boards 30cm high and 20cm wide which were turned through 90°, free space being left at each side of the boards. The set of temperature measurements given in figure 5, obtained for 7W dissipated (typical of the results obtained) shows that there is no winglet effect. The integrated temperature (this time on vertical strips) is the same on both the rough and smooth sides.

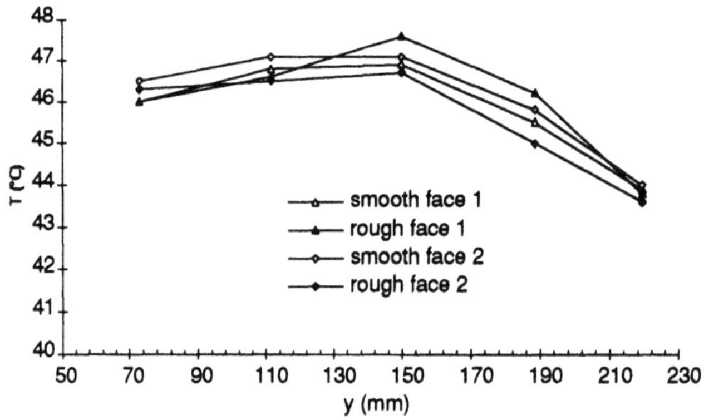

Figure 5. Temperature profiles on each face. Components in vertical rows and surfaces face to face.

Despite doubling the heat exchanger surface, no significant temperature change is observed. Viscosity effects are sufficient to reduce velocity in the small vertical ducts. Once again, the phenomena that arise in the region of the asperities are such that we have overall smooth face behavior. The slight dissymmetry observed between the temperatures on the right and left sides of the circuit boards is normal in this type of experiment where perfect symmetry would demand a miracle.

To show that our observations were not the result of badly designed circuit boards and, in particular, of excessive thermal conduction in the insulating material sandwiched in the boards which would have contributed to the same temperature at all levels on both sides of the board, we carried out the following trial : a circuit board was placed horizontally, smooth face down on insulating material in an isothermal medium (23°C). For a given amount of heat dissipated, temperatures measured on the smooth and rough sides are quite different (figure 6).

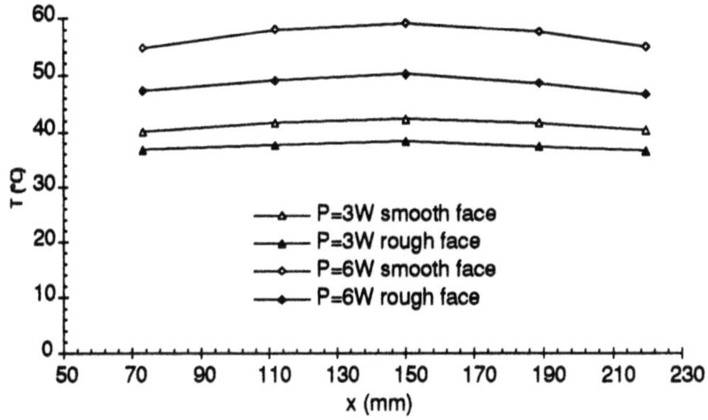

Figure 6. Temperatures monitored on a circuit board placed horizontally on an insulating surface, component side up.

This clearly demonstrates that different modes of transfer (natural convection on the upper face, conduction through an insulating material on the other) are associated with different temperatures. Moreover, if the heat dissipated on the circuit board is doubled, temperature variations are also approximately doubled (within radiation correction), which indicates that metrology is not just qualitative

4. Conclusion

Natural convection flow and transfer in airtight enclosures containing electronic circuit boards display complex, interacting mechanisms. When convection is a significant parameter (heat sources below heat sinks), good knowledge of local phenomena is necessary for the calculation of heat transfer on the boards. Using a qualitative analysis of the fluid dynamics around components mounted on cooled circuit boards, we have shown that a wide diversity of flow regimes can be encountered locally. According to their position on the circuit board or of the board itself in the enclosure, the components, (considered as obstacles) or the spaces between them (considered as cavities), can be the seat of phenomena linked to pure forced convection or pure natural convection, passing through all intermediate stages.

It is interesting to note that this wide diversity in dynamics is not accompanied by differences in thermal behavior – in particular no winglet, obstacle or cavity effects can be shown for components mounted in horizontal, or even vertical rows, i.e. respectively perpendicular or parallel to the main flow. It is therefore possible to use the laws of convective transfer on the smooth faces when simulating the thermal and dynamic behavior of this type of circuit board.

This claim can in no way be taken as universally valid and demands further investigation using different configurations (asperity height, cavity width, etc.) to determine up to what point asperities can be considered negligible.

Acknowledgements The authors would like to thank the Centre National des Télécommunications (C.N.E.T.), LANNION, FRANCE for their financial support without which this study would not have been possible. We would also like to thank PAUL WASS for translating the original French text into English.

References

1. Penot F., Pimont V., Rifi A., Viault A., cadre M.: Experimental and numerical investigations of heat transfer in enclosures containing heat dissipating plates. *Proceeding of 9th International Heat Transfer Conference, Jerusalem* (1990) 139-143.
2. Afrid M., Zebib A.: Natural convection air cooling of heated components mounted on a vertical wall. *temperature/fluid measurements in electronic equipment* ASME HTD 89 1987 17-24
3. Jaluria Y.: Mixed convection flow over localized multiple thermal sources on a vertical surface. *Phys. Fluids* 29 (1986) 934-940.
4. Khalilokkahi A., Sammakia B.G.: Transient mixed convection in air adjacent to discrete uniform flux heat sources. *Numerical simulation of convection in electronic equipment cooling* ASME HTD 121 (1989) 55-60
5. Ortega A., R.J. Moffat: Buoyancy induced convection in a non uniformly heated array of cubical elements on a vertical channel wall. *Heat transfer in electronic equipment* ASME HTD 57 (1986) 123-134
6. Perng C.Y., Street R.L.: A domain decomposition technique for solving geometrically complex flow problems. *Numerical simulation of convection in electronic equipment cooling* ASME HTD 121 (1989) 61-68
7. Quintela D.A.A., Borges A.R., Penot F.: Convective heat transfer from surface mounted bluff bodies: A model study. *Experimental Heat transfer, fluid mechanics and thermodynamics Dubrovnik* (1988) 1221-1228
8. Dalbert A.M., Penot F., Peube J.L.: Convection naturelle laminaire dans un canal vertical chauffé à flux constant. *Int. J. Heat Mass Transf.* 24 N°9 (1981) 1463-1473
9. Rifi A. Convection naturelle en enceintes fermées fortement partitionnées. Application aux cartes électroniques en boîtiers étanches. *Thèse de l'Université de Poitiers France* (1990)
10. Penot F.: Numerical calculation of two-dimensional natural convection in isothermal open cavities. *Numerical heat transfer* 5 (1982) 421-435
11. Shakerin S., Bohn M., Loerkhe R.I.: Natural convection in an enclosure with discrete rougness elements on a vertical heated wall. *Int. J. heat Mass Transf.* 31/7 (1988) 1423-1430

A parametric study of heat spreading through multiple layer substrates

J. PUNCH & M. DAVIES
Department of Mechanical Engineering, University of Limerick, Limerick, Ireland

Abstract. A double Fourier series technique is used to develop an analytical expression for the three-dimensional temperature field within a thin multiple layer substrate cooled by convection and heated by sources mounted on one side. The temperature distribution, in a dimensionless form, is a function of non-dimensional parameters: the Biot number, aspect ratios and conductivity ratios of each layer, and the ratios of Fourier coefficients which describe the sources. A parametric study is presented to illustrate the influences of these parameters for a practical electronics cooling problem. Equivalent series and parallel orthotropic conductivity components are derived which can be used to model multiple layer structures as single layers.

NOMENCLATURE

a, b	Substrate dimensions, m	x, y, z	Cartesian co-ordinates
a_{mn}	Fourier coefficients	λ	Wave number, 1/m
D_{imn}	Factor in solution	θ	Dimensionless temperature
h	Heat transfer coefficient, W/m^{2}K		
H	Layer thickness, m	Subscripts	
I	Number of layers		
K	Thermal conductivity, W/mK	i	Layer index
q	Heat flux per unit area, W/m^2	m, n	Indices of Fourier terms
Q	Heat flux, W	s, p	Series, parallel
T	Temperature, K	T	Total

1. Introduction

Thin rectangular multiple layer substrates carrying heat sources are common in electronic systems. Printed circuit boards (PCBs) are the most prominent example. Various analytical models have been derived to describe the conductive heat spread from the sources through the substrate to the convective sink. These models are useful to system designers because components' junction temperatures can be related to the substrate surface temperatures by junction-to-casing thermal resistance data. A literature review pertinent to these models has been compiled by Punch and Davies (1993). The purpose of this paper is to develop, using an analytical method, a parametric study of the heat spread in multiple layer substrates.

The analytical method uses a double Fourier series to describe the temperature distribution within a multiple layer substrate. A dimensionless temperature field is derived which is a function of non-dimensional parameters. The parametric study illustrates the influence of these parameters for a practical cooling problem. A novel method is developed which saves computational effort by approximating a multiple layer structure as a single layer.

C. J. Hoogendoorn et al. (eds.), Thermal Management of Electronic Systems, 159–168.

2. Analytical Model

A double Fourier cosine series approach is used to develop an expression for the three-dimensional steady-state Laplacian temperature field within the multiple layer structure shown in figure 1. The substrate is comprised of I layers, each layer of thickness H_i and of orthotropic conductivity components K_{ix}, K_{iy} and K_{iz}.

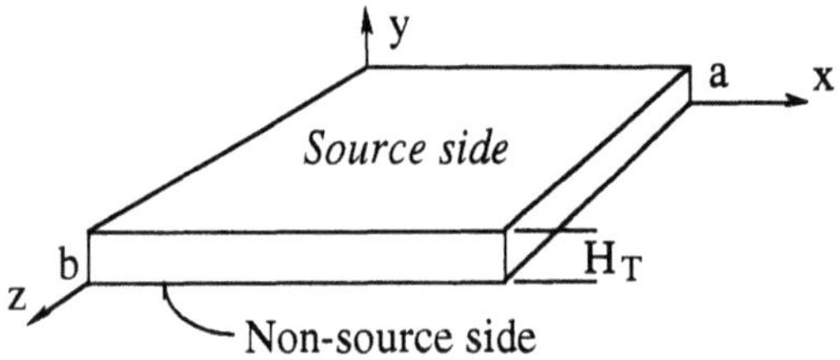

Fig. 1 Co-ordinate axes and dimensions of domain

The edges of the substrate are adiabatic, and cooling from the source and non-source sides is by convection of constant heat transfer coefficient, h, and sink temperature, T_∞. Sources are defined by the heat flux applied to the source side of the substrate:

$$q(x,z) = \sum_{m=0}^{\infty} \sum_{n=0}^{\infty} a_{mn} \cos\left(\frac{m\pi x}{a}\right) \cos\left(\frac{n\pi z}{b}\right) \tag{1}$$

The Fourier coefficients are found from the applied flux by the expressions of Carslaw and Jaeger (1959). In each layer, a Fourier series describes the variation of temperature in the xz-plane, and the decay in the y-direction is governed by exponential terms. Continuity of temperature and conservation of heat flux provide the relationships between layers. The method of solution is listed by Punch and Davies (1993). A dimensionless temperature field is derived which is a function of these non-dimensional parameters: the Biot number, aspect ratios and conductivity ratios of each layer, and the ratios of Fourier coefficients:

$$Bi_i = \frac{hH_i}{K_{iy}} \qquad Bi_T = \sum_{i=1}^{I} Bi_i \tag{2}$$

$$Asp_{ix} = \frac{H_i}{a} \qquad Asp_{iz} = \frac{H_i}{b} \tag{3}$$

$$K_{ix}^{+} = \frac{K_{ix}}{K_{iy}} \qquad K_{iz}^{+} = \frac{K_{iz}}{K_{iy}} \tag{4}$$

$$a_{mn}^{+} = \frac{a_{mn}}{a_{00}} \tag{5}$$

The aspect ratios and conductivity ratios are referred to as the spreading ratios which, through the wave numbers, influence the spread of heat in the xz-plane of the substrate:

$$\left(\lambda_{imn}^{+}\right)^2 = K_{ix}^{+} (Asp_{ix}\, m\pi)^2 + K_{iz}^{+} (Asp_{iz}\, n\pi)^2 \tag{6}$$

Heat flux generated by the sources is, by definition:

$$Q_s = a_{00}\, ab \tag{7}$$

Convective heat losses from the source and non-source sides, Q_1 and Q_2 respectively, are:

$$\frac{Q_1}{Q_s} = \frac{1 + Bi_T}{2 + Bi_T} \qquad \frac{Q_2}{Q_s} = \frac{1}{2 + Bi_T} \tag{8}$$

Of most interest is the temperature distribution on the source side, normalised by the area-averaged source side temperature:

$$\theta(x,z) = \frac{T(x,z) - T_\infty}{\overline{T(x,z)} - T_\infty}$$

$$= 1 + \sum_{m=0}^{\infty} \sum_{n=0}^{\infty} a_{mn}^{+} \cos\left(\frac{m\pi x}{a}\right) \cos\left(\frac{n\pi z}{b}\right)\left[\frac{2 + Bi_T}{1 + Bi_T}\right]. \left[\frac{Bi_1\left(e^{\lambda_{1mn}^{+}} + D_{1mn}\, e^{-\lambda_{1mn}^{+}}\right)}{\left(\lambda_{1mn}^{+} + Bi_1\right) e^{\lambda_{1mn}^{+}} - D_{1mn}\left(\lambda_{1mn}^{+} - Bi_1\right) e^{-\lambda_{1mn}^{+}}}\right] \tag{9}$$

The Fourier sum m = n = 0 is not evaluated. Equation 10 can be used iteratively to relate the factors D_{1mn} of the source side layer to those of layer I, defined by equation 11:

$$\frac{Bi_i}{\lambda_{imn}^{+}}\left(\frac{1 + D_{imn}}{1 - D_{imn}}\right) = \frac{Bi_{(i+1)}}{\lambda_{(i+1)mn}^{+}}\left(\frac{e^{\lambda_{(i+1)mn}^{+}} + D_{(i+1)mn}\, e^{-\lambda_{(i+1)mn}^{+}}}{e^{\lambda_{(i+1)mn}^{+}} - D_{(i+1)mn}\, e^{-\lambda_{(i+1)mn}^{+}}}\right) \tag{10}$$

$$D_{Imn} = \frac{\lambda_{Imn}^{+} - Bi_I}{\lambda_{Imn}^{+} + Bi_I} \tag{11}$$

The dimensionless temperature distribution on the source side of a multiple layer substrate consists of two components: a constant term, unity, which expresses the area-averaged temperature, and a spreading term which is a function of the Biot number, aspect ratios and conductivity ratios of each layer, and the ratios of Fourier coefficients which define the source shape, position and flux distribution.

There are several limitations in the application of the analytical model to practical substrates. The edges of real substrates are not strictly adiabatic, but leakage is usually small. Flux distributions beneath practical components are never completely known, and must be estimated by the designer. Heat sources in the model are considered to be planar, and this ignores the local variations in the convective conditions which occur around real components. Radiation cooling cannot be described by the model. Contact resistance between layers and between the sources and the substrate is not considered. Despite these limitations, the model reveals the important parameters which influence a cooling problem.

3. Parametric Study

The influence of the non-dimensional parameters on the dimensionless temperature is illustrated here for a default example of a power transistor surface-mounted on a PCB cooled by natural convection. Practical multiple layer substrates are also investigated. A method is derived which approximates a multiple layer structure as a single layer.

3.1. DEFAULT COOLING PROBLEM

The default cooling problem is a TO-220 power transistor centrally mounted on a 0.1m square PCB, 1.6 mm thick, of isotropic conductivity 0.3 W/mK. Cooling from both sides of the board is by convection. Source dimensions are 0.01 m x 0.015 m, the overall heat dissipation is 1W and the flux distribution is assumed to be uniform over the source footprint. Dimensionless temperature is calculated at the centroid of the footprint of the source, a suitable point for the estimation of junction temperatures by the use of junction-to-casing thermal resistance data. Non-dimensional parameters, evaluated by the default values, are varied singly, or in pairs, over practical ranges. Cooling problems can be scaled by these dimensionless parameters.

3.2. SINGLE LAYER SUBSTRATE

The influences of two sets of non-dimensional parameters are illustrated here. Firstly, the parameters defined by the board and the convective conditions: the Biot number, the aspect ratios and the conductivity ratios. Secondly, the factors which define the sources: the flux distribution shape, and the source footprint dimensions and position on the substrate.

3.2.1 *Biot Number and Spreading Ratios.* Dimensionless temperature is shown in figure 2 as a function of the Biot number and the spreading ratios. These parameters are defined in equations 2 - 4 by the board material and dimensions, and by the convective conditions.

The Biot number is the most important of the non-dimensional parameters. For most practical cooling problems, the range of Biot numbers is 10^{-5} - 10^{0}. Dimensionless temperature at a point, for a particular applied flux distribution, is bound by two limits, the 'thermally thick' and the 'thermally thin' limits. The upper limit, the thermally thin limit, is reached at high Biot numbers where the internal conductive resistance is much greater than the external convective resistance. Large gradients of temperature can exist within the substrate. No net heat transfer occurs through the substrate and all heat loss is by source side convection. The dimensionless temperature at a point is directly proportional to the flux density at that point. Lower Biot numbers, however, produce the thermally thick limit, unity. The conductive resistance is negligible compared to the convective resistance and the substrate becomes isothermal. Heat is then lost in equal amounts from both sides.

For a constant value of spreading ratio there is a regime of cooling in which the temperature is a strong function of the Biot number. Within this range, reducing the dimensionless temperature reduces peaks of temperature within the substrate. However, the designer must ensure that by minimising peaks of temperature, the area-averaged source side temperature is not greatly increased.

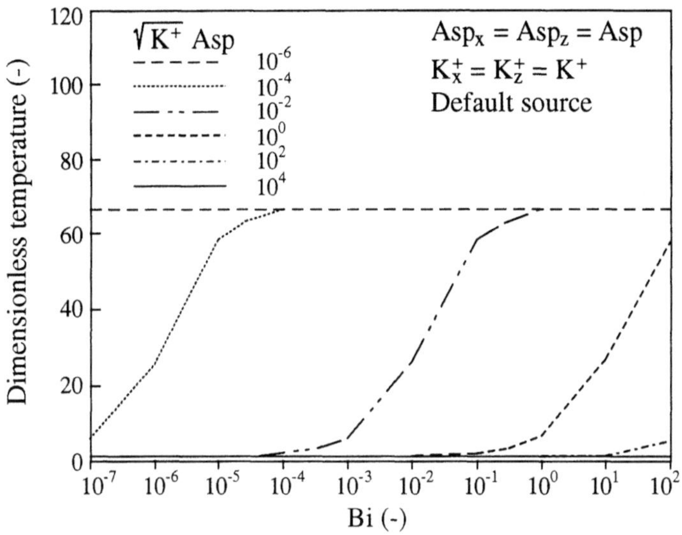

Fig. 2 Dimensionless temperature as a function of Biot number and spreading ratios

For a square substrate, $Asp_x = Asp_z = Asp$, with equal conductivity ratios, $K_x^+ = K_z^+ = K^+$, the wave numbers, equation 6, can be expressed as:

$$\lambda_{mn}^{+} = \left(\sqrt{K^+}Asp\right)\sqrt{(m\pi)^2 + (n\pi)^2} \tag{12}$$

Figure 2 shows that the larger the aspect ratios or conductivity ratios, the greater the spread of heat in the plane of the substrate and the closer the dimensionless temperature is to the thermally thick limit. For unequal values of the aspect ratios or the conductivity ratios, the heat spreads preferentially in the direction of the larger ratios.

3.2.2. *Source flux distribution, size and position.* The ratios of Fourier coefficients, a^+_{mn}, describe the source shape, size and position, and the flux distribution beneath the source.

Flux distributions beneath practical components are rarely fully known. To evaluate substrate temperatures, and to relate these to the junction temperatures of the components, the designer should know the approximate modes of heat transfer between the source and the substrate, and the surface areas through which they occur. For small components with an integral heat spreader, such as the TO-220, the flux distribution is likely to be uniform or parabolic in shape. The dimensionless temperature at the centroid of the source footprint is shown in figure 3 as a function of the Biot number for these distributions. For low Biot numbers, the distribution shape is immaterial but at higher Biot numbers, the thermally thin limit is approached and the temperature at a point is directly proportional to the applied flux. Hence, the temperature of the centroid is much greater under the parabolic distribution.

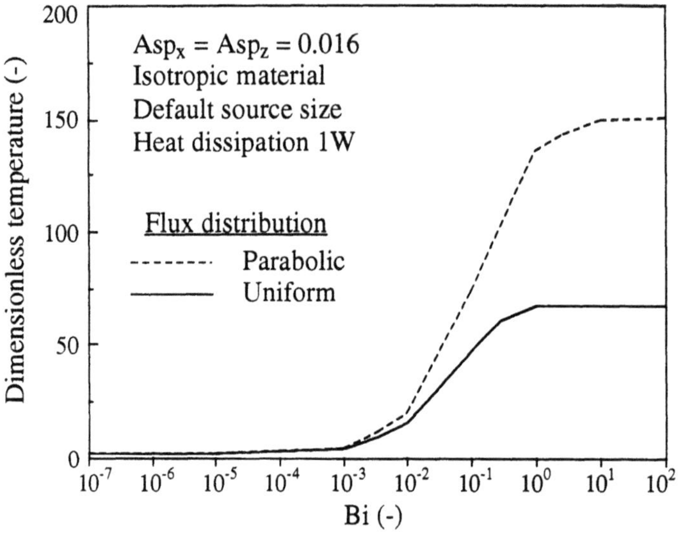

Fig. 3 Dimensionless temperature as a function of Biot number and flux distribution shape

Dimensionless temperature at the centroid of the source footprint is shown in figure 4 as a function of the source side dimension for a square source. For a constant Biot number, the smaller the source footprint, the greater the temperature. Source dimension represents, in a qualitative manner, the packing density of sources on the substrate surface. The example of smallest source represents a sparsely populated substrate, where only a small area is covered. Temperature beneath the source is high. As the source dimension increases, the substrate becomes more densely populated and the source side becomes more isothermal.

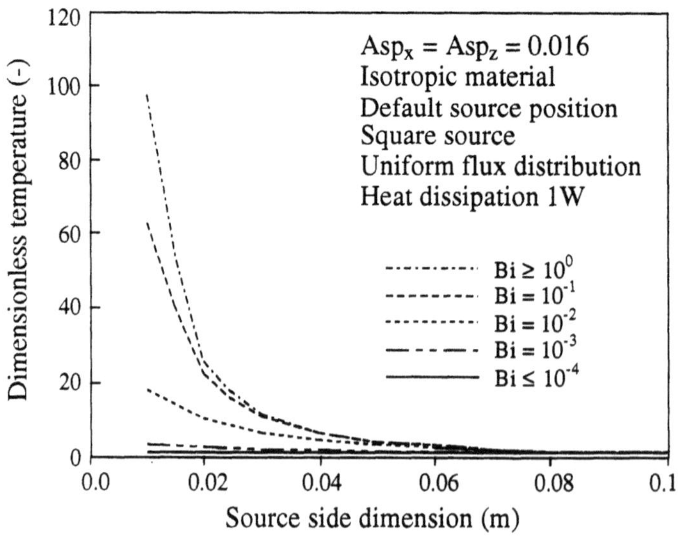

Fig. 4 Dimensionless temperature as a function of Biot number and source dimension

Source position on a substrate is important in two ways: the proximity of a source to other sources, and to the adiabatic edges of the substrate. Dimensionless temperature at a source centroid is shown in figure 5 as a function of Biot number and source spacing for two sources. Both are positioned on the lateral centreline of the substrate and are equi-spaced from the other centreline. Proximity to the substrate edge or to another source causes an increase in temperature for certain Biot numbers. If the component is close to the substrate edge or an adiabatic boundary created by another source, then heat spread is inhibited in that direction and the temperature increases. For high and low Biot numbers respectively, the thermally thin and thick limits have been reached and source position is immaterial.

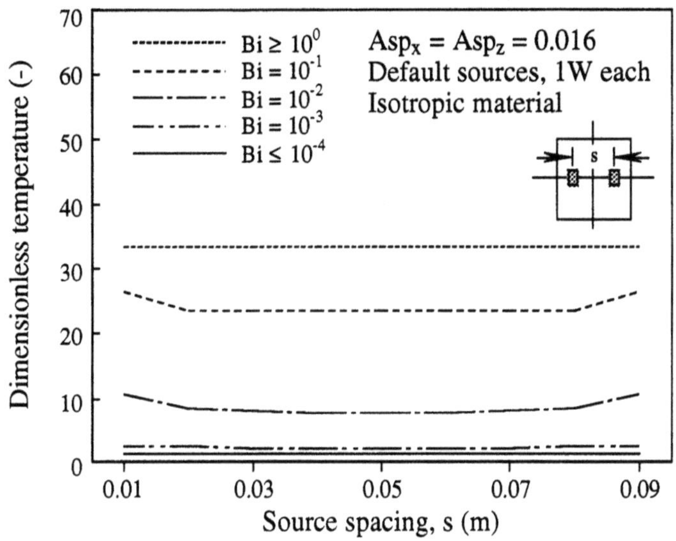

Fig. 5 Dimensionless temperature as a function of Biot number and component spacing

3.3. MULTIPLE LAYER SUBSTRATES

In this section, the effects of heat spreader layers and the behaviour of thermal substrates are demonstrated for the default cooling problem. Equivalent series and parallel orthotropic conductivity components are derived to model multiple layer substrates as single layers.

Figure 6 shows dimensionless temperature as a function of heat transfer coefficient for three different multiple layer boards: the default PCB substrate; the default PCB with a two-ounce copper heat spreader placed on the source side (2oz - PCB), in the mid-plane (PCB - 2oz - PCB) and on the non-source side (PCB - 2oz); and a commercially-available 'thermal' substrate, which is detailed in a publication by the Bergquist company (1991). The substrate consists of a one-ounce copper layer on the source side, 0.076 mm of a dielectric material of conductivity 3 W/mK and a 1.6 mm thick aluminium layer.

Clearly, the addition, in any position, of the copper layer to the PCB greatly reduces the dimensionless temperature at the source. The thermally thick limit is reached only at large values of the heat transfer coefficient. For the most effective use of the heat spreader layer, it should be placed as close to the source side as possible. The thermal substrate produces the lowest temperatures of all the boards due to the high conductivity of the structure.

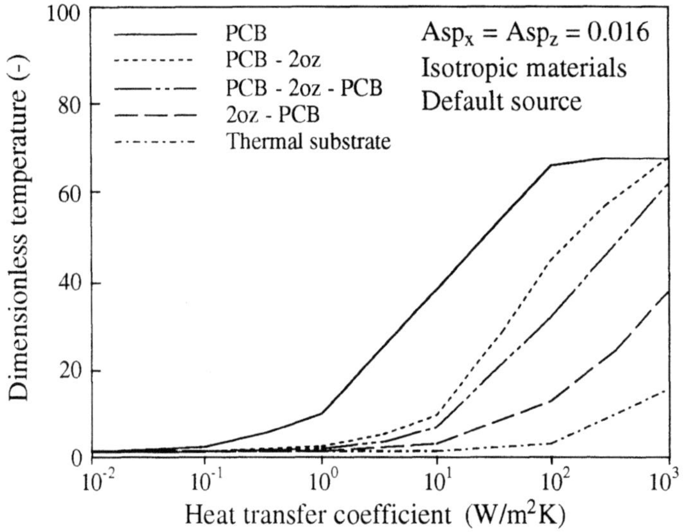

Fig. 6 Dimensionless temperature as a function of heat transfer coefficient for different multiple layer boards

Series and parallel equivalent conductivities were reported by Lemczyk et al (1991) which represent the conductivities for one-dimensional heat flow through, and in the plane of, the substrate respectively. Accounting for orthotropic conductivities in the layer materials:

$$K_{sy} = \left(\sum_{i=1}^{I} \frac{H_i}{K_{iy} H_T} \right)^{-1} \qquad K_{px} = \sum_{i=1}^{I} \frac{H_i K_{ix}}{H_T} \qquad K_{pz} = \sum_{i=1}^{I} \frac{H_i K_{iz}}{H_T} \tag{13}$$

Lemczyk et al (1991) used the harmonic average of the series and parallel conductivities to approximate a multiple layer structure as an isotropic single layer. Greater accuracy, however, can be obtained by using the equivalent orthotropic series and parallel conductivities to evaluate the conductivity ratios K_x^+ and K_z^+ which describe an orthotropic single layer. The accuracy of these approximations is tested in figures 7 and 8 as functions of the heat transfer coefficient and source dimension respectively for the default PCB with the two-ounce copper layer.

The figures show that the orthotropic conductivity model closely approximates the PCB with the copper layer at mid-plane. However, the orthotropic conductivity model cannot account for the position of the copper layer within the structure and it overestimates and underestimates the temperatures produced by the 2oz - PCB and the PCB - 2oz structures respectively. The harmonic model predicts very high temperatures.

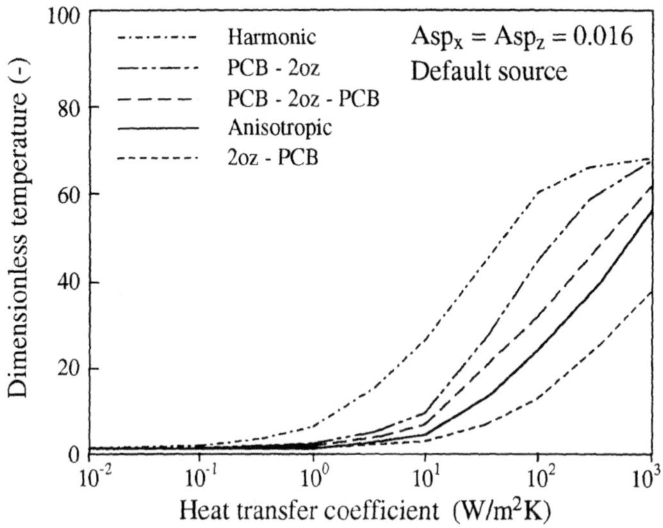

Fig. 7 Dimensionless temperature as a function of heat transfer coefficient for different multiple layer approximations

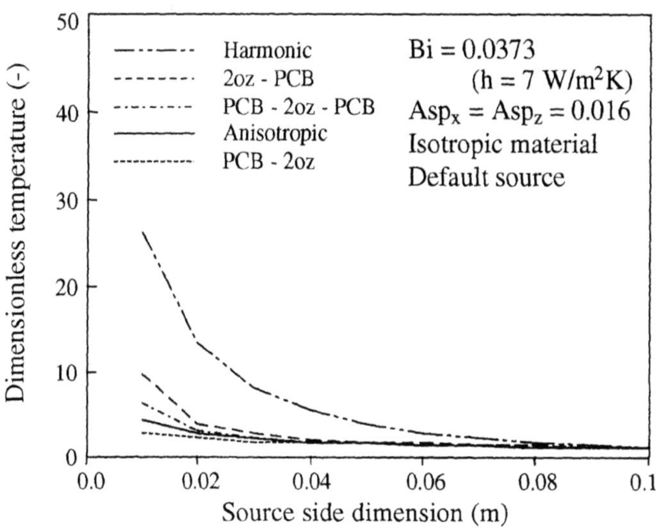

Fig. 8 Dimensionless temperature as a function of source dimension for different multiple layer approximations

These two parametric tests do not constitute a comprehensive assessment of the accuracy of the equivalent orthotropic model, however it is clear that it is a useful simplification. Even though the default PCB with a single two-ounce copper layer is only one example of a multiple layer substrate, it is perhaps a worst-case because of the large disparities in the conductivities and in the thicknesses of the layers.

4. Conclusions

The dimensionless temperature field within a multiple layer substrate is a function of a Biot number and two each of aspect ratios and conductivity ratios of each layer, and ratios of Fourier coefficients which define the source flux distribution, size and position.

The dimensionless temperature at a point on the source side of a substrate is bound by two limits: the 'thermally thin' limit, at which large temperature gradients in the substrate are sustained and all heat loss occurs by convection from the source side; and the 'thermally thick' limit, unity, at which the source side becomes isothermal and heat is lost in equal proportions from the source and non-source sides. These thick and thin limits are characterised by low and high values of Biot number respectively.

Aspect ratios and conductivity ratios govern the spread of heat in the plane of the substrate. Large ratios cause dimensionless temperatures to approach the thermally thick limit.

The dimensionless temperature at a point on the source side of a substrate outside of the thermally thick limit can be a strong function of the applied flux density at that point.

The dimensionless temperature on a substrate densely and uniformly populated by sources is, by definition, close to the thermally thick limit.

For substrates not within the thermally thick or thin cooling regimes, the dimensionless temperature beneath the centroid of a source is proportional to the proximity of the source to the adiabatic edge of the substrate, or to another source. Source spacing has little effect.

For optimum heat spreading, a spreader layer must be as close as possible to the sources.

Equivalent orthotropic series and parallel conductivity components can be used to model a multiple layer structure as a single layer.

ACKNOWLEDGEMENTS

The authors gratefully acknowledge the support of Eolas through the Power Electronics Program in Advanced Technology.

REFERENCES

Bergquist company, 1991, "The Thermal Clad Design Guide," available from Bergquist, 5300 Edina Industrial Boulevard Mpls., MN 55435, p. 6.

Carslaw, H.S., and Jaeger, J.C., 1959, *Conduction of Heat in Solids*, Clarendon, Oxford. pp. 180-182.

Lemczyk, T.F., Mack, B., Culham, J.R., and Yovanovich, M.M., 1991, "PCB Trace Thermal Analysis and Effective Conductivity," *Proceedings of the 7th IEEE SEMI-THERM Symposium*, pp. 15-22.

Punch, J., and Davies, M., 1993, "Three-Dimensional Heat Conduction Through Rectangular Multiple Layer Plane Substrates," accepted for presentation at the ASME National Heat Transfer Conference, August.

Thermal drift effects in power microwave hybrid circuits

ALKHOJA J. [1]., RAULY D.[2], GRANDJEAN F.[3]

[1] *LET - URA 1403 ENSMA - 86034 POITIERS CEDEX FRANCE*
[2] *LTTM - IUT - 03107 MONTLUCON CEDEX FRANCE*
[3] *LAHC - UNIVERSITE DE SAVOIE - 73376 LE BOURGET DU LAC*

Abstract. The analysis of transient thermal transfers in a power microwave thick film resistor is performed by a coupled Boundary Element/Nodal Method. The thermal nodal network is then coupled to an equivalent electrical circuit, experimentally extracted by an iterative procedure. A step-by-step time domain simulation exhibits the non-linear behaviour of the scattering parameters during an indicial power excitation at 1 GHz.

1. Introduction

In recent years, thermal management of hybrid microelectronics devices have taken an increasing part, focusing on their reliability in power applications. Advanced packaging techniques have been proposed, in order to meet multiple requirements such as efficient heat evacuation or spreading, fast electrical response and low thermomechanical stresses [1],[2].

In the field of microwave circuits, particular specifications are usually imposed, concerning the scattering parameters Sij, and consequently the propagation and losses characteristics of the structure. The Suspended Strip technology offers appreciable performances [3], but accurate modeling of these structures requires heavy tools in both electrical and thermal domains. In the former, the propagation characteristics must be computed in a highly dissipative line, while in the later it is requested to analyse transient heat transfers with conductive and radiative modes. Moreover, this kind of structure is composite, with strongly unequal extension in 3D space, and it should be advisable to use a numerical method with a complicated grid

The heaviness of modeling is still increased when thermal and electrical problems must be coupled. An attractive alternative should be the use of an unique modeling software, operating with interactive thermal and electrical equivalent networks. Simulator softwares in personal computer version, like SPICE or ESACAP, are suitable for this application. In return, the difficulty is then transferred on the elaboration of the equivalent networks.

C. J. Hoogendoorn et al. (eds.), Thermal Management of Electronic Systems, 169–178.

In the present paper, this method is applied to the simulation of a Suspended-strip Microwave Resistor at 1GHz, carried out by means of ESACAP. The equivalent circuit is based on previous rigorous analysis in both thermal and electrical domains.

2. Studied Microwave Device

The studied device is a microwave thick film resistor, inserted in a shielded Suspended Stripline (see figure 1), with a resistive ink (DUPONT QS 872) characterised by a 100 Ω/❑ sheet resistance. The operating frequency ranges up to 5 GHz and the typical power density within the resistive material is around 10^{10} W/m^3.

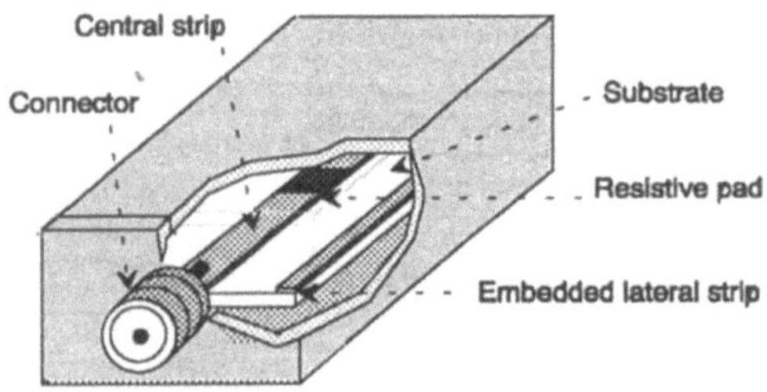

Figure 1-a - Hybrid Microwave Thick Film Resistor, with its case.

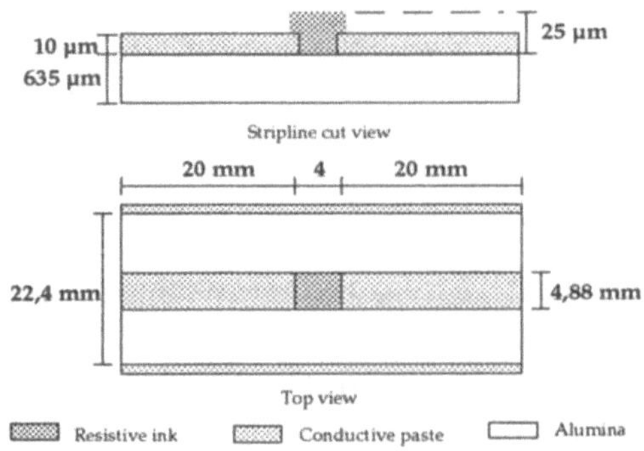

Figure 1-b - Stripline structure

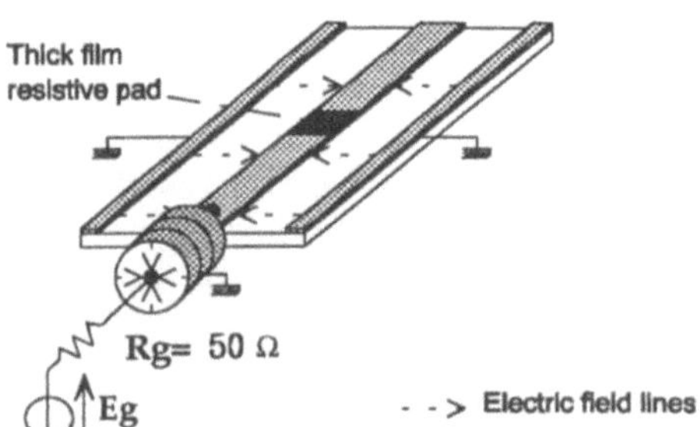

Figure 2-a - Set up for Launching the Electromagnetic Wave

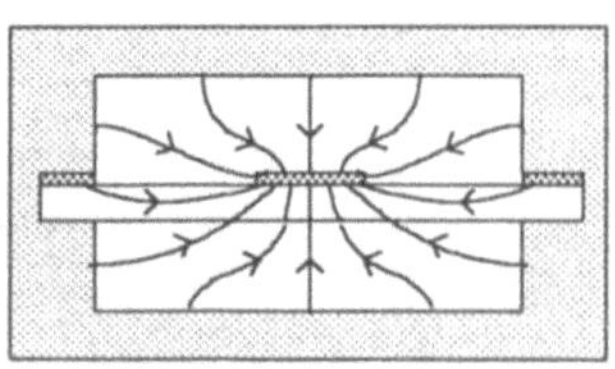

Figure 2-b - Electric Field Lines Pattern

The electromagnetic wave is launched from a coaxial line, with the central conductor connected to the central strip, while the external conductor is relied to the metallic case. This kind of connection excites a quasi-TEM (Transverse Electro-Magnetic) mode in a lossless line, which corresponds to the field lines pattern represented in figure 2. In fact, the resistive pad modifies this schematic representation, inducing losses and dispersive propagation. The detailed electrical analysis is related in next section.

3. Electrical Model

The equivalent electrical circuit is determined by an iterative extraction procedure [4]. It consists in measuring the Scattering parameters Sij with a vectorial network analyser (WILTRON 360) in frequency domain, and comparing them to the simulated values, resulting from an initial equivalent circuit. The interesting parameters are for a symmetrical structure S_{11} and S_{21}, which represent respectively the input reflection and forward transmission coefficients with the output matched.

An error function is then calculated and the elements of the circuit are consequently corrected. The initial equivalent circuit and the form of the error function governess the rapidity of the convergence. An other requirement for the resistive equivalent elements is to meet the static value.

The chosen error function is indicated below:

$$.F=\frac{1}{n}\sum_{i=1}^{2}\sum_{k=1}^{n}\left|S_{ij}(f_k)\right|^2(1-K_i^2) \quad \text{where} \quad K_i=\frac{\sum_{k=1}^{n}\left|S_{i1}(f_k)\right|.\left|\hat{S}_{i1}(f_k)\right|}{\sqrt{\sum_{k=1}^{n}\left|S_{i1}(f_k)\right|^2.\sum_{k=1}^{n}\left|\hat{S}_{i1}(f_k)\right|^2}}$$

n is the number of optimisation in the frequency band,
f_k is the operating frequency for simulated and measured points,
S_{i1} and $\hat{S}_{i1}$ are the measured and simulated parameters respectively.

Good agreement between measured and simulated Sij parameters have been found and reported in [4]. The extracted equivalent circuit (see figure 3 below) exhibits reactive and resistive elements. These are assumed to undergo a - 200 ppm/°C thermal drift for ulterior simulation at the operating frequency 1 GHz.

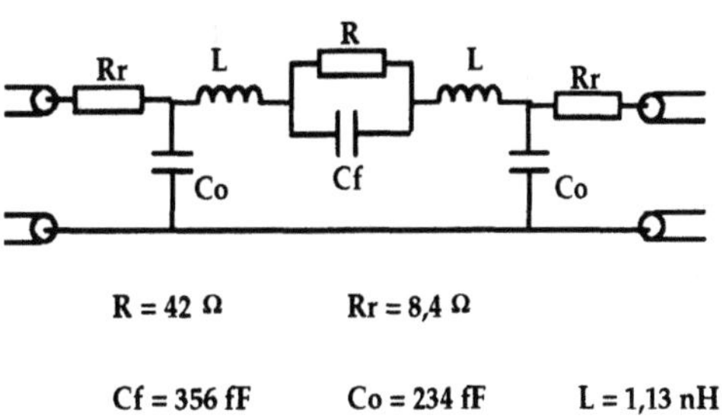

Figure 3 - Equivalent Circuit of the Thick Film Suspended-Strip Resistor

The resistor R represents the central part of the resistive pad, while Rr modelises the paste overlap on the stripline. The capacitor Cf is caused by the gap between access lines and also by the inner capacitance of the paste, while Co is the end capacitor of the access line. The self L modelises the electrical delay due to the length of the component

4. Thermal Diffusion Model

The thermal modeling of the device is performed by means of a combined Boundary-Element (BEM)/Nodal method .The first step is a 2D BEM analysis in several critical cuts of the central plate (alumina substrate with the conductive strip and the resistive pad). The second step achieves a 3D nodal network, with the nodes localisation based on the precedent analysis. Moreover, the thermal surroundings of the substrate, i.e. the different heat transfers modes towards the case and the case/air interface,. are taken into account.

4.1 Boundary Element Analysis

BEM Formulation:

Given an isotropic domain Ω, closed by a boundary Γ, and characterized by the parameters:

λ: thermal conductivity $[W.K^{-1}m^{-1}]$
a: thermal diffusivity$[m^2.s^{-1}]$
Q: internal generated heat source $[W.m^{-3}]$

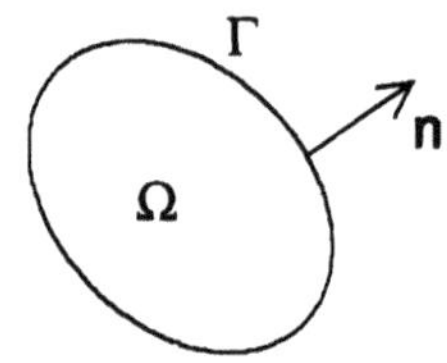

The FOURIER differential equation, governing the temperature θ within Ω is:

$$\nabla^2\theta - \frac{1}{a}\frac{\partial\theta}{\partial t} + \frac{Q}{\lambda} = 0 \qquad t\rangle 0 \quad \text{where } \nabla^2 \text{ is the Laplace's operator} \quad (1)$$

The domain problem can then be reduced to a boundary problem by applying the GREEN's theorem:

$$\int_0^t\int_\Omega (U.\nabla^2\theta - \theta.\nabla^2 U)\, d\Omega\, d\tau = \int_0^t\int_\Gamma (U\frac{\partial\theta}{\partial n} - \theta\frac{\partial U}{\partial n}) d\Gamma\, d\tau \qquad (2)$$

with n defined as the outer unit normal to Γ.

The function U must be selected to satisfy the heat equation (3), which represents the response of the system and where a unit pulse source is applied at time τ at co-ordinates $\overline{X}, \overline{Y}$.[7]:

$$\nabla^2 U + \frac{1}{a}\frac{\partial U}{\partial\tau} + \delta\langle X, Y | \overline{X}, \overline{Y}\rangle \delta\langle t|\tau\rangle = 0 \quad \tau\langle t \quad \text{where } \delta \text{ is the Dirac function} \quad (3)$$

The so-called "fundamental solution" [5] of equation (3) is for a two dimensional problem:

$$U = \frac{\exp(-r^2/4a(t-\tau))}{(4\pi a(t-\tau))} \qquad (4)$$

r is the distance between the source point and the considered response point.

For a short time interval $[t_1,t_2]$, the basic equation (2) becomes after simplification [6]:

$$C_i\theta_i(t_2)+\frac{1}{2\pi}\int_{\Gamma}\left(\theta(t_2)\exp(-\beta/r^2)\frac{\partial r}{\partial n}\right)d\Gamma = \frac{1}{4\pi}\int_{\Gamma}\frac{\partial\theta}{\partial n}(t_2)\,E_1(\beta)\,d\Gamma + \int_{\Omega}(U\,\theta(t_1))\,d\Omega + a\int_{\Omega}\int_{t_1}^{t_2}(U\,Q/\lambda)\,d\Omega\,d\tau \qquad (5)$$

Ci = 1 for a point within the domain
Ci = 1/2 for a point on the boundary
$\beta=(r^2/4a\,\Delta t)$ with $\Delta t=t_2-t_1$

$E_1(\beta)$ is the exponential integral. [10]

$$U=\frac{\exp(-r^2/4a\,\Delta t)}{4\pi a\,\Delta t}$$

Both kinds of conditions, i.e. DIRICHLET and NEUMANN, may be found on the boundary. The later is then divided in N elements, taking in consideration its geometry and discontinuities on θ or $\partial\theta/\partial n$. The temperature and flux density are considered uniform on each element and the time step is fixed constant.

This discretization applied to equation (5) leads to the matricial equation, to be resolved step by step in temporal domain.:

$$[D]\,[X]=[R] \qquad (6)$$

where $[D]$ is the coefficient matrix, $[X]$ is the unknown matrix and $[R]$ is the known parameters matrix (initial conditions, generated heat, ...). Internal temperatures must be re-evaluated for each time step.

2D BEM results

Figure 4-a shows the boundary conditions simulated on an half stripline cut view (see fig.1b). It must be noted that the 20°C condition simulates a perfect contact with the case, which improves the role of lateral thermal drain usually played by the substrate. Only the alumina substrate is studied, and the heat generated within the resistive pad is consequently described as a surface heat density φ.

The isothermal lines are plotted figure 4-b for different instants. It can be observed that the steady-state is reached in about ten seconds. A thickness-directed temperature gradient exists only within the portion located just under the heating pad. The heat rate is consequently one-dimensional in the other direction. This result leads to a first choice concerning the nodal network, namely a unique node in the thickness direction.

Figure 5-a represents the simulation conditions on the quarter top view of the composite substrate (see fig. 1-b). The heat source within the resistive paste is here 3.10^7 W.m^{-3}, and isothermal lines are shown on figure 5-b, for different

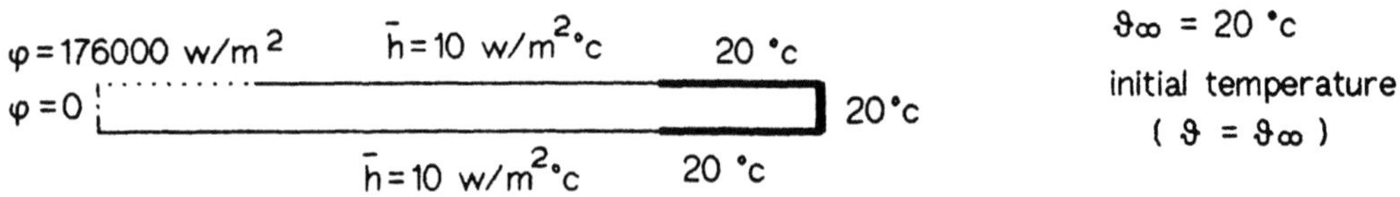

Figure 4-a - Initial temperature and boundary conditions

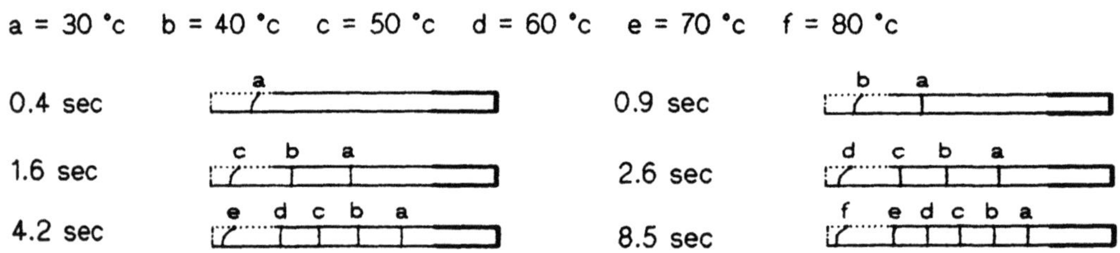

Figure 4-b - Temperature distributions

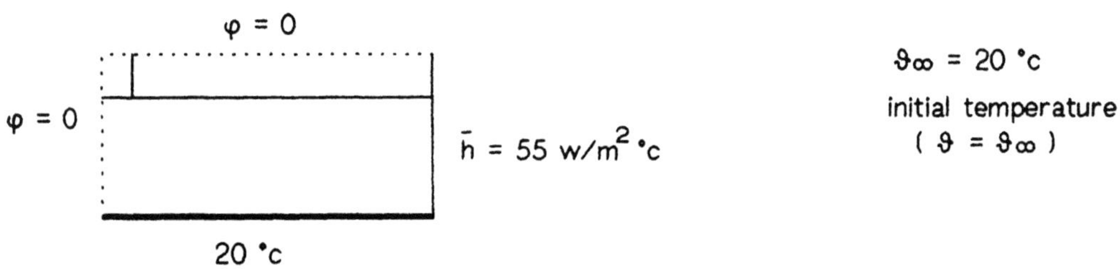

Figure 5-a - Initial temperature and boundary conditions

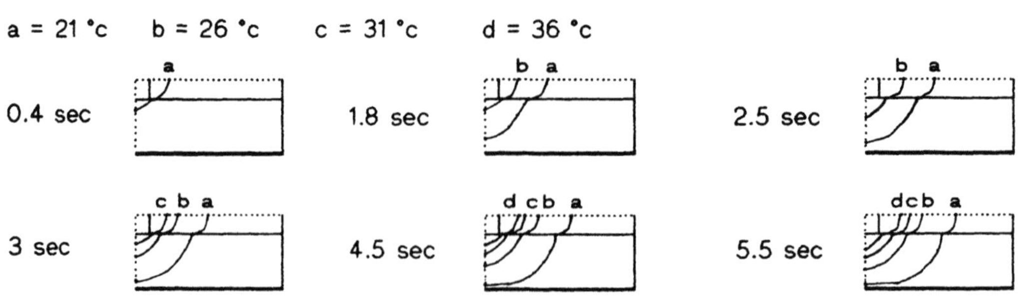

Figure 5-b - Temperature distributions

time steps. It appears that the temperature is almost uniform within the heating paste, which constitutes a favourable condition for electrical operation. An other consequence is the possibility of placing an unique node for the resistive pad volume in the nodal network.

In other respects, the isothermal lines are lightly broken at the metal/alumina interface, with a heat wave front progressing more quickly in the metallic strip than in the substrate. Nevertheless, it must be noted that the 2D modeling doesn't take into account the low 10 µm thickness of the central strip. This effect should be less important in the reality.

Finally, the temperature gradient is lower in the strip direction than in the perpendicular one because of the geometry and the boundary conditions. This will influence the number of nodes in both corresponding directions of the nodal network.

4.2 3D Nodal Equivalent Network

The nodal network, represented on figure 6, has been built with the classical thermal/electrical analogy, the number of node in the substrate, the metallic and resistive deposits being determined from the BEM analysis. All the conductances are orthogonal, except whose which connect the resistive overlap to the substrate.

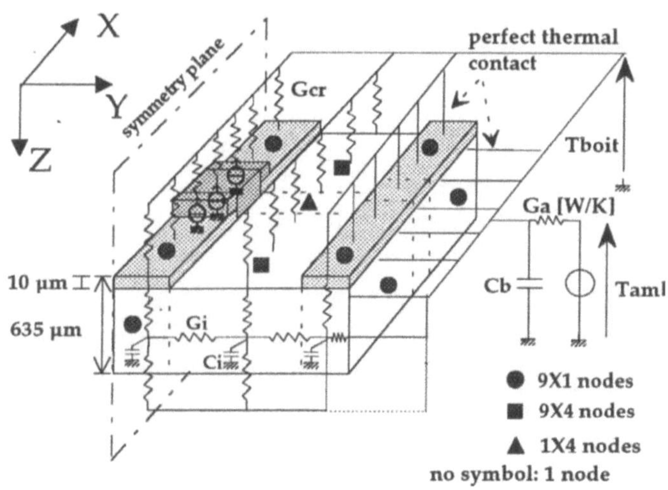

Fig. 6: Schematic nodal network of the device, modeling the thermal transfers within the central plate and between plate and case. Each layer contains one node in Z-direction. The number of nodes in X and Y directions are indicated on each sub domain.
Additional nodes have been placed in the surelevated part of the resistive pad. The three visible power sources correspond to the electrical power dissipated in resistors Rr and R in the equivalent electrical circuit (Fig. 3)

Next, the heat is transferred from the plate toward the case with a conductive mode via the embedded lateral strips, but also with surface exchanges. These are due to the conductivity of air and the radiation. No convection can occur in the enclosed air gap because of its low 3 mm thickness. Indeed, the RAYLEIGH number, calculated with extreme conditions (T_{plate} = 160 °C and T_{case} = 20°C) has been found equal to 530 [8], while the minimum value allowing convection is 1700. [9].

Since the case can be assumed isothermal in first approximation [8], the conductances Gcr (see fig.6) are calculated from the surface exchange coefficient $h_{ar} = h_a + h_r$ [$W.K^{-1}m^{-2}$], where:

$$h_a = \frac{\lambda_{air}}{g_{air}} \approx \frac{0.026}{2\times10^{-3}} \approx 13\ W.K^{-1}.m^{-2} \qquad (7)$$ modelises the conduction in air

$$h_r = \frac{\sigma}{S_p} \times \frac{\varepsilon_p\varepsilon_c}{\varepsilon_p + \varepsilon_c - \varepsilon_p\varepsilon_c} \times \frac{\theta_p^4 - \theta_c^4}{\theta_p - \theta_c} \qquad (8)$$ modelises the radiation

The temperature and the emissivities are designed θ and ε, while subscripts p and c relate to the plate and the case respectively. S_p is the global plate surface. The used equation (8) describes a radiative system with two infinite plane surfaces [9]. Indeed, the radiation between vertical and horizontal internal faces of the case can been neglected, since the former are much less the later. This comes to consider the global internal face as plane, but receiving the totality of the emitted flux from the plate.

5. Results

Both equivalent electrical (fig. 3) and thermal (fig. 6) circuits have been coupled and implemented in the software ESACAP (E.C. ,Hoersholm, Denmark). The temperature coefficient of the equivalent resistors has been assumed to the typical value -200 ppm/°C. A time domain, step-by-step, simulation has been perform with an incident high level (70V amplitude, with 50 Ω internal resistance), 1GHz wave, modulated by a step, as electrical source condition. The step duration was 50 ms from 0 to 2 s, .5s from 2 to 10s, 5s from 10 to 100s and 10s from 100 to 200s.

The simulated temperature and power evolutions in the resistances and the consequent parameters drift are plotted on figures 7 to 9 below. The major part of evolution is accomplished in 20 s, but a slow drifting subsists up to 200 s and the steady state is not rigorously reached at this time. This slow effect is due to the case thermal capacity, interfaced with air by the thermal conductance of the heat-sink (0.2 $K.W^{-1}$).

The final temperature of the central resistive pad (≈115°C) is greater than the overlap's one (≈95°C), although the microwave power density is 0.2 $W.mm^{-2}$ in the central pad, and 1.1 $W.mm^{-2}$ in the overlaps. In fact the overlap volume is very low ($4.88\times10^{-3} \bullet 150\times10^{-6} \bullet 15\times10^{-6}\ m^3$) and easy heat evacuation is offered by the underlying metallic strip.

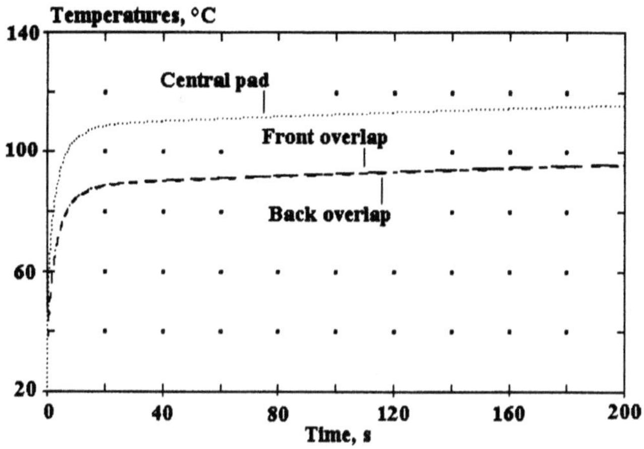

Figure 7-

Evolution of resistive central pad and both overlap temperatures, in response to a microwave power step, corresponding to a high level, 1GHz, incident signal (70V generator with a serial 50 Ω resistance).

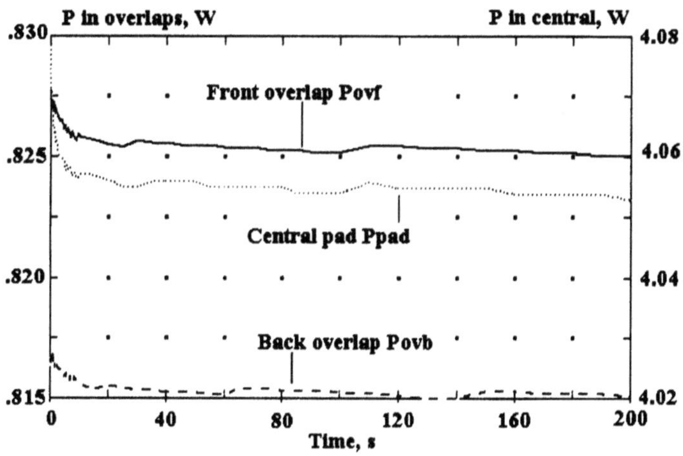

Figure 8 -

Evolution of the microwave power dissipated in the resistive central pad (modelised by R in fig.3) and both overlaps (modelised by Rr), with the same incident signal. The corresponding power densities are around 8.3 GW.m^{-3} within central pad and 75 GW m^{-3} within overlaps.

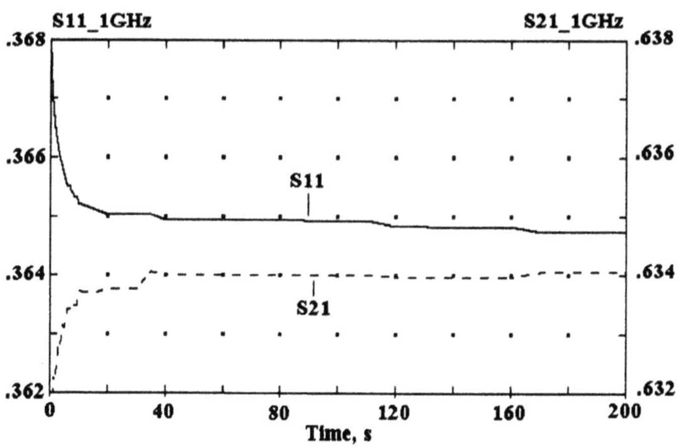

Figure 9 -

Thermal drift of S_{11} (reflection) and S_{21} (transmission) parameters, induced by the overheating of the resistive paste, with negative temperature coefficient. The global serial resistance is diminished and the reflection is consequently lowered, while transmission is increased.

The overheating of the resistive paste diminishes the serial resistance of the microwave device, which leads to lower S_{11} (reflection coefficient with matched output) and to increase S_{21} (transmission coefficient with output matched). This is because the resistive pad must be considered as a serial obstacle embedded in the Suspended-strip line. Other effects could be expected for a parallel resistive obstacle.

Regarding the energetic balance, the microwave power dissipated in the device is slightly reduced, which contributes here to stabilize the component. Once more, other designs concerning the localisation of pads could lead to other behaviours.

6. Conclusion

A time domain, step-by-step simulation of a non-linear power-hybrid microwave component has been carried out. It reveals a light thermal drift of its electrical parameters, which doesn't here produce irreversible effects because of the serial configuration of the resistive pad.

Other particular geometries and excitation modes, like pulsed signals, must be studied. It requires extension of the method towards multiple pads designs, which represents an ulterior work. Different patterns and materials should be proposed in order to meet the new thermal and electrical requirements.

References

[1] N.B. NGUYEN:" Adoption of Advanced Materials in Hybrid Packaging Techniques".*Proceedings of the 1992 Internat. Symp.on Microelectronics, San Fransisco, USA*, pp 567-570,october 19-21, 1992

[2] J. REDOUTEY, D. LAFORE, JM LI:" A Very Fast Power Hybrid Module in the Thyristor Mode". *Proceedings of EPE 91, Florence*, vol 2, pp 439-441, sept. 1991

[3] A. RONG:" Transmission characteristics of Suspended Striplines with Finite Metallisation Thickness and Substrate Supporting Grooves". *Electron. Lett.* , Vol.26, n°7, pp431-432 (29 march 1990)

[4] F. GRANDJEAN, G. ANGENIEUX, B. FLECHET:" Extraction of Equivalent Circuits of Hybrid Microstrip Resistor by Harmonic Measurements". *Proceedings of Internat. Symp. on Microelectron., ISHM'92*, pp 168-172 (oct 19-21, 1992)

[5] H.S. CARSLAW, J.C. JAEGER:"Conduction of Heat in Solids". *Clarendon Press, Oxford*,1959

[6] J. ALKHOJA, N. MARILIER, J.J. VULLIERME:"Utilisation de la Méthode des Intégrales de Frontières pour la Détermination des Champs Thermiques en Milieu Isotrope ou Anisotrope". *Journées d'Etude SFT, Paris*, 5 june 1991

[7] J.C.F. TELLES, L.C. WROBEL, C.A. BREBBIA:" Boundary Element Techniques". *Springer-Verlag*, 1984

[8] D. RAULY, N.E. SIFI:"Dérive Thermique des Paramètres d'une Charge Adaptée de Puissance". *4ème Colloque L'Electronique de Puissance du Futur, Marseille* (18-20 novembre 1992)

[9] D.R. PITIS, L.E. SISSOM:"Theory and problems of Heat Transfers". *Mc Graw Hill Book Company*

[10] M. ABRAMONITZ, I.A. SEGUN:"Handbook of Mathematical Functions". *National Bureau of Standards, Washington*, 1972

Computational modelling of wafer cooldown in a staging module

K. NEDERVEEN

ASM Microelectronics Technology Centre, P.O. Box 100, 3720 AC Bilthoven, The Netherlands

Abstract. It is investigated how long it will take to cool down a batch of wafers from 600 to 70 °C in a so-called staging module. Different approaches are followed to estimate the time scale of the cooldown process, from simple analytical approximations to a numerical grid-based model. The numerical model describes both unsteady-state and steady-state heat transfer in an axisymmetric geometry, taking into account conduction and radiation. Water cooling and free (convective and radiative) cooling have been considered.

1. Introduction

In semiconductor production equipment, typically LPCVD (low pressure chemical vapour deposition) processes take place. I.e., chemically reacting gases are deposited as a thin layer on a silicon surface (the wafer). These processes take place at temperatures in the order of 500 to 1200 °C. In ASM's Advance 600/2 cluster tool several batch-reactors (modules) are gathered around a central wafer-handling robot. This implies that a batch of wafers may undergo a sequence of process steps without having to leave the vacuum controlled environment [1]. In Figure 1 schematic pictures of a cluster tool are shown. The modules are vertically positioned, with the wafers horizontally lying in a quartz boat.

One of the possible modules of the Advance 600/2 is a staging (or: cooldown) module. Its function is to cool down a batch of (typically) 75 wafers that has returned from a process module to an acceptable temperature level, prior to transfer to the outside world. For reasons of throughput the cooldown time has to be as low as possible.

The aim of the investigation described here is to determine the time to cool down a batch of 75 wafers from 600 to 70 °C. The staging module has a stainless steel wall. Water cooling and free (convective and radiative) cooling is considered. The non-stationary energy transport equation is solved in cylindrical coordinates, taking into account conduction and radiation, using a discretization scheme based on the finite-volume approach.

2. First approximation

The spectral emissivity ε_λ of silicon is presented in [2] for temperatures between 270 and 800 °C. For temperatures above 800 °C Si acts as a grey body, since its emissivity is a constant of about 0.7 for all wavelengths. From [2] the *total* emissivity is estimated to lie between 0.2 and 0.7.

C. J. Hoogendoorn et al. (eds.), Thermal Management of Electronic Systems, 179–187.

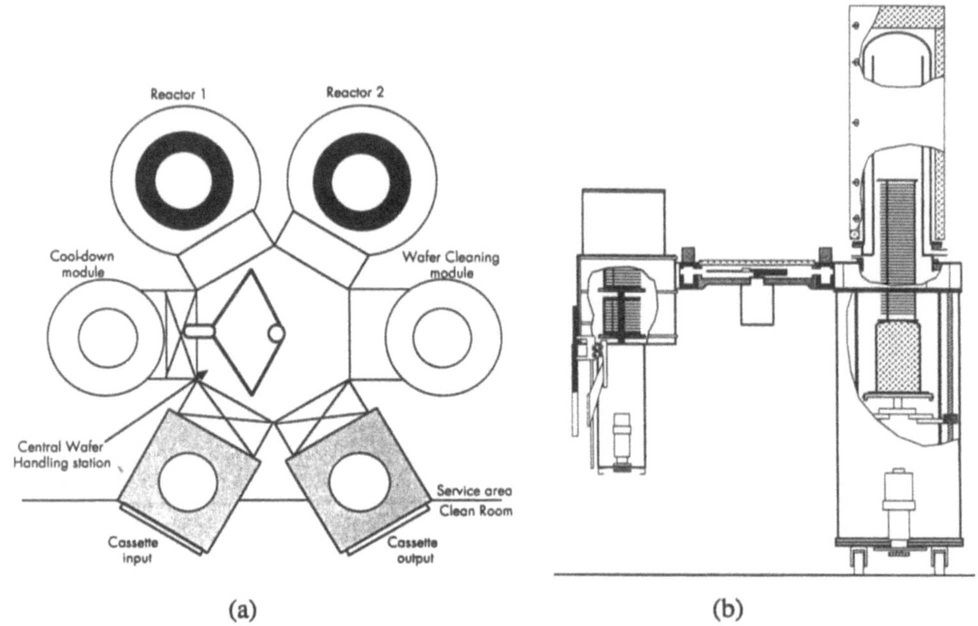

Figure 1: *Schematic arrangement of a cluster tool: (a) top view, (b) side view.*

All the wafers in the batch certainly have reached a certain temperature when the *middle* wafer has reached this temperature. Consider two coaxial wafers with similar radiative characteristics and equal radius. The radiative flux leaving a surface that faces the other wafer is given by, assuming a zero transmissivity,

$$q = \frac{\varepsilon\sigma T_1^4 (1 - F)}{1 - \rho F} + \frac{\varepsilon^2 \sigma F \left(T_1^4 - T_2^4 \right)}{1 - \rho^2 F^2} \tag{2.1}$$

where
F = view factor
q = radiative flux [W/m^2]
T_1 = temperature of the wafer under consideration [K]
T_2 = temperature of the opposite wafer [K]
ε = emissivity
ρ = reflectivity = 1 - ε.

The first term describes the energy escaping from the gap between the wafers. This term is not dependent upon the temperature of the opposite wafer. The second term describes the energy exchange between the two surfaces. The total heat flow leaving the surfaces of one wafer with T_1 between *two* others of T_2, is found by multiplication of the total term of eq. (2.1) by the total area of the wafer (i.e., taking into account top and bottom side). Note that the contribution of the rim is neglected here.

When F ~ 1, as is the case here, the first part q_1 of eq. (2.1) is approximately equal to

$$q_1 = \sigma T_1^4 (1 - F) \tag{2.2}$$

This is independent upon the emissivity of the wafer. When emissivity is low a relatively small amount of radiation leaves the gap between the wafers directly. On the other hand, then there is a relatively large contribution leaving the gap by multiple reflection. For a high emissivity the reverse holds.

In order to assess how the number of wafers influences the outcome a quick numerical iteration procedure is set up to solve for the wafer temperatures using eq. (2.1). For an infinite number of wafers the influence of neighbouring wafers should vanish and the outcome should equal that obtained upon using eq. (2.2).

Results are shown in Table 1 for 15, 75 and ∞ wafers. The latter is obtained with eq. (2.2). The cooldown time is defined as the time needed for the middle wafer to reach 70 °C, starting from 600 °C. The radius of the wafer is 200 mm, the centre-to-centre wafer distance 6.35 mm, and the thickness 0.785 mm.

Table 1: *Cooldown time (min.) for 15, 75 and ∞ wafers, as a function of Si emissivity.*

ε_{Si}	Δt (15) (min.)	Δt (75) (min.)	Δt (∞) (min.)
0.2	31.9	31.9	32.0
0.3	29.0	29.1	29.2
0.4	27.2	27.6	27.7
0.5	25.9	26.6	26.6
0.6	24.8	25.8	25.9
0.7	23.7	25.1	25.2

For small emissivities there is no difference between 15 and 75 wafers. Even for emissivity 0.7 the difference is small. The difference between 75 and ∞ wafers is insignificant. It is concluded that 15 wafers suffice to calculate the cooldown time within an accuracy sufficient for ASM purposes.

Considering the favourable approximations, the values obtained here for the cooldown time are a lower limit. For a pitch of 6.35 mm this lower limit for the cooldown time lies around 25 to 30 minutes. Given the reactor height and the number of wafers the pitch practically cannot be varied. Note that the temperature of the outside world (wall of the reactor module) is not regarded in eq. (2.1). Neither is conduction, which becomes important at low temperatures. The actual value of emissivity depends not only on temperature, but also on pre-treatment of the wafer and the kind of deposition that has taken place [3].

3. Numerical model

The Fourier equation for unsteady-state heat transfer is solved in cylindrical coordinates

$$\frac{\partial T}{\partial t} = \frac{1}{r}\frac{\partial}{\partial r}\left(\alpha\, r\,\frac{\partial T}{\partial r}\right) + \frac{\partial}{\partial z}\,\alpha\,\frac{\partial T}{\partial z} + S \tag{3.1}$$

where
α = thermal diffusivity [m^2/s]
S = source term [K/s].

For the discretization of eq. (3.1) the control-volume approach is followed. The resulting finite-difference equations are solved line-by-line using the Tri-Diagonal Matrix Algorithm [4]. From Figure 2 the orientation of the coordinate system can be read.

3.1 Source term

The source term S is determined as follows. The surfaces of those control volumes (computational cells) in the grid which are defined as a solid and which are not neighboured by a 'solid' surface are collected. The respective view factors for each possible pair of surface elements are calculated [5].

Then, for each radiative surface the net heat flux has to be determined, and is included in the source term of eq. (3.1). The radiative flux leaving surface *i* is described by

$$W_i = \varepsilon_i \sigma T_i^4 + \rho_i H_i + \tau_i H_i^* \tag{3.2}$$

where
H_i = incident radiative flux [W/m^2]
$H_i{}^*$ = incident radiative flux on surface *i**, i.e. the surface opposite to *i*, on the same solid [W/m^2]
ε_i = emissivity of surface *i*
ρ_i = reflectivity of surface *i*
σ = Boltzmann's constant
τ_i = transmissivity of surface *i*.

The incident flux on surface *i* is given by

$$H_i = \sum_{j=1}^{N} F_{ji} W_j \tag{3.3}$$

where
F_{ij} = view factor from surface *i* to *j*
N = number of surfaces present in the system.

Using matrix notation this can be combined into

$$\mathbf{q} = \sigma\,(\mathbf{I} - \mathbf{F})\,(\mathbf{I} - \mathbf{RF})^{-1}\,\mathbf{E} \tag{3.4}$$

and
$\mathbf{q}$ = vector containing the N net heat fluxes $W_i - H_i$
$\mathbf{E}$ = vector containing the N εT^4 terms
$\mathbf{F}$ = NxN matrix containing the view factors
$\mathbf{I}$ = unity matrix
$\mathbf{R}$ = an NxN matrix with the reflectivities on the diagonal positions and transmissivities on (i,i^*) positions.

All surfaces are assumed to be grey, so the r.h.s. of eq. (3.4) needs to be calculated once. Multiplication of each element of $\mathbf{q}$ by the corresponding area yields the power contribution to the source term for those grid points which are associated to free radiative surfaces.

The algorithms which have been developed to determine the view factors are general in the sense that any number of discs contained in a cylinder can be treated in this way, provided the geometry is axisymmetrical. The dimensions of the discs, both thickness and radius, can be chosen arbitrarily and different from each other. Naturally, there is an upper limit in the number of grid points and free radiating surfaces, due to computational constraints.

3.2 Thermal boundary conditions

Boundary conditions can be either isothermal or free-convective and radiative combined. See Figure 2 for clarification. The first possibility speaks for itself. The radiative heat transfer coefficient is easily calculated from the current wall temperature under the assumption that the staging module is contained in a infinitely large container of temperature 20 °C, i.e. the heat transfer coefficient is found from

$$h = \frac{\varepsilon\sigma\left(T^4 - T_\infty^4\right)}{\left(T - T_\infty\right)} \qquad (3.5)$$

The free-convective Nusselt number (Nu) for heat transfer from a plate to fluid (i.c. air) is given by

$$Nu = C\,(Gr\,Pr)^{0.25} \qquad (3.6)$$

where
Gr and Pr have their usual meaning [6].

For a vertical plate C = 0.59, and the expression is valid for $10^4 < GrPr < 10^9$. For cooled horizontal plates facing downwards C = 0.54, for $10^5 < GrPr < 2x10^7$. Finally, for cooled horizontal plates facing upwards C = 0.27, for $3x10^5 < GrPr < 3x10^{10}$. With L = 0.6 m, for air of room temperature, the above conditions for GrPr are met. From Nu the heat transfer coefficient is easily found.

4. Results

A grid has been set up for 15 wafers contained in a cylinder. The number of grid points is 91 in axial and 15 in radial direction. As has been seen section 2, the cooldown time reached at for 15 wafers is very similar to that for 75 wafers. The limitation to 15 wafers has computational benefits. In Figure 2 the geometry is depicted.

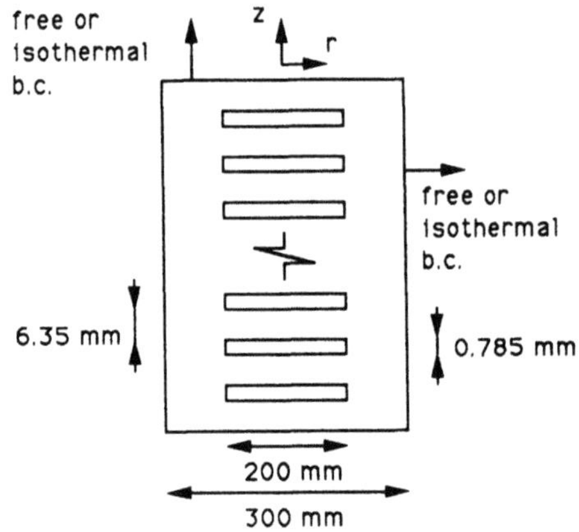

Figure 2: *Schematic picture of the geometry and the boundary conditions.*

The relevant dimensions are as follows:

diameter wafer : 200 mm — height module : 600 mm
diameter module: 300 mm — thickness wafer: 0.785 mm
interwafer distance (centre -to-centre) : 6.35 mm.

The gas between the wafers is nitrogen. The temperature dependence of the thermal conductivity of the nitrogen is taken into account. The thermal conductivity of the wafers is set to 35 W/m/K. In the Table 2 a few results obtained for various parameter combinations are presented, where the boundary condition is isothermal at 20 °C.

Table 2: *Cooldown time for the middle of 15 Si wafers in a stainless steel (s.s.) cylinder from 600 to 70 °C, as a function of the emissivities (ε), reflectivities (ρ) and transmissivities (τ). The wall is isothermal at 20 °C.*

	Si			s.s.		Δt
ε	ρ	τ	ε	ρ	τ	(min.)
0.7	0.3	0	0.4	0.6	0	43
0.2	0.3	0.5	0.4	0.6	0	46
0.7	0.3	0	1	0	0	34
0.2	0.3	0.5	1	0	0	38
0.1	0	0.9	0.4	0.6	0	43
0.1	0.9	0	0.4	0.6	0	70

These results are compiled in Figure 3 together with those obtained for free (convective and radiative) cooling on a wall with thickness of 3 mm. From the scarce data available, the reflectivity is considered to have a constant value of 0.3 [7].

In Figure 4 the cooldown time is shown as a function of wall thickness for wafer emissivities of 0.2 and 0.7, when free cooling boundary conditions are applied. The thicker the wall, the lower the cooldown time. This effect is reflected in Figure 5, where for various wall thicknesses the temperature development of the wall is shown. The thicker the wall, the lower the temperature increase.

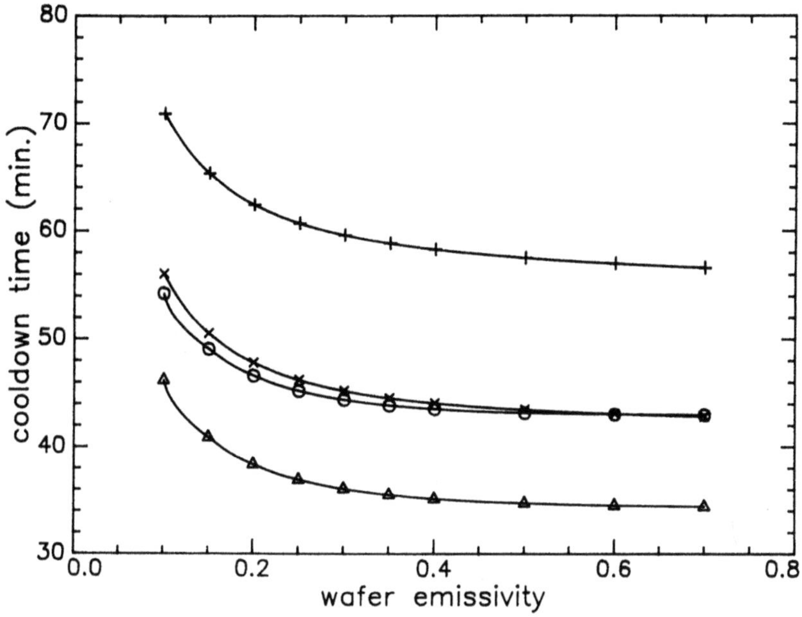

Figure 3: *Cooldown time for the middle of 15 wafers as a function of wafer emissivity. The reflectivity is a constant of 0.3. The boundary conditions are isothermal (Δ and o), and free (x and +). The wall emissivity is 0.4 (o and +), and 1.0 (Δ and x).*

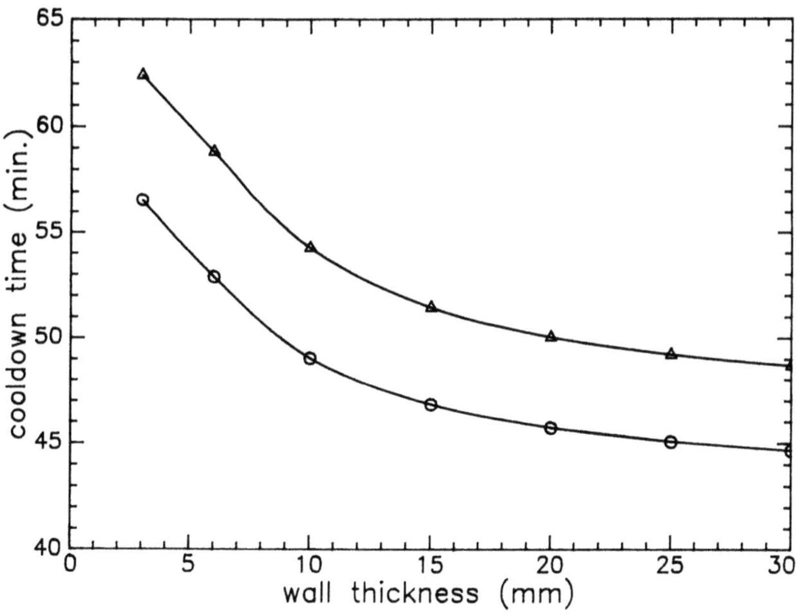

Figure 4: *Cooldown time for the middle of 15 wafers as a function of wall thickness when free cooling boundary conditions are applied. Wall emissivity is 0.4, wafer emissivity is 0.2 (Δ) and 0.7 (o).*

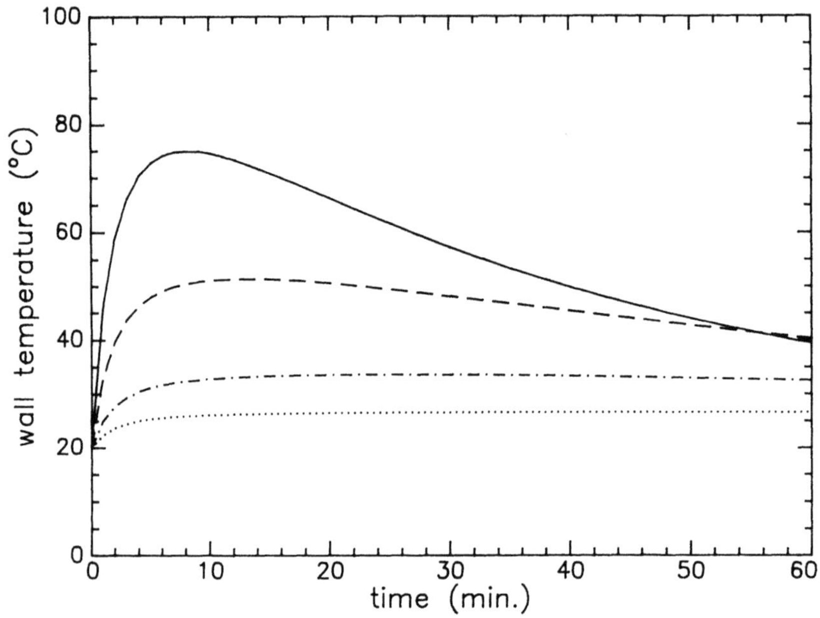

Figure 5: *Wall temperature vs. time half way up the wall, for wall thicknesses of 3 (—), 6 (- -), 15 (- · · -) and 30 mm (···). Free cooling boundary conditions are applied. Wall emissivity is 0.4, wafer emissivity is 0.7.*

5. Discussion and conclusions

The most important source of heat loss of the wafers is by radiation to the wall, and not so much to neighbour wafers of different temperature. Therefore, the cooldown time is not directly proportional to the emissivity of the wafer.This is concluded from both the analysis in section 2 and the numerical results shown in Figure 3. Thus, the fact that the wafer emissivity is not exactly known does not prevent a satisfactorily accurate prediction of the cooldown time range.

A constant value of 0.3 for the wafer reflectivity is considered, while the value of the emissivity is between 0.2 and 0.7. Then, the staging time achieved lies between 43 and 46 minutes for a typical wall emissivity of 0.4. In the idealized case with wall emissivity 1, the staging time lies between 34 and 38 minutes. This is a lower limit. The difference between the lower limits found analytically in section 2 is caused by the inclusion of the interaction with the wall. When the emissivity becomes lower than 0.2, which however is not likely, the cooldown time rapidly increases. Preliminary experimental observations have indicated that the cooldown time indeed lies in the order of 45 minutes.

Free convection cooling, instead of water cooling, of the staging module is in principle feasible. However, the applicability depends on the exact demands for the cooldown time. With the actual wall thickness (3 mm) and wall emissivity 0.4 cooldown times are expected to be about 30% higher than when the wall is water cooled. When the wall thickness is greater than 15 mm the cooled situation is approached. However, a module with a stainless steel wall thicker than 3 mm is not easily manufactured, apart from other drawbacks such as costs and weight.

Furthermore, it has to be realized that consecutive batches are to be cooled in the staging module. The wall temperature has to be back at 20 °C, when the next batch is entered. When this is not the case a higher cooldown time will arise. Consecutive batches might continuously heat up the wall. This will only partly be compensated for by a higher heat transfer coefficient. Therefore, for typical operating conditions water cooling cannot be disposed of.

In the approach followed in this paper free convective heat transfer between the wafers is neglected. This is justified by the fact that the Grashof number based on the interwafer distance, given the fact that temperature differences between neighbouring wafers (apart from the end wafers) are in the order of degrees, remains well below 10^3. In that case conduction dominates free convection between the wafers [6].

As mentioned in section 3, the applicability of the numerical model described here is broad in the sense that any number of discs contained in an axisymmetric geometry can be dealt with. E.g., the model has also been applied in the design of single wafer based heating and cooling stations.

References

1. Werkhoven, C.J., Westendorp, J.E.M., Huussen, F. and Granneman, E.H.A.: Poly-emitter device manufacturing in a cluster tool. *Semiconductor International* May (1991) 228 - 230.
2. Sato, T.: Spectral emissivity of silicon. *Japanese Journal of Applied Physics* 6 (1967) 339 - 347.
3. Vandenabeele, P. and Maex, K.: Emissivity of silicon wafers during rapid thermal processing. *SPIE Rapid thermal and related processing techniques* **1393** (1990) 316 - 336.
4. Patankar, S.V.: *Numerical heat transfer and fluid flow*. New York: McGraw-Hill (1980).
5. Hottel, H.C. and Sarofim, A.F.: *Radiative transfer*. New York: McGraw-Hill (1967).
6. McAdams, W.H.: *Heat transmission, 3rd ed.* Singapore: McGraw-Hill (1958).
7. Sala, A.: *Radiant properties of materials*. Amsterdam: Elsevier (1986).

5. MEASUREMENT TECHNIQUES

Recent innovations in thermal technology instrumentation applied to materials used in electronics applications

R. P. TYE[1] & R. L. GARDNER[2]

[1]*ULVAC/SINKU-RIKO, INC., Andover, MA USA*

[2]*ULVAC Technologies, Inc, Andover, MA USA*

Abstract. Development, production and characterization of materials and components for applications in emerging technologies has generated needs for new and improved thermal processing and properties measurement instrumentation. For electronics in particular, many materials and components now require fabrication under very precisely controlled rapid annealing and related processing conditions in non-contaminating environment. Materials are often used or available in limited amounts and special forms such as thin films and sheets which require new or modified relevant, absolute, and/or comparative measurement techniques in order to determine reliable thermal property valves.

Applications of infrared gold image furnace systems and several new and modified property measurement techniques will be discussed with reference to these requirements. Example of the use of the furnaces will include:

- Rapid thermal annealing and other heat treatments to provide materials having the required uniformity of form and optimum performance properties.
- Combination furnace/microscopes to study surface and related behavior at simulated fabrication and application conditions to high temperatures.

New and modified thermoanalytical techniques, thermal properties measurement methods and associated fabrication and characterization instrumentation are reviewed.

Specific property measuremen techniques will include:

- Modified flash diffusivity utilizing a two-dimensional analysis combined with off-center temperature measurements to investigate the homogeneity and uniformity of property of "large" wafer type specimens.
- Modified ac calorimetry for thermal diffusivity and device thermal conductivity.

1. Introduction

The continuing growth in materials development and use has produced substantial quantities of new and improved but often expensive products. In many cases the materials have to or can only be produced in special forms and combinations having optimum properties based on specific purities, inhomogeneties, heat treatments, etc. These factors require that fabrication and characterization be undertaken under special conditions, including especially clean environments and much faster controlled uniform heating and

C. J. Hoogendoorn et al. (eds.), Thermal Management of Electronic Systems, 191–200.

cooling rates than those used more conventionally. In addition, the properties of materials in the form of thin films can be distinctly different to those of the bulk material and can also be application dependant.

As a result of these issues, there has been a need for development of new thermal technology based instrumentation for fabrication and subsequent analysis of materials including new and modified techniques for determining quantitative thermal transport properties performance relevant to the application. This has required that measurement be made more rapidly on special forms and under conditions that cannot necessarily be satisfied using appropriate classical methods. Electronic packaging requires the use of a wide variety of materials, many having distinctly different properties used separately and in combination, and direct measurement of property becomes essential for the application.

The critical issues are now the numbers and types of materials requiring investigation. Results are required on smaller specimens, more rapidly and with high precision. This has been the stimulus for development of methods based on transient techniques for transport properties.

The present paper describes two tools developed especially for use in fabrication and analysis of these new materials together with a discussion of the essential principles of two newer methods for measurement of thermal diffusivity and controlling thermal transport property for many applications and especially for the ever increasing high power density packages.

A. Instrumentation

1. Infrared Gold Image Furnace

Many semiconductors, superconductors and composites require fabrication in very clean high vacuum or gaseous environments often at high temperatures. In some cases, such as annealing after ion implation to decrease resistance, for recrystallization, alloying and sintering purposes, precisely controlled temperatures and/or heating, cooling or cycling rates are required. In addition, these applications require heating rates that are well in excess of those attainable with conventional resitively heated furnaces and at very precise controlled rates that cannot be attained with other rapid heating processes.

A development to address these critical requirements is the infrared gold image furnace. This consists essentially of one or more high power density tungsten filament, argon or halogen sealed, quartz tube or banks of tubes together with one or more very high reflective, gold coated water cooled, elliptical, parabolic, or planar gold surface reflectors. These surround an appropriate size quartz tube to contain the wafer or artifact being treated. The result is a compact, low-mass, energy efficient, maintenance free system that allows different sizes, types, and forms of specimen to be heated and cooled at controlled rates from the conventional 1 to 100°C/minute to 100°C/s and higher. High precision of control and uniformity of temperature is attained by use of the instrumented specimen itself as the temperature sensor.

These individual furnaces and larger systems based upon them are used for many applications that include the development and production of semiconductor materials and components, composites, rapid or continuous processing and annealing and simulation of processes to develop materials with optimum properties. The shorter time heating technique provides the means to irradiate large area specimens more uniformly with less contamination potential such that it can be used for diffusion of implanted impurities, for formation of uniform oxide and nitride films, for control of interface reactions and of wafer electrical resistance. The advantage of more precise temperature control makes the technique most suitable for more uniform crystal growth.

2. High Temperature Surface Observation Furnace

A more recent extension of the furnace is its incorporation with a standard metallurgical microscope, CCD color camera and monitor. This allows continuous visual observation and analysis of materials to very high temperatures. In contrast to a "hot stage", no special optics or mechanical conversion are required.

The lamp acts simultaneously as the heating and light sources and the compact system can be attached directly and simply such that the specimen remains focused continuously as shown in Figure 1. The furnace/microscope combinations have been used extensively in the examination of semiconductor materials and electronic packages. Examples include studies of the failure mechanisms of soldered and mechanical joints, surface related behavior including phase transitions, melting and solidification, and oxide growth, sintering and deposition processes, thermal shock, and surface adhesion. For improved material characterization, the furnace microscope system can now be combined with thermal analysis instrumentation such that simultaneous visual behavior with change in a thermoanalytical parameter is possible.

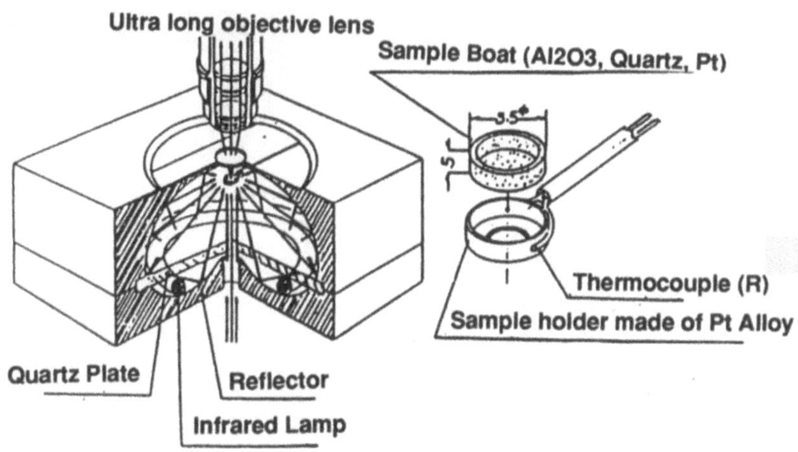

Figure 1 High Temperature Furnace Microscope System

B. Thermal Properties Measurement

1. Flash Thermal Diffusivity

For well over 20 years, the well-known transitory temperature "flash" method has been the most widely used method to measure this property on all types of solid homogeneous materials. The specimen is a small flat 10-12mm diameter, uniform disc having a thickness of 4mm or less. The principle is extremely well known and documented since it was first developed by Parker et al.[1]

The front face of a specimen is subjected to a short burst of radiant energy for 1ms or less. The temperature history of the rear surface is measured and the thermal diffusivity, D, in the normal to surface direction is computed from values of the half time (time to reach 50% of maximum) and square of thickness. Other percentage times can be used depending on the analytical model used, see e.g. Taylor and Maglic.[2] With certain modifications, it can also be used to measure specific heat on the same specimen. Its advantages are relative simplicity of concept, speed of measurement and small specimens.

The technique, while revolutionary is not a panacea for all applications. The common problem has been that of the mistaken application of deriving thermal conductivity, λ, directly from the relationship relating D to specific heat, C, and density, ρ for all materials and applications. The relationship holds only for homogeneous materials, where the prime mode of heat transmission is by solid conduction. For inhomogeneous, layered, and anisotropic materials and for transparent specimens, where radiation occurs, special care needs to be applied both in the measurement and the interpretation of results by use of special analytical procedures.

In general, the method has been used to determine absolute values of D in order to derive λ. However, it has been found very useful for a number of qualitative and quantitative applications. In particular, during the development of different materials such as AlN for use as substrates or other components all for electronics packaging applications the method has been particularly less time consuming for measuring the many specimens necessary to study relationships between processing parameters and the absolute or optimum property value and the uniformity of thermal property. Such factors as increased density, reduction of impurity, increased grain size, eliminating or minimizing imperfections and grain boundaries have all contributed to continuing improvements in property of such materials.

More recently, a development has been made Azumi and Kishi [3], to apply the method to measure D in the parallel to surface direction in order to study the uniformity and homogeneity of thin wafers such as silicon, AlN and CVD diamond produced for electronic components. It is based on a two-dimensional analysis such that larger specimens than the normally used 10 to 12mm diameter can be evaluated.

The principle is illustrated in Figure 2. The temperature history at point r, off-center can be derived analytically by assuming that the specimen is an infinite plate: the thickness is very much less than the distance r, and that there are no heat losses. The technique assumes that for normalizing and analyzing of the temperature response at the usual $t_{1/2}$ condition

$$Dt_{1/2} = K_{1/2}$$

where $K_{1/2}$ is a constant depending on the uniformity of the incident energy and various geometrical parameters including specimen profile, aperture diameter, applicable area of temperature measurement and the distance between the incident light and temperature measurement positions. The analysis is somewhat complicated and experimentally $K_{1/2}$ can be determined by use of one or more homogeneous specimens of known properties. Providing r/r_0 is greater than 1.8, for thicknesses less than 1mm reliable quantitative results can be obtained as shown in Figure 3. This illustrates some results obtained on a 40mm x 30mm x 0.6mm AlN wafer. However, the method has significant potential as a quality control tool when used in a qualitative mode.

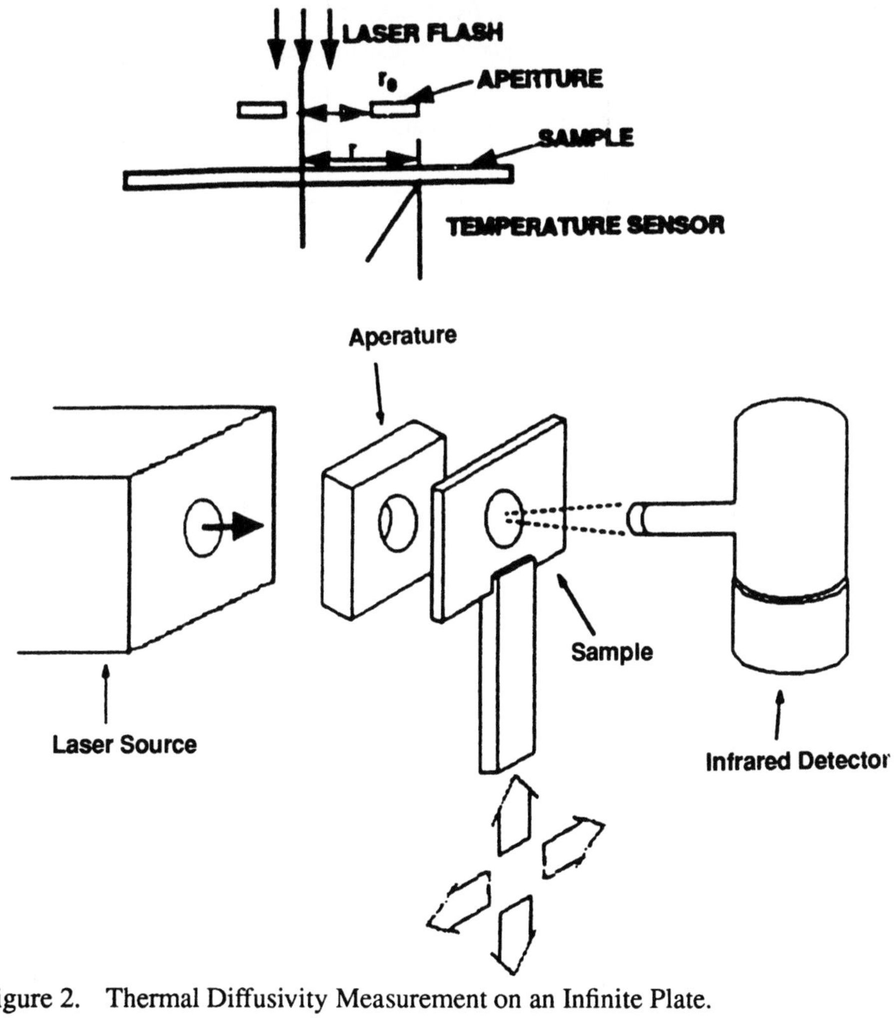

Figure 2. Thermal Diffusivity Measurement on an Infinite Plate.

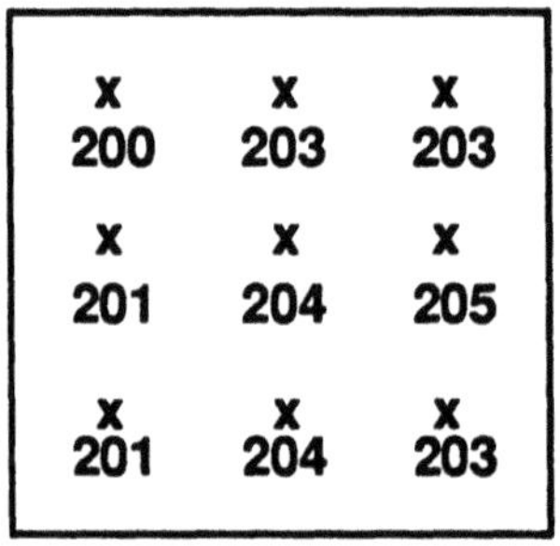

Figure 3. Uniformity of Thermal Conductivity (W/m•K) of a 0.6mm thick AlN wafer.

2. Modified ac Calorimeter

A specific method developed to measure much thinner films and foils including layered composites, using a simple "mixtures" relationship, is a modification of the original ac calorimeter technique developed to measure specific heat of small specimens Kraftmakher.[4] In the latter technique, the front surface of a specimen is irradiated by light from a halogen lamp, chopped at frequency, f (usually between 0.01 and 30Hz). By using a specially designed very precise lock-in amplifier, the small temperature amplitude $|T_{ac}|$ is measured on the rear surface of the specimen.

The heat capacity C is given by:

$$C = \frac{Q}{(\omega\, |T_{ac}|)}$$

Where Q is the heat flow during one cycle, ω is the frequency, and T_{ac} is an amplitude of temperature variation.

Absolute values can be obtained if T_{ac} values are measured for a specimen(s) of known specific heat, e.g. sapphire, and the slopes of the plots of $1/T_{ac}$ vs. frequency compared. However, due to its high precision in the "comparative mode", the technique is most valuable for determining very small changes in heat capacity e.g. due to phase transitions, especially second and third order, which occur for example in the ceramic oxide superconductors, over a very narrow temperature range. Since the actual rise in temperature during a measurement is only milliKelvins, the condition may be described as stationary or quasi-adiabatic.

For the measurement, a small specimen of the order of a maximum of 4mm x 4mm x 0.1mm 0.3mm for absolute values and 2mm x 2mm x 0.01mm for relative values is supported inside the furnace by two 25μm thermocouples wires welded together and attached with a silver paste. Helium gas is introduced in the specimen cell to maximize heat transfer between the specimen and surroundings. For this measurement and that of thermal diffusivity, D, the thickness of the specimen must be less than the thermal diffusion length which is given by $(2D/\omega)^{1/2}$.

The essential features of the modified technique to obtain D, are quite similar, see Hatta etal.[5] The optimum specimen size is now a film 10mm x 4mm with a maximum thickness of 0.35mm (up to 1mm or more for CVD diamond). However, as shown in Figure 4, only part of the surface is subjected to the periodic thermal energy. The remainder is masked from the chopped light irradiation. The ac temperature at a position of the specimen under the mask is measured as in the ac calorimeter case. The thermocouple does not affect the accuracy (see Hatta et al)[6].

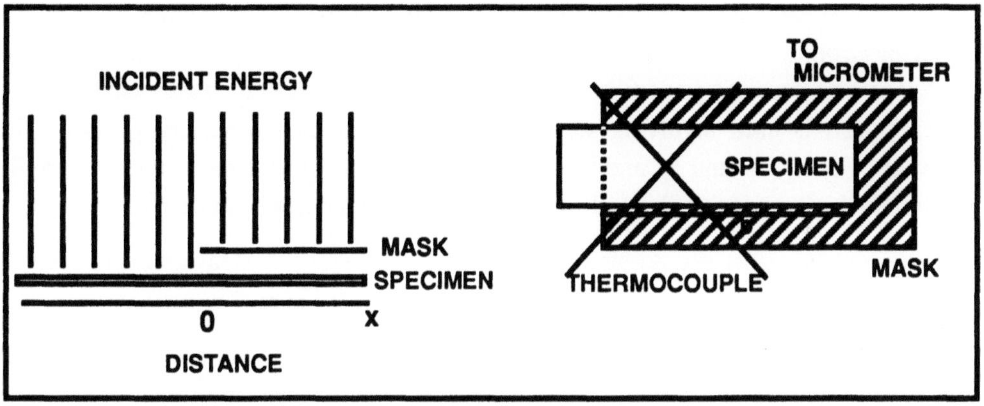

Figure 4. Modified ac calorimetry (schematic) a) cross-sectional view, b) view of rear surface l is change of distance as mask is moved incrementally.

The thermal diffusivity, D, in the direction parallel to the surface, is obtained by measuring the ac temperature, T_{ac} as a function of distance x between the position of the sensor and the edge of the mask. In the instrument, the mask is moved with respect to the fixed specimen.

The relation Tac vs. x is given by:

$$|Tac| = \frac{Q \exp(-kx)}{2 \omega Cl}$$

Where Q is the amplitude of the applied energy flux for unit area, is the angular frequency of the periodic heating, C the heat capacity of the specimen, 1 the specimen thickness, and k is the thermal decay constant $(\omega/2D)^{1/2}$ or $(\pi f/D)^{1/2}$ where is $(\omega/2\pi)$ the ac frequency. D can thus be obtained from the decay constant. An advantage of the method is that the specific heat can also be determined separately on a smaller part of the same specimen.

In practice, to ensure uniformity of heat flow, the front surface of the specimen is coated with a very thin layer of carbon, bismuth or some other highly energy absorbent material. This is particularly necessary for transparent materials. It has been demonstrated analytically and experimentally that effects of such layers less than 10nm in thickness can be neglected even on specimens some 10μm thick, see Hatta et al[7].

Measurements can be carried out in air on many materials. However, for very thin films (<25μm), especially those having a very low diffusivity materials, e.g. polymers, it has been found necessary to undertake measurements in vacuum, see Gu and Hatta[8]. If measurements are undertaken in air, the results have to be analyzed in terms of a two layer composite consisting of the specimen plus an air layer having a thickness proportional to the air diffusion length.

The method has proven valuable for evaluating the properties of all types of materials used in electronics applications not only in the form of free standing films but also composite layers consisting of a film and supporting substrate. For the latter application, a simple analysis is used involving separate measurements on the substrate and on the composite. For λ_1,C_1, and l_1, and λ_2,C_2, and l_2 are the thermal conductivities, specific heats and thicknesses of the film and substrate respectively, it has been found.

That reliable results are possible providing $\lambda_1 l_1 \geq \lambda_2 l_2$. Measurements are still possible but with decreasing precision as $\lambda_1 l_1$ approaches $0.1\lambda_2 l_2$.

One of the more recent applications of the method is for studying both the effect of processing parameters on the thermal properties of CVD films deposited on substrates and the degree of homogeneity of property of the free standing CVD diamond. Since diamond is both a very good electrical insulator and has extremely high thermal conductivity, it has many potential uses as heat spreaders and heat sinks in high power density systems. Diamond films now are being developed in increasing quantities by various processes (plasma CVD, sputtering, etc.). A variety of processing parameters can influence the properties of CVD diamond films. These include crystal size, numbers of grain boundaries, thickness, uniformity, anisotropy, etc. As one example, Figure 5 illustrates the effect of crystallite size where it is seen that high, moderate and low thermal diffusivity values correspond to decreasing crystal size of one type of CVD diamond film grown at different rates.

In one investigation, the effects of methane concentration on the properties of the diamond film were studied. In this silicon wafers 95 10μm thick were coated with plasma CVD diamond films from 5 to 30μm thick. As shown in Figure 6, as the methane concentration is reduced, the crystalline structure becomes closer to that of the diamond (very high thermal diffusivity). For the specimen produced with the lowest methane concentration, a value of derived thermal conductivity of approximately 1225W/m•K was obtained.

It should be pointed out that thermal diffusivity and thermal conductivity are very sensitive to structure differences. Thus, the evaluation technique is not only useful for evaluating the effects of processing but also can be used for studying the existence of local variations of film quality over larger areas of a specimen.

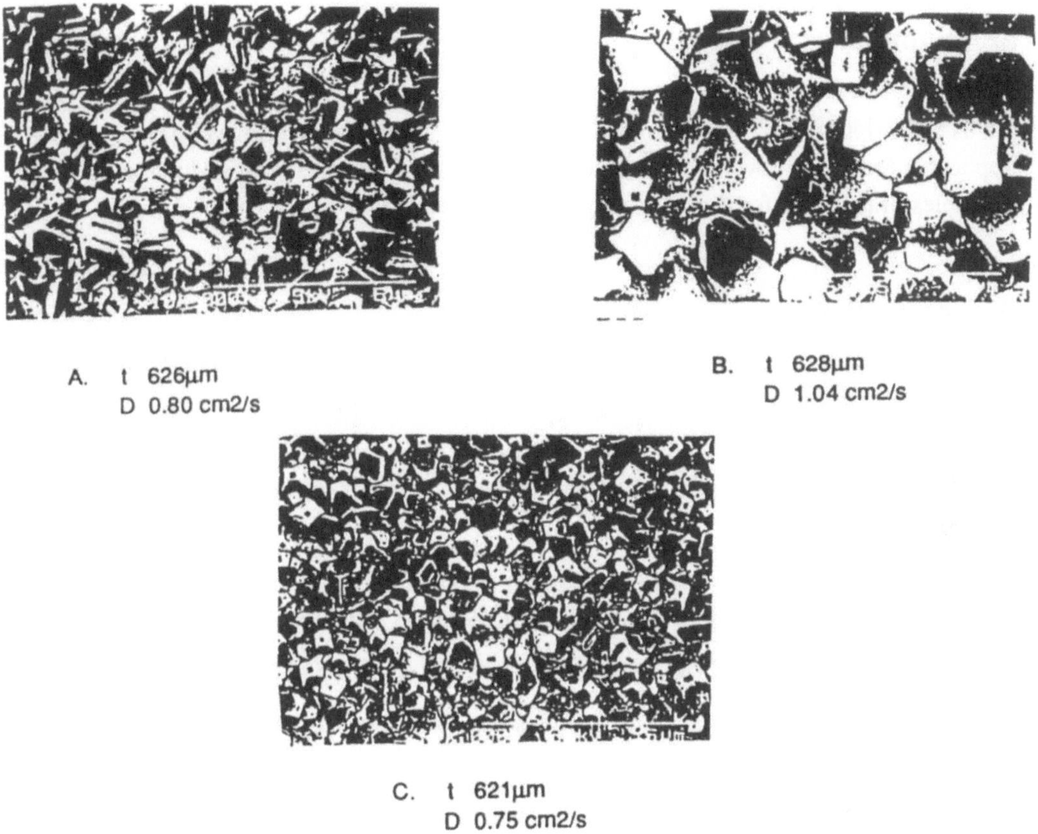

Figure 5. Effect of crystal size on thermal diffusivity of CVD diamond film (approx. 620 µm thick) grown on silicon substrates.

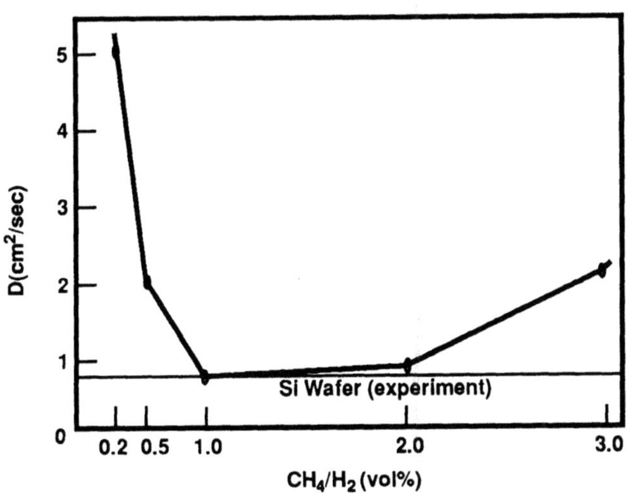

Figure 6. Influence of methane concentration upon thermal diffusivity of diamond film.

2. Summary

A description of several new or modified thermal technology based systems applicable for use in developing and characterizing materials has been given together with examples of their use for different electronics applications.

3. References

1. W.J. Parker, R.J. Jenkins, C.P. Butler and G.L. Abbott, *Journal of Applied Physics.*, 32 (1961) 1679.
2. R.E. Taylor and K.D. Maglic,*Compendium of Thermophysical Property Measurement Methods,* Eds K.D. Maglic, A Cezairliyan, and V.E. Peletsky, Plenum Press, New York, 1984, pp 305-336.
3. T. Azumi and A. Kishi, *Proc. 6th Japanese Symposium of Thermophysics* , Japanese Thermophysical Properties Society, Tokyo, 1985, pp 201-203.
4. Y.A. Kraftmakher, Zh. Prikl. Mekh, and Tekh. Fiz., 5 (1962) 176.
5. I. Hatta, Y. Sasuga, R. Kato, and A. Maesono, Rev. *Scientific Instrumentation.*, 56(8) (1985) 1643.
6. I. Hatta, R. Kato, and A. Maesono, *Japanese Journal of Applied Physics,* 25(6) (1986) 493.
7. I. Hatta, S. Hiraiwa, and H. Yao, *Proc. Second Asian Thermophysical Property Conference.*, Chinese Physical Society, Beijing 1989, pp 543-546.
8. Y. Gu and I. Hatta, *Japanese Journal of Applied Physics,*, 30(6), 1991, 1295.

AN EXPERIMENTAL METHOD TO DETERMINE THE EFFECTIVE THERMAL CONDUCTIVITY OF PRINTED CIRCUIT BOARDS.

H.J.L. VANLAER and C.J.M. LASANCE
Philips N.V.
Centre for Manufacturing Technology
P.O. Box 218
5600 MD Eindhoven
The Netherlands

ABSTRACT. An experimental method to determine the effective thermal conductivity is proposed for flat plates such as Printed Circuit Boards (PCB), using an analytical model with constant boundary conditions. The centre temperature of a heated, horizontally placed (anisotropic) PCB is a function of the thermal conductivities of the various materials constituting the board. By fitting the thermal conductivity of the isotropic analytical model until the centre temperature is equal to the measured centre temperature ensures that this thermal conductivity can be interpreted as the effective thermal conductivity of the considered PCB. Because the analytical solution requires constant and uniform boundary conditions, special attention has been paid to construct the same set of boundary conditions in the experimental set-up. Packing the PCB with a material of which both the thermal conductivity λ and the thickness c are well known guarantees that the unknown and non-uniform heat transfer coefficient α is replaced by a known and uniform quotient λ/c. Packing once more the packed PCB between two aluminium plates, kept at a constant temperature, forces the reference temperature to a constant value. Both the heat dissipation and the temperature measurement were performed via contactless means. The calibration and validation of the device showed a good agreement with known data for PTFE and Glass G291 test samples. The results have been compared with literature data.

Nomenclature

Symbol		SI unit
A	square area, cross section	m^2
a,b,c	length, width and thickness of a board	m
Bi	Biot number	-
E_{lm},F_{lm}	integration constants	
h	ratio of heat transfer coefficient and heat conduction coefficient	m
l,m	number of Fourier terms	
q	heat flux	W/m^2
R	thermal resistance	°C/W
T	temperature	°C
x,y,z	cartesian coordinates	
Δx,Δy	length and width of the heated area on the board	m
α	heat transfer coefficient	$W/m^2{}^\circ C$
ε	emission coefficient	-
Θ	dimensionless thermal resistance	-
λ	thermal conductivity	W/m°C
σ	$5.67.10^{-8}$ Stefan-Boltzmann's constant	W/m^2K^4
ϕ	power dissipation	W

C. J. Hoogendoorn et al. (eds.), Thermal Management of Electronic Systems, 201–210.

1. Introduction

Together with increasing demands on reliability, one of the major problems in the design of modern electronics is how to control the heat that is produced by an increasing number of components in a decreasing volume of space. Because the reliability is a strong function of the temperature, the accurate knowledge of thermal behaviour of an electronic apparatus is of great importance. The modern hardware designer can use a variety of software programs, which help him or her avoid thermal problems at component, Printed Circuit Board (PCB) or system level. As an input, this software requires thermal coefficients for the PCB such as the thermal conductivity λ. The values for the thermal conductivities of PCBs cited in literature are mostly determined via standardised measurement methods, that produce values for the direction normal to the PCB. However, these values are useless in practice because:

a) the thermal conductivity of the boards insulating material (glass fibre epoxy) can be highly anisotropic (Savar et al. (1990), Agarwal et al. (1991)), so the in-plane thermal conductivity is much more important from a practical point of view.
b) the very complex patterns of metal tracks and layers influence the thermal conductive behaviour considerably.

For an analysis at system level, and to a certain extent also on board level, average values for the thermal conductivities of PCBs are required. Therefore, an experimental technique has been devised for determining the *"effective thermal conductivity"* of PCBs.

A three-dimensional single layer model of the device was constructed by solving the three dimensional Laplace equation analytically over an isotropic rectangular plate with no internal heat generation, but heat dissipation on a small area in the centre, and heat loss from the top and bottom surfaces according to Newton's cooling law (the heat loss from the edges is assumed negligible). The steady state analytical solution provides the temperature distribution over the surface as a function of the thermal conductivity λ. The analytical centre temperature of an isotropic plate is equated to the experimentally measured centre temperature of an anisotropic PCB by varying the thermal conductivity λ. The resulting thermal conductivity of the model can be interpreted as the *"effective thermal conductivity"* in the experiments.

For a one to one translation of the analytical solution to the experimental situation, it is mandatory that the boundary conditions are equal in both cases. This is why a simple experimental set-up of a PCB, horizontally positioned on some insulating material, powered by foil heaters and measured with thermocouples encounters three major problems. The first problem is that heating a horizontally placed PCB causes free convection flow over the surfaces. Free convection on horizontal plates is generally being characterised by a variable and unknown heat transfer coefficient α, thereby rendering the boundary conditions in the experiments unreliable (see the paper by Lasance ('93) in this volume for an overview of natural convection on horizontal plates). A special treatment of these experimental boundary conditions is required in order to link the experiments to the model. The applied method makes use of the analogy between the heat convection coefficient α, the ratio of the thermal conductivity λ and the thickness c of a material. By packing a PCB in a material (foam) of which both the thermal conductivity λ and the thickness c are well known, one ensures that the unknown and non-uniform α is translated into a known and uniform quotient λ/c.

The second problem comes from the reference temperature. As a result of a temperature gradient in the packed PCB, the reference temperature of the device is not constant. By packing the packed PCB once more between two aluminium plates, kept at constant temperature by heated water, the reference temperature is forced to a constant value.

The third problem is how to minimise the disturbing influences of the heat generating and temperature measuring equipment in order to obtain accurate results. This idea is realised with the use of a lamp for the contactless heat dissipation in the centre of the PCB and the use of a Heimann Infrared pyrometer type KT4 for the contactless temperature measurement in the centre of the PCB.

The experimental device was calibrated and validated using samples with a known, independently measured, thermal conductivity. The experimentally measured thermal conductivities showed a good agreement with the literature values for PTFE and Glass G291 test samples. Furthermore, the accuracy of estimated effective thermal conductivities found in literature (Dickerson (1967), Dean (1985)) was tested. These estimated thermal conductivities lie within 15% of the experimentally found effective thermal conductivities.

2. The Analytical Model

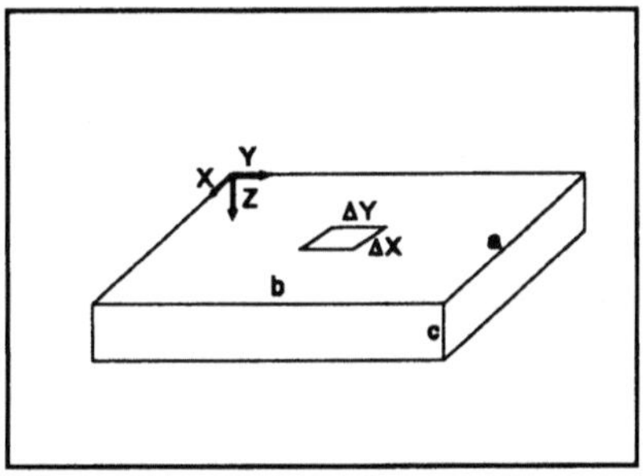

Figure 1 *Board geometry*

The three-dimensional Laplace equation, derived from the energy balance of a cubical element

$$\frac{\partial^2 T}{\partial x^2}+\frac{\partial^2 T}{\partial y^2}+\frac{\partial^2 T}{\partial z^2}=0 \tag{2.1}$$

is solved analytically (e.g. Estes (1988), Carslaw and Jaeger (1978), Chatenier (1987)) over a rectangular board (figure 1) with heat dissipation on a small area $\Delta x, \Delta y$ in the centre.

It is assumed that the heat loss from the edges is negligible

$$\frac{\partial T}{\partial x}\Big|_{x=0,a}=\frac{\partial T}{\partial y}\Big|_{y=0,b}=0 \tag{2.2}$$

and that the top and bottom surfaces lose heat by Newtonian cooling, i.e., linearised convection

$$\left[\frac{\partial T}{\partial z}-h_t(T_t-T_{amb})\right]\Big|_{z=0}=-g(x,y)\ ;\ \left[\frac{\partial T}{\partial z}+h_b(T-T_{amb})\right]\Big|_{z=c}=0 \tag{2.3}$$

in which

$$h_t=\frac{\alpha_t}{\lambda}\ ;\ h_b=\frac{\alpha_b}{\lambda}\ ;\ g(x,y)=\frac{q(x,y)}{\lambda} \tag{2.4}$$

with q (x,y) the heat flux in W/m^2.

The three-dimensional Laplace equation is solved by the usual technique for separation of variables, in which the coordinate dependent temperature is separated in three different coordinate dependent quantities, each being dependent on only one coordinate:

$$T(x,y,z)=X(x)Y(y)Z(z) \tag{2.5}$$

Substituting equation 2.5 in equation 2.1 provides three differential equations, which are solved separately by using the boundary conditions. Particular solutions, also satisfying the Laplace equation are added to the homogeneous solutions, resulting in the general solution for a single layer board, which is valid for every layer in a (multilayer) board, only with different values for E_{00}, F_{00}, E_{lm} and F_{lm}.

$$T=F_{00}+E_{00}z+\sum_{l=0}^{\infty}\sum_{m=0}^{\infty}F_{lm}\cos\frac{l\pi x}{a}\cos\frac{m\pi y}{b}[E_{lm}\sinh(\nu(c-z))+\cosh(\nu(c-z))] \quad (2.6)$$

Because the specific interest is the centre temperature on the top-surface (z=0), E_{00} is not required in this case and will not be taken into account. F_{00}, E_{lm} and F_{lm} follow from the boundary conditions on the top and bottom surfaces of the board.

$$F_{00}=\frac{g_{00}(1+h_b c)}{h_t+h_t h_b c+h_b} \quad (2.7)$$

$$E_{lm}=\frac{h_b}{\nu}\ ;\ F_{lm}=\frac{g_{lm}}{E_{lm}[\nu+h_t\tanh\nu c]+[h_t+\nu\tanh\nu c]}\frac{1}{\cosh\nu c} \quad (2.8)$$

The source is positioned between $x_1<x<x_2$, $y_1<y<y_2$ and is expanded in Fourier series

$$g(x,y)=\sum_{l=0}^{\infty}\sum_{m=0}^{\infty}g_{lm}\cos\frac{l\pi x}{a}\cos\frac{m\pi y}{b} \quad (2.9)$$

from which it can be calculated that:

$$g_{00}=g\frac{(x_2-x_1)(y_2-y_1)}{ab}\ ;\ g_{lo}=\frac{2g}{\pi bl}(y_2-y_1)\left[\sin\frac{l\pi x_2}{a}-\sin\frac{l\pi x_1}{a}\right] \quad (2.10)$$

$$g_{0m}=\frac{2g}{\pi am}(x_2-x_1)\left[\sin\frac{m\pi y_2}{b}-\sin\frac{m\pi y_1}{b}\right]\ ;\ g_{lm}=\frac{4g}{lm\pi^2}\left[\sin\frac{l\pi x_2}{a}-\sin\frac{l\pi x_1}{a}\right]\left[\sin\frac{m\pi y_2}{b}-\sin\frac{m\pi y_1}{b}\right] \quad (11)$$

Let us define a dimensionless thermal resistance Θ as the ratio of the thermal centre reference resistance to the thermal resistance of an 'ideal' board, i.e. a board with an infinite thermal conductivity. Let us also define, for the general case, two Biot numbers which are measures of the relative importance of the internal resistance of the board (the resistance in the board) to the external resistance (the resistance from the surface to the reference temperature). (top: Bi_1, bottom: Bi_2).

$$\theta=\frac{T_c-T_{amb}}{\phi}\alpha_t 2ab \; ; \; Bi_1=\frac{\alpha_t c}{2\lambda} \; ; \; Bi_2=\frac{\alpha_b c}{2\lambda} \tag{2.12}$$

substituting equation 2.7 and 2.8 in 2.6 gives the total solution for the centre temperature $T_c=T(a/2,b/2,0)$:

$$\theta=2Bi\left[\frac{1+2Bi_2}{Bi_1+Bi_2+2Bi_1Bi_2}+2\sum_{l=1}^{\infty}\frac{\sin\left(\alpha_l\frac{\Delta x}{a}\right)}{\left(\alpha_l\frac{\Delta x}{a}\right)}\frac{\alpha_l\frac{c}{a}+Bi_2\tanh\left(2\alpha_l\frac{c}{a}\right)}{\alpha_l\frac{c}{a}(Bi_1+Bi_2)+\left[Bi_1Bi_2+\left(\alpha_l\frac{c}{a}\right)^2\right]\tanh\left(2\alpha_l\frac{c}{a}\right)}\right.$$

$$+2\sum_{m=1}^{\infty}\frac{\sin(\beta_m\frac{\Delta y}{b})}{(\beta_m\frac{\Delta y}{b})}\frac{\beta_m\frac{c}{b}+Bi_2\tanh(2\beta_m\frac{c}{b})}{\beta_m\frac{c}{b}(Bi_1+Bi_2)+\left[Bi_1Bi_2+(\beta_m\frac{c}{b})^2\right]\tanh(2\beta_m\frac{c}{b})}$$

$$\left.+4\sum_{l=1}^{\infty}\sum_{m=1}^{\infty}\frac{\sin(\alpha_l\frac{\Delta x}{a})\sin(\beta_m\frac{\Delta y}{b})}{(\alpha_l\frac{\Delta x}{a})(\beta_m\frac{\Delta y}{b})}+\frac{\gamma_{l,m}+Bi_2\tanh(2\gamma_{l,m})}{\gamma_{l,m}(Bi_1+Bi_2)+[Bi_1Bi_2+\gamma_{l,m}^2]\tanh(2\gamma_{l,m})}\right] \tag{2.13}$$

in which

$$\alpha_l=l\pi \; ; \; \beta_m=m\pi \; ; \; \gamma_{l,m}=\left[\alpha_l^2\left(\frac{c}{a}\right)^2+\beta_m^2\left(\frac{c}{b}\right)^2\right]^{\frac{1}{2}} \tag{2.14}$$

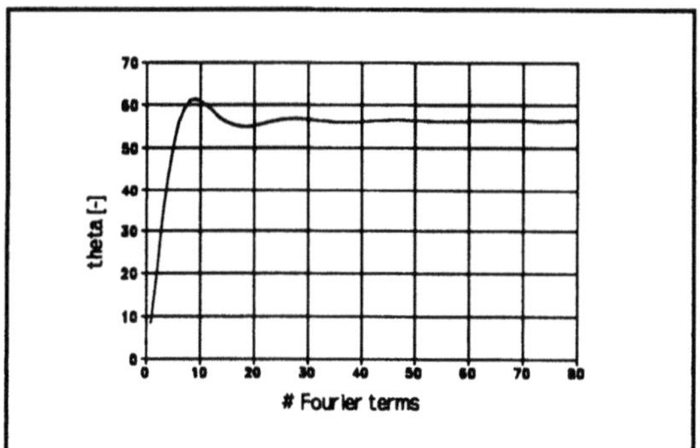

Figure 2 *Convergence analysis*

The accuracy of the solution given in equation 2.13 depends on the number of Fourier terms taken into account. The exact analytical solution requires an infinite number of Fourier terms. However, after a limited number, the solution for Θ approaches its final value (figure 2). The analysis of the experimental set-up requires about 50 Fourier terms to reach the final solution for the worst case situation (minimum value of λ).

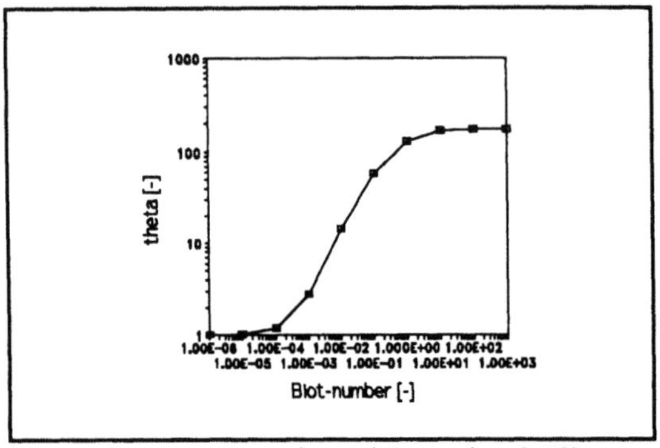

Figure 3 Θ *against Bi (equation 2.13)*

The relationship between the Biot number and Θ is shown in figure 3. The lower limit of the model $\lambda \approx 25$ W/m°C (Bi = 10^{-4}) and the upper limit $\lambda \approx 0.025$ W/m°C (Bi = 1.0) restrict the area in which the use of the model is recommended.

Because the shape of the curve varies with the geometry of the board, these limits are different for every board geometry. The limits of figure 3 represent the worst case situation. For other practical board geometries, the restrictions are less severe.

3. The experimental set-up

A small area of the top surface of a horizontally placed PCB is contactlessly heated by a halogen lamp. The dissipated power ϕ is measured with an OPHIR Laser Power Monitor type 30ADGHH(P). This power monitor provides up to the maximum power of 30 W (continuous) an inaccuracy of ±3%. The temperature of the top surface centre T_c during heating is determined by the pyrometer which is capable of measuring up to 300 °C with an inaccuracy of 1.5 °C, and has a focal diameter of 4 mm.

As pointed out in the introduction, the experimental set-up (figure 4) had to be designed in such a way that reproducible boundary equations ensure the comparison with the analytical analysis. The proposed replacement of α in the analytical model and λ/c in the experiments comes from the thermal resistance of the free convection case ($1/\alpha A$) which is equated to the thermal resistance of the packed case ($c/\lambda A$). Unfortunately, the translation is not one to one. The lateral heat conduction in the foam is responsible for a deviation from the analytical boundary conditions, that are constant and uniform for the whole surface. This influence was investigated numerically with a software package specially developed for the analysis of PCBs, based on Finite Difference techniques (THERMAX from CADENCE). For the worst case - a high lateral temperature gradient in the foam due to a poorly heat conducting board (λ = 0.25 W/m°C) - a deviation of 4% was calculated. Hence the translation of the single layer model with constant boundary conditions to a multiple layer problem (PCB+foam) is considered to be valid.

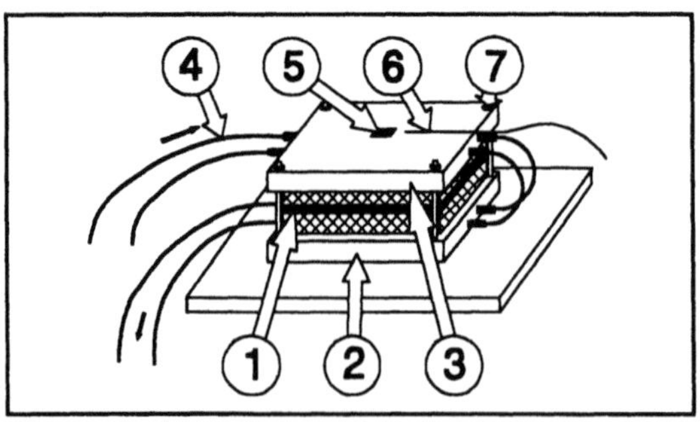

Figure 4 *The experimental device*

Another problem in the experimental setup is found in keeping the reference temperature of the device constant, as is required in the analytical model. By packing the foam packed PCB (1) once more between two aluminium plates (120 x 120 x 8 mm) (2+3), kept at a constant temperature by heated water (4), the ambient temperature is forced to a constant value. The upper aluminium plate (3) contains a central square hole (8.0 x 8.0 mm) (5), through which the generated heat from the lamp reaches the PCB. The heated water comes from a Tamson TC tank and has a temperature variation of less than 0.5 °C at a level of 50 °C . The temperature of the aluminium plates is measured with a thermocouple (6). Controlling the distance between the two aluminium plates by screws (7) ensures a reproducible value for c_{foam} within 5%.

The influence of the contact resistances that occur at the two foam interfaces has been investigated by changing the foam thickness c_{foam}. The total thermal resistance between the PCB and the Aluminium follows from

$$R_{tot}=R_{c_1}+R_{foam}+R_{c_2}=\frac{T_{PCB}-T_{Al}}{\phi} \quad (3.1)$$

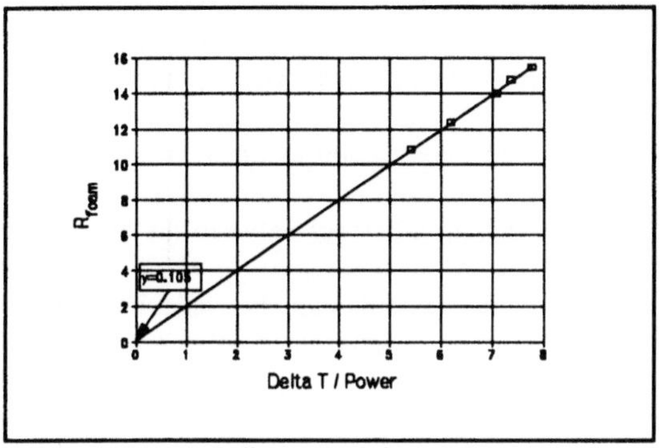

Figure 5 *Influence of the contact resistances*

A plot of the total foam resistance R_{foam} against $\Delta T/\phi$ is given in figure 5. Extrapolating the linear regression line to $\Delta T/\phi = 0$ provides the influence of the contact resistances as the remaining value on the vertical axis. As this value (0.105) is negligible to the foam resistance (R_{foam} >11.1), the contact resistances are neglected in the subsequent analysis.

4. Comparison of the experiments with the analytical model

4.1 CALIBRATION OF THE PACKING MATERIAL

Rewriting formula 2.12, the conductivity of the foam λ_{foam} follows from:

$$\lambda_{foam}=\frac{\phi}{\Delta T}\frac{c_{foam}\Theta}{2ab} \quad (4.1)$$

in which all the parameters can be derived from the experiments, except for the dimensionless thermal resistance Θ. This Θ comes from the exact analytical solution in which the heat transfer coefficient α in the Biot number is replaced by the quotient $(\lambda/c)_{foam}$. The problem is that the foam conductivity is not known. The following calibration procedure has been used to determine its value. The use of a solid copper board sets the Biot number for the calculation of Θ below the lower limit of figure 3 (Bi ~ 10^{-5}). In this region, Θ is almost independent of the Biot number. The estimated Biot number is calculated with a foam conductivity, calculated from equation 4.1 in which Θ is equated to 1.00. In this way, an accurate value for Θ is used in the calculation of the thermal foam resistance, and the heat conductivity of the foam was experimentally measured, resulting in 0.050 W/m°C for a foam thickness of 4.0 mm.

4.2 ESTIMATION OF THE HEAT LOSSES

The heat losses of the board occur at the edges of the plate and through the heating hole. In the experiments a possible heat-loss from the edges will be accounted for by involving the edge area in the heat transferring top and bottom surfaces. Since the boundary conditions on the edges are about similar to the surface boundary conditions (some foam also covers the edges), this correction method gives a small error.

The effective heat loss through the heating hole is calculated as the difference between the total heat loss through the hole and the heat loss in case the whole board is foam packed. The total heat loss through the heating hole consists of:

- heat loss by air conduction $\phi_\lambda = A(T_{board} - T_{ref})\lambda_{air}/c_{air}$ (no convection assumed in the heating hole)
- heat loss by radiation ϕ_σ, calculated with use of the radiation balance between the board, the foam and the heating source.

The heat loss in case the whole board is foam packed can be calculated with $\phi_{foam} = A\Delta T\lambda_{foam}/c_{foam}$. The correction factor for the heat loss through the heating hole is now: $\phi_{loss} = \phi_{foam} - (\phi_\lambda + \phi_\sigma)$. The power dissipation on the board is finally corrected with this factor. The measured centre temperature has to be corrected because the analytical model provides the maximum centre temperature, while the pyrometer reads an average value over the focal spot. This correction, being a function of the thermal conductivity, has been calculated by using the proprietary package HEAT3D, based on the same mathematical techniques as the model, but allowing for multiple layers and the output of average temperatures over a certain area.

4.3 THE "EFFECTIVE THERMAL CONDUCTIVITY" OF A PCB.

To arrive at a value for the effective thermal conductivity, the first step is to calculate Θ from the output of the experiments

$$\Theta = \frac{T_c - T_{amb}}{\phi}\frac{\lambda_{foam}}{c_{foam}}2ab \tag{4.2}$$

Substituting this experimentally found dimensionless thermal resistance < in the analytical model provides the corresponding Biot number. The effective thermal conductivity of the PCB is then calculated as:

$$\lambda_{board} = \frac{\lambda_{foam}}{c_{foam}}\frac{c}{2Bi} \tag{4.3}$$

5. Results

The thermal conductivity of several materials was determined using the described method. First of all, a calibration and validation was carried out, using an isotropic glass G291 plate (75 x 75 x 2 mm) and three PTFE (teflon) plates of different sizes (75 x 75 x 3.9 mm ; 50 x 50 x 3.9 mm and 25 x 25 x 1.0 mm). The thermal conductivity of the glass G291 plate was independently measured by a transient hot strip method (Philips Research Labs. (1992)) and given as 0.95 W/m°C. The experimentally measured thermal conductivity resulted in 0.92 W/m°C. For PTFE, the thermal conductivity found in literature (Savar et al. (1990)), is given as 0.26 W/m°C. The experimentally measured thermal conductivity was 0.24 W/m°C for the two largest boards, and 0.21 W/m°C for the smallest board. Clearly, the values are very close to the literature values, so these results give confidence in the reliability of the method. However, it is difficult to draw conclusions on the PTFE results because no research was focused on the determination of the variance in the literature data. The discrepancy between the values of the larger and smallest PTFE plates can be attributed to the increasing role of the heat losses.

Next, the effective thermal conductivities of two standard 50 x 50 mm FR4 epoxy/glass PCBs covered with copper on both sides were determined, with an overall thickness of 1.0 and 1.6 mm respectively, and 17.5 μm of copper cladding.

	board thickness mm	λ_{board} W/m°C
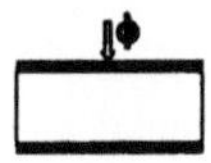	1.0	11.0
	1.6	6.8
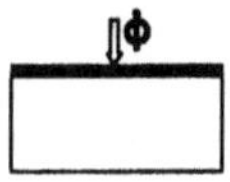	1.0	7.4
	1.6	4.8
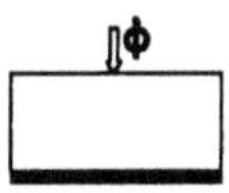	1.0	1.3
	1.6	0.76
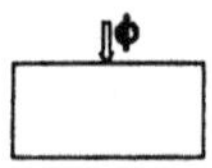	1.0	0.32
	1.6	0.32

Table 1 Effective thermal conductivity for various copper-epoxy combinations

According to Dickerson, the effective thermal conductivity of the 1.6 mm thick single copper layer board is 4.1 W/m°C. With Dean's Empirical Formula, the effective thermal conductivity is calculated as 4.5 W/m°C, while our measurement gives λ=4.8 W/m°C. The results compare reasonably well; the lower values obtained by the estimated methods could be explained by their assumption of no thermal influence of the epoxy board in case of a solid conductor sheet.

In case of no copper cladding, the thermal conductivity of the pure FR4 epoxy/glass board was measured as λ_{exp}=0.32 W/m°C. This result is contrary to the anisotropy assumption, as described by Savar et al. (1990), who concluded that - under vacuum - the thermal in-plane conductivity ($\lambda_{//}$=1.06 W/m°C) is about three times the thermal conductivity normal to the plane (λ_n=0.34 W/m°C). Using THERMAX, imposing their anisotropic values, resulted in temperatures way from our measured temperatures. No explanation has been found to clarify the observed discrepancy, and further research is recommended.

A sensitivity analysis was carried out to determine which parameters of the model require special attention in order to obtain more accurate results. For this reason, a 1% variation was added to all parameters of interest, resulting in the following percentile variation of λ, found with the analytical solution of equation 2.13):

1% variation in parameter:	% variation in λ	1% variation in parameter:	% variation in λ
T_c	8.25	c_{foam}	1.21
ΔT	2.19	a,b	0.89
ϕ	2.21	c	0.94
λ_{foam}	1.23	$\Delta x, \Delta y$	0.30

Table 2 Sensitivity analysis

From this table it can be concluded that the centre temperature T_c, the temperature difference between the board and the reference ΔT and the dissipated power ϕ require special attention in order to obtain an accurate result. For a different parameter configuration, the sensitivity on λ will be different, however, the values in table 2 are based on the worst case situation.
Note: The term 'effective thermal conductivity' could raise confusion, because it is not a genuine material property. It is also dependent, however slightly, on the boundary conditions and the dimensions.

6. Conclusions

An experimental device has been devised to measure the effective thermal conductivity of flat plates such as Printed Circuit Boards. Special attention has been paid to ensure uniform and constant boundary conditions in order to enable the calculation of the thermal conductivity by means of an analytical solution. Calibration against known thermal conductivities (PTFE and Glass G291) showed excellent agreement. Because of the fairly laborious correction procedures necessary to correctly apply the analytical solution, improvements could be expected from the combination of an optimisation algorithm with a Finite X software package.

References

AGARWAL, R.K., DASGUPTA, A. et al.: 1991;" Prediction of PWB/PCB Thermal Conductivity", Int. J. for Hybrid Microelectronics Volume 14 no3.
CARSLAW, H.S. and JAEGER, J.C.: 1978; "Conduction of Heat in Solids", Oxford Press, pp. 176-186.
CHATENIER du, F.: 1987; Philips Internal Rapport, CTB 87.34.134.
DEAN, D.J.: 1985; "Thermal Design of Electronic Circuit Boards and Packages", Electrochemical Publications, pp. 151-152.
DICKERSON, P.: 1967; "Convenient Thermal Analysis Techniques for PCB Assemblies" pp.43, Proc. Nepcon.
ESTES, R.C.: 1988; "Thermal Design Considerations for COB Applications", Hybrid Circuits vol.17 pp. 49-55.
LASANCE, C.:1993; "Thermal Management of Air-Cooled Electronics Systems: New Challenges for Research", This Volume.
PHILIPS RESEARCH LAB 1992, Personal Communication.
SAVAR, F.,POOLE, N.J., and WITTING, P.A.: 1990; "PCB Glass-Fibre Laminates: Thermal Conductivity Measurements and Their Effect on Simulation" Journal of Electronic Materials vol.19 no.12. pp.1345-1350.
SIEGEL, R. and HOWELL, J.R.: 1972; "Thermal Radiation Heat Transfer", mcGraw-Hill, pp. 18-26.

Air Velocity and Temperature Measurements Around a Naturally Convecting Rectangular Fin Array

P. Rodgers and M. Davies
Department of Mechanical Engineering
University of Limerick
Limerick
Ireland

Abstract.

A calibration method is presented that allows hot wire anemometry to be used to evaluate the inlet and outlet temperature and velocity fields for a rectangular naturally convecting fin array. Contour plots of these flows clearly show the salient features ; the most important of which is the large scale ingestion of flow in through the fin tip region. A simple theory predicts the two dimensional core flow well.

Introduction.

Extended surfaces or fin arrays are one of the most common methods used by electronic designers to enhance heat transfer rates. Such arrays come in a wide variety of geometries and sizes and they may be used in either free or forced convection. This paper is concerned with understanding the fluid flow and heat transfer from a rectilinear array in free convection. Such arrays are commonly found in electronic applications, this particular study was motivated by the problem of cooling electronic power supplies. The results presented herein will also be of considerable interest to the study of free convection parallel plate heat transfer such as that found in arrays of printed circuit boards.

The effectiveness of a particular fin array design is usually accessed by using a U channel or parallel plate Nusselt number correlations such as those presented by Van De Pol & Tierney (1974) and Bar Cohen & Rohsenow (1984) respectively. The parallel plate correlations work well when the height of the fins above the base plate is relatively large. Both are global predictions which give little insight into the passage flow. It is usual to accept the ensuing radiative heat flux as there is little the designer can do geometrically to enhance this.

The are many unanswered questions as to the nature of this type of fluid flow. In particular, and importantly, the detailed nature of the flow is not known nor is it known what is the most accurate numerical modelling technique. Many of the secondary questions that arise out of these problems concern the nature of the boundary conditions. Generally the questions that are asked are as follows :

C. J. Hoogendoorn et al. (eds.), Thermal Management of Electronic Systems, 211–220.

(i) What is the shape of the inlet flow velocity and temperature profiles ? Martin et al. (1991), Naylor et al. (1991),Chappidi and Eno (1990).

(ii) What is the significance of heat conduction to the incoming and outgoing flow ? Martin et al. (1991), Ramanathan and Kumar (1991).

(iii) What effect does the fin thickness and the temperature of the edges have on the flow ? Naylor et al. (1991), Aihara (1970).

(iv) How much is the flow effected by the proximity of the enclosure walls ? O'Meara and Poulikakos (1987), Buller and Martinek (1978).

Although this investigation did not set out to understand these phenomena explicitly many of these questions will be addressed for the particular fin array examined. The primary aim of this work was to develop an experimental technique such that the inlet and outlet flow fields could be investigated. With such a technique in hand any improvements made by novel designs such as fin stagger, Guglielmini et al. (1987), or fin inclination, Praught (1982), could be understood and hence improved upon.

Hot wire anemometry provides a method whereby the temperature and velocity of a stream of fluid may be measured. For a general view of this technique see, Jaluria (1980). The measurement principles are that if the hot wire is operated in a constant current mode it may be calibrated to measure temperature and if operated in constant temperature mode it may be calibrated to measure velocity in a known constant temperature flow.

A standard hot wire probe typically used for either temperature or velocity measurements is a Dantec 55P11 5 μm diameter platinum plated tungsten wire probe with the sensor having a length diameter ratio of 250 and a maximum operating temperature of 300°C. In the present investigation this probe is not suitable for velocity measurement because of its poor sensitivity at low velocities due to excessive heat conduction down the supports. To enhance velocity sensitivity length to diameter ratio and operating temperature should be increased. This was achieved by rewiring a Dantec 55P01 probe with 10 μm diameter platinum rhodium wire giving a length diameter ratio of 300 and an operating temperature of 700°C. For optimum temperature measurement a Dantec 55P31 platinum resistance thermometer probe with a wire diameter of 1 μm and a sensor length of 0.4 mm was used.

In this experiment the upper and lower opening of a fin passage were traversed, firstly for temperature then for velocity. The velocity calibration was performed over a range of velocities in a constant temperature flow. This calibration was then repeated for three other temperatures. The initial temperature traverse of a fin passage therefore allowed for the correct velocity calibration isotherm to be used.

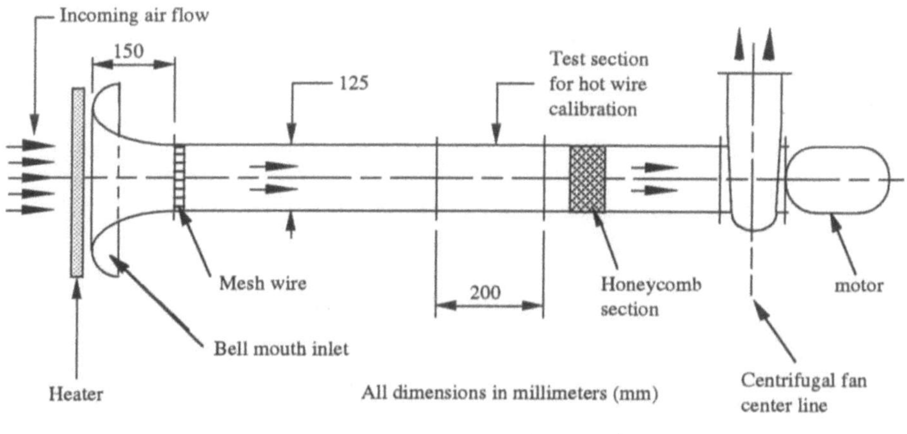

Figure 1 : Hot wire calibration facility shown horizontally.

Constant Current and Constant Temperature Calibration.

The most accurate method of calibration is to calibrate a transducer in an exactly similar environment as it is to be used in. The hot wires in these tests were to be used in a flow with spatially varying temperature at a low velocity in a direction approximately perpendicular to the surface of the earth. It was therefore required that this configuration be repeated in the calibration process.

A heated tunnel, was placed vertically in the gravitational field, figure 1, with a preset velocity and air temperature. The wire was placed in the tunnel so that its thermal plume rose in the same direction as the tunnel flow. A modified 55P01 probe was calibrated over a range of velocities and at four air temperatures, Figure 2. The offset between the calibration lines is nearly constant therefore linear interpolation was used for points falling between these lines. The velocities in the calibration were measured by a pitot static tube connected to a micromanometer measuring to within 8 % accuracy at 0.1 m/s. The zero velocity reading was taken in a small isothermal enclosure. Air temperature measurement was made by using a Radionics Type T thermocouple with an accuracy of +/- 1°C. It was constructed of foil material, 0.05 mm thick, butt bonded at the junction. It should be noted that the temperature and velocity measurements in the tunnel work section were made in the same horizontal plane as the hot wire probe and within the uniform portion of the temperature and velocity profiles.

A 55P31 probe was then calibrated in a constant current mode in a temperature controlled oven against the same Type T thermocouple as used above. The calibration is shown in Figure 3. This probe was then ready to measure the inlet and outlet temperature field.

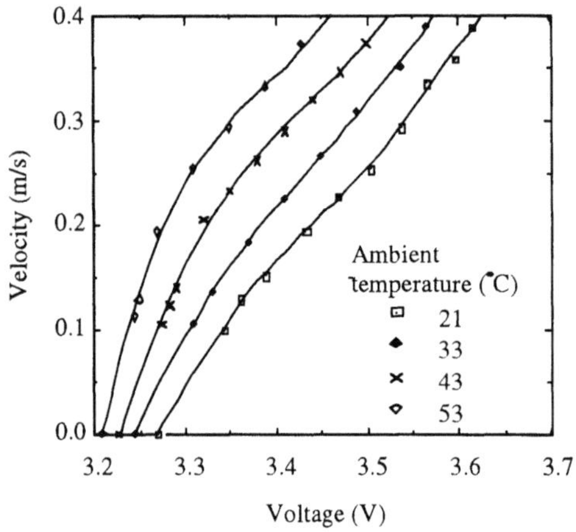

Figure 2 : Hot wire velocity calibration.

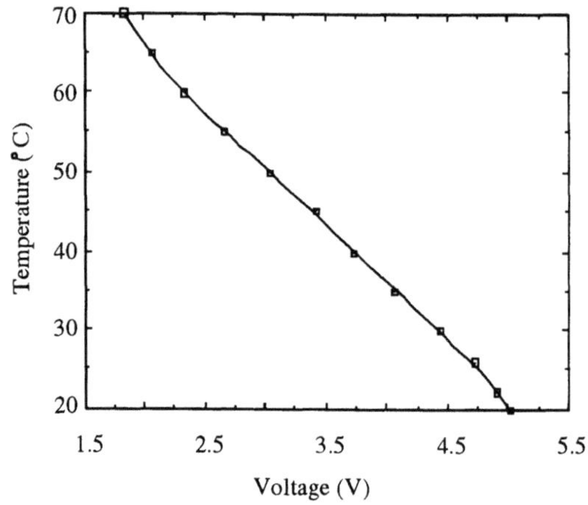

Figure 3 : Hot wire temperature calibration.

Experimental Description.

The aluminium fin array to be tested is shown in figure 4. It was chosen because it is simple, relatively long finned with a relatively large fin spacing so the probe would not significantly block the passage. It is typical of designs used to cool power supplies in 19" racks. A double sided design was chosen so that the fins were symmetrical and, consequently, so that the base plate stagnation line was well defined. This makes the numerical analysis of such an array somewhat easier, and it ensures that the heat transfer rate is known. The array was heated by three electrical cartridge heaters in the base plate and the temperature was monitored by three baseplate thermocouples. The fins are isothermal to within 1 °C. The emissivity of the surfaces is less than 0.1 and using a gray body U channel radiative analysis, as developed by Ellison (1979), it was calculated that radiative dissipation would account for less then 8% of the total heat dissipated. Therefore radiative heat transfer may be ignored.

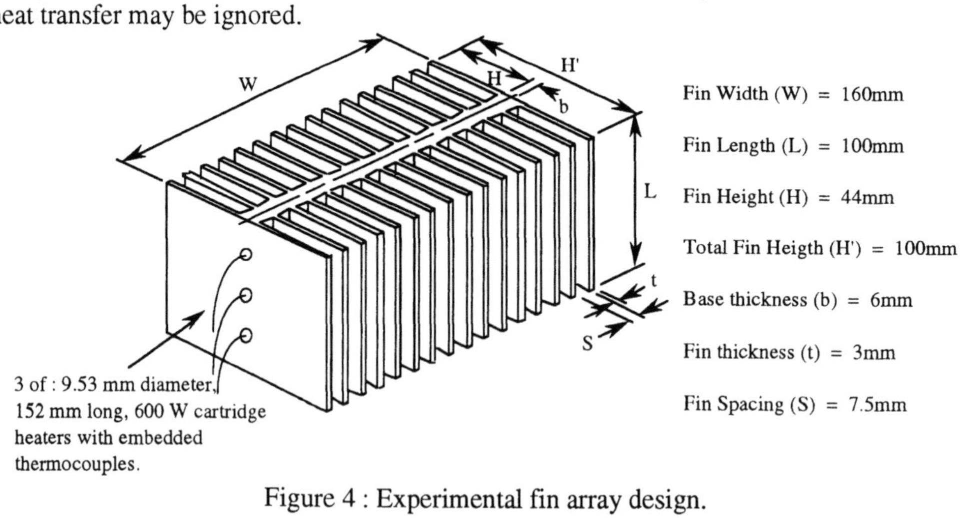

Figure 4 : Experimental fin array design.

The array was suspended in the centre of a perspex enclosure to exclude drafts. The enclosure dimensions of height, width and depth were 0.58, 0.54 and 0.46 meters respectively which are typical of an industry standard electronic cabinet. The enclosure wall thickness was 10 mm. The heat transfer system is then no longer an isolated fin array but an interactive system of laboratory, enclosure and array. The convection conditions in the enclosure were a function of the air temperature in the laboratory, the level of insulation or cooling of the outside of the enclosure and the power level of the fin heaters. All of these were adjusted to give the required convective conditions. The total power input was held constant at 80 W and the fin temperature at 70 °C. For this steady state condition the outside of the enclosure was cooled by a 10 m/s stream of air at 20°C.

The traverses of the inlet and outlet flows were achieved by attaching the hot wire probe to a two dimensional stepper motor driven traverse gear. 208 points were taken per scan 1mm above and below a single passage. Data was taken when the probe was stationary. The array of points relative to the fin passage is shown in figure 5 and the complete measurement system is shown in figure 6.

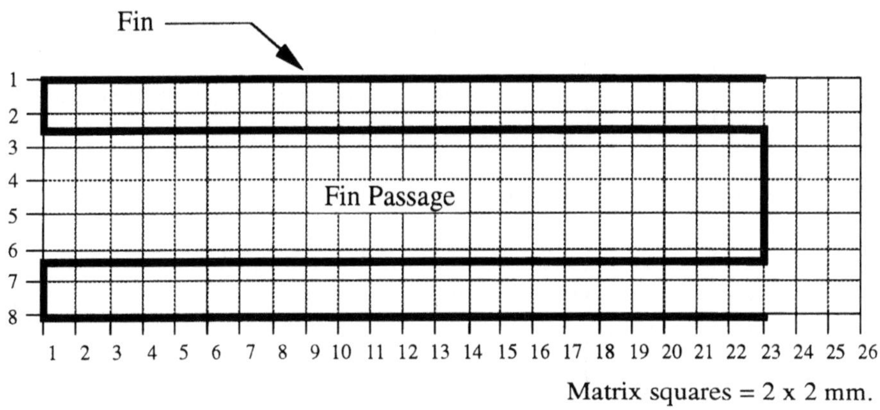

Figure 5 : Matrix scan of channel at inlet and outlet.

Results and Discussion.

Figures 7 and 8 shows the temperature fields at inlet and outlet respectively. The air entering the passage is preheated by two mechanisms : Firstly by thermal conduction down through the thermal boundary layers as can be clearly seen in figure 7, where the centre passage flow is significantly cooler than the flow in the region of the fins. The second mechanism is the heating of the air as it flows over the base of the fins. This causes the highest temperatures in the inlet flow and, again, it can be clearly seen in figure 7.

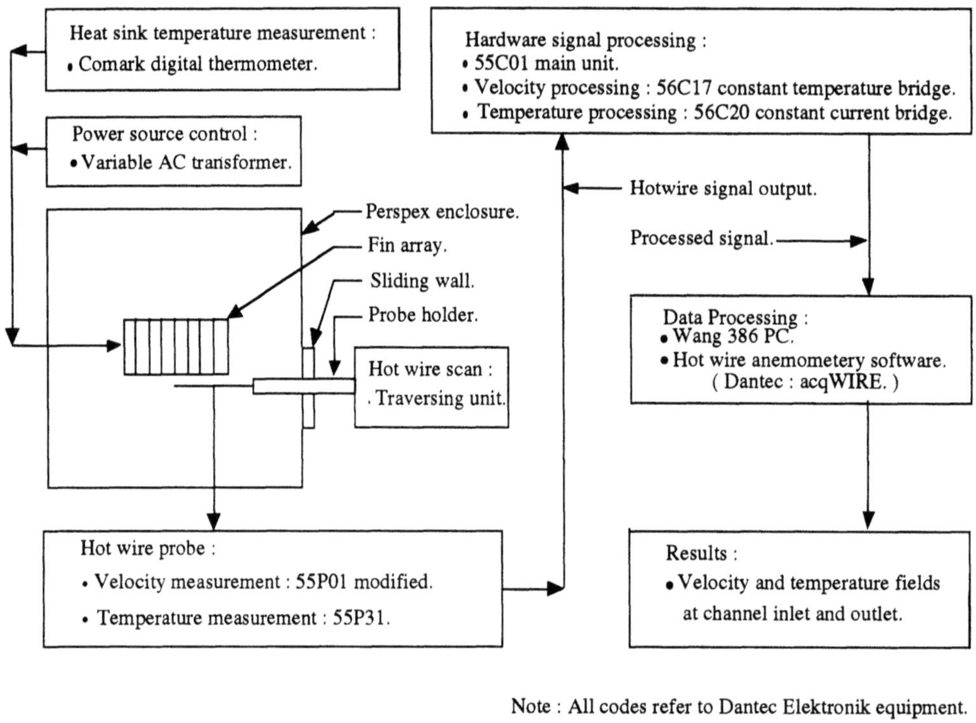

Figure 6 : Experimental instrumentation schematic.

At outlet from the passage the temperature distribution shows a quite different character. The dominant gradients here are caused by the interaction between the passage flow and the cool air at the fin tip. This has an effect over some one third of the fin passage. Despite the length of the fins there seems to be a considerable ingestion of air in through the fin tip region. This phenomena has been seen in flow visualisation by Aihara (1970).

Figures 9 & 10 shows the velocity fields at inlet and outlet respectively. The fields are quite different in magnitude; the maximum velocity is almost three times higher at outlet than at inlet. The mean velocity is more than twice as high. This increase in velocity cannot be accounted for by a decrease in density due to the heating of the air, the density drops by no more than 10%. The extra mass flow must therefore enter through the fin tip region and this flow must play a major part in the heat transfer process. It's effect can be clearly seen in the outlet temperature traverse as stated above.

It might be expected that the flow in the tip outlet region would be fast in comparison with the core flow. This is not the case. Smoke flow visualisation experiments showed that the tip ingested flow was drawn into the core flow and very little of it emerged in the tip outlet region. This is in agreement with the velocity traverse data and hence gives a coherent description of the flow.

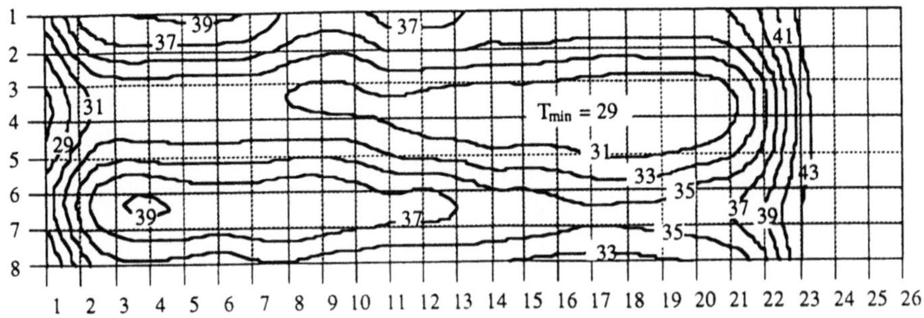

Figure 7 : Temperature distribution at inlet (°C).

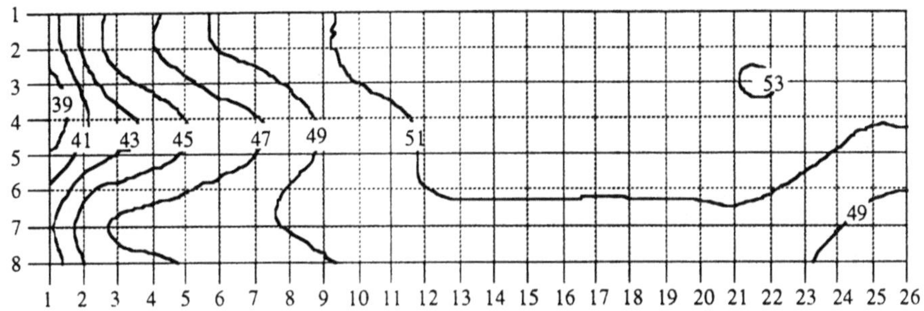

Figure 8 : Temperature distribution at outlet (°C).

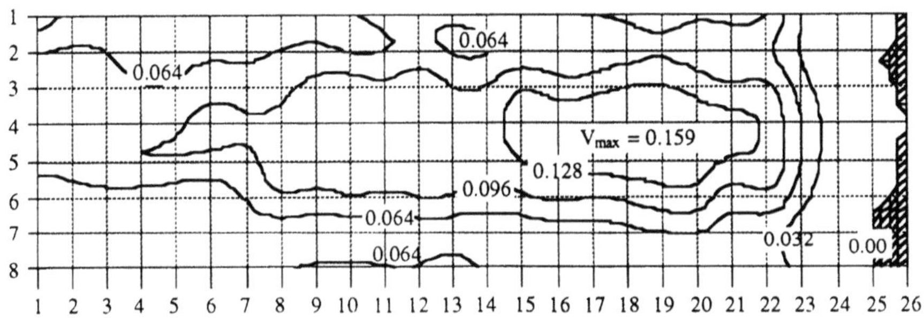

Figure 9 : Velocity distribution at inlet (m/s).

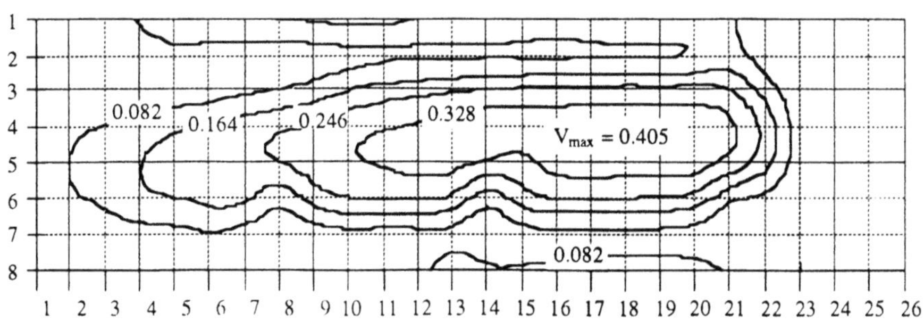

Figure 10 : Velocity distribution at outlet (m/s).

Analysis.

Inlet and outlet flows are given by :

$$\dot{m}_i = \int_A \rho\, u\, dA = 3.8 \times 10^{-5}\ \text{kg/s} \tag{1}$$

$$\dot{m}_o = \int_A \rho\, u\, dA = 1.0 \times 10^{-4}\ \text{kg/s} \tag{2}$$

Where ρ and u are density and velocity respectively and area A is the area between the fin stagnation lines. The density changes over an area are small so equations (1) and (2) are essentially velocity area averages times a mean density.

The ingested mass flow is therefore :

$$\dot{m}_{ing} = \dot{m}_o - \dot{m}_i = 6.2 \times 10^{-5}\ \text{kg/s} \tag{3}$$

A full energy integral analysis is not possible because there is no detailed information of the tip region ingested flow. The total heat dissipated may be approximated in the following way. The heat flux per fin passage is :

$$Q = \dot{m}_o\, C_P \left(\overline{T_o} - T_\infty\right) \tag{4}$$

Where T_∞ is the coldest air temperature in the container and $\overline{T_o}$ is the area averaged exit temperature. There are 15 channels and two end fins, therefore the total heat dissipated per fin array is :

$$Q_T \approx 48\ \text{W} \tag{5}$$

The experimental input power was 40 watts per array. The above calculation is very dependent on the correct assessment of T_∞. It is expected that some of the air will not have been cooled to the minimum temperature before it re-enters the fin passage, hence the overestimation of the heat flux.

A simple analytical method derived by Bejan (1984) was used for a comparison with the measured velocities. For infinite parallel plates:

$$\frac{\partial^2 u}{\partial x^2} = \frac{g\, \beta \left(T - T_\infty\right)}{\nu} \tag{6}$$

Approximating : $T - T_\infty = T_f - T_\infty$ where T_f is the fin temperature allows integration to the following two dimensional velocity distribution perpendicular to the fins :

$$u = \frac{Gr_L \nu}{8L}\left(\frac{S}{L}\right)^2\left[1-\left(\frac{x}{S/2}\right)^2\right] \quad (7)$$

Where Gr_L is the Grashof number based on fin length L and ν is the kinematic viscosity.

For the section marked 17 in figure 10 the two dimensional velocity profile is compared with this prediction in equation (7) with good effect in Figure 11. Showing both that the velocity profile is parabolic in the middle of the passage outlet and that the velocity distribution is quite accurately predicted under these circumstances.

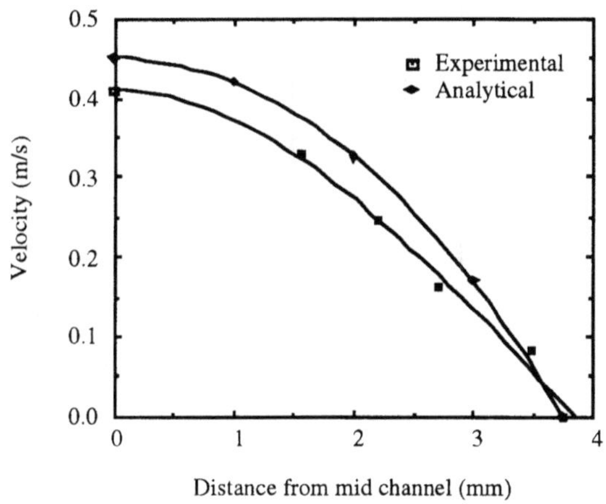

Figure 11 : Comparison of analytical and experimental velocity profiles at channel outlet.

Conclusions.

* The results show that over 50 % of the fin mass flow is ingested in through the fin tip region.

* The preheating of the air by conduction before it enters the passage is clearly seen.

* The outlet velocity profile in the centre of the passage is parabolic and well predicted by simple theory.

* The velocity varies strongly with fin height at outlet, with very little flow in the tip region.

Acknowledgements.

The authors gratefully acknowledge the support of Eolas through the Programme of Advanced Technology (PAT) in Power Electronics.

References.

Aihara, T., 1970, "Natural Convection Heat Transfer From Vertical Rectangular-Fin Arrays (Part 2, Heat Transfer From Fin Edges)", *Bulletin of Japan Society of Mechanical Engineers*, Vol. 13, No. 64, pp. 1182-1191.

Bar-Cohen, A., and Rohsenow. W., 1984, "Thermally Optimum Spacing of Vertical, Natural Convection Cooled, Parallel Plates", *ASME Journal of Heat Transfer*, Vol. 106, pp 116-123.

Bejan, A., 1984, *Convection Heat Transfer*, Wiley, New York, pp. 139.

Buller, M., and Martinek, F., 1978, "Experimental Study of the Inflow Effects on a Natural Convection Heat Sink", *ASME Paper 78-WA/HT-30*, pp. 1-5.

Chappidi, P., and Eno B., 1990, "A Comparative Study of the Effect of Inlet Conditions on a Free Convection Flow in a Vertical Channel ", *ASME Journal of Heat Transfer*, Vol. 112, pp 1082-1085.

Ellison, G., 1979, "Generalised Computations of the Gray Body Shape Factor for Thermal Radiation from a Rectangular U-Channel", *IEEE Trans. on Components, Hybrids, and Manufacturing Technology*, Vol. 67, pp 621-631.

Guglielmini, G., Nannei, E., Tanda, G., 1986, "Natural Convection and Radiation Heat Transfer From Staggered Vertical fins", *Int. J. Heat Mass Transfer*, Vol. 30, No. 9, pp. 1941-1948.

Martin, L., Raithby, G., Yovanovich, M., 1991, "On the low Rayleigh Number for Natural Convection Through an Isothermal, Parallel Plate Channel", *ASME Journal of Heat Transfer*, Vol. 113, pp. 899-905.

Jaluria, Y., 1980, *Natural Convection Heat and Mass Transfer*, Pergamom pp. 300

Naylor, D., Floryan, J., Tarasuk, J., 1991, "A Numerical Study of Developing Free Convection Between Isothermal Vertical Plates", *ASME Journal of Heat Transfer*, Vol. 113, pp. 620-621.

O'Meara, T., and Poulikakos, D., 1987, "Experiments on the Cooling by Natural Convection of an Array of Vertical Heated Plates With Constant Heat Flux", *The International Journal of Heat and Fluid Flow*, Vol. 8, pp. 313-319.

Praught, R., 1982, "Inclined fins for a more efficient heat sink", *Proc. IEEE Int. Telecomm. Energy Conf.*, CH1818-4/82, pp. 158-164.

Ramanathan, S., and Kumar, R., 1991, "Correlations for Natural Convection Between Heated Vertical Plates", *ASME Journal of Heat Transfer*, Vol. 113, pp. 97-107.

Van De Pol, D., and Tierney, J., 1974, "Free Convection Heat Transfer from Vertical Fin Arrays", *IEEE Trans. Parts, Hybrids and Packaging*, Vol. PHP-10, No. 4, pp 267-271.

6. LIQUID COOLING OF ELECTRONIC DEVICES

Application and Performance of Si-Microcooling Systems for Electronic Devices

Bernd Gromoll

Siemens AG, R & D Laboratories , Erlangen

Synopsis: A silicon-etched type of microcooler has been developed which, it is hoped, will form the basis of an integrated design of microcooling systems for electronic components. A number of different versions of the microcooler have been made to suit different coolants and different applications. The performance that has been achieved so far using air and water and employing natural and forced convection is described. The initial results of using the microcooler as the evaporator of a thermosyphon circuit show promise for other possible applications.
The practical application of the microcooler with air as the cooling medium is described, taking a Type 486 microprocessor as an example. Assuming the power loss of the processor to be 4 W, its operating temperature can be reduced by 30 K; which allows the clocking frequency to be increased by 30%. The microcooler that does this is extremely small, being only 0.5 cm^3 in physical volume.

1. Introduction

In the field of electronics, figures of power-loss density per unit volume are continuing to rise ever more sharply. Over the past 5 years, for example, the power density of semiconductor packaging systems, such as those employed in mainframe computers, has risen from 1 kW/litre to 4 kW/litre at present (Fig.1). A Power Density of 20 kW/litre is predicted for the year 2000. The same applies to other fields, such as PC's and WS´s,

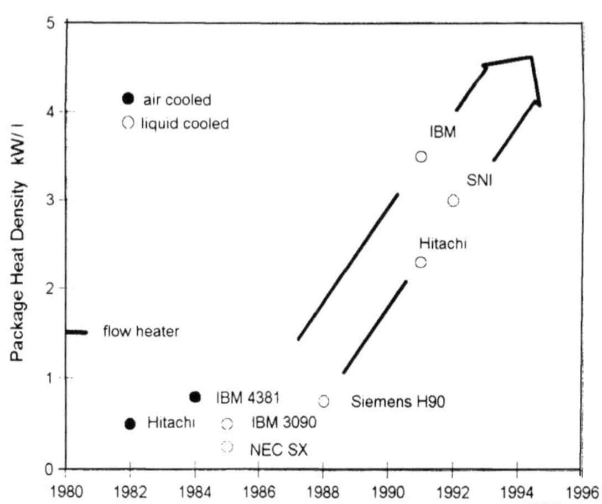

Fig.1: Package power density -Trend in semiconductor packages for mainframe systems

C. J. Hoogendoorn et al. (eds.), Thermal Management of Electronic Systems, 223–232.

where, at processor level, the power figures have increased from 3 - 5 W to 15 - 20 W. The trend in computer systems is from 2D to 3D architecture employing appropriate interconnection techniques. In fact, the arrangement of power electronic devices in equipment cubicles being used at present is already a close approximation to a 3D layout. Microcooling systems with an integrating capability are needed so that the power losses from such arrangements can be dissipated without restricting the performance of the electronic devices themselves [1]. Apart from the

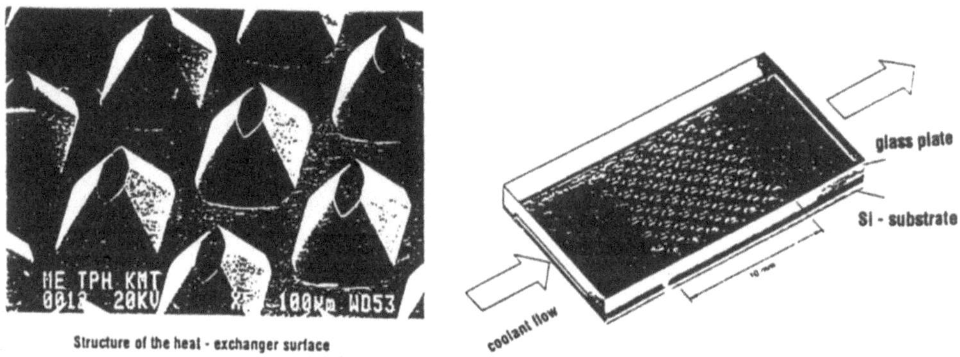

Structure of the heat - exchanger surface

Fig. 2: Microcooler, Si-etched with glass plate 300 W/cm², ΔT = 25 K (Gromoll/Stephani)

physical volume of the cooling system itself, the manner in which the system is coupled to the heat source and the type of cooling medium used are of the utmost importance. Development work in Siemens' R & D laboratories has produced a microcooler suitable for very high values of heat flow density per unit volume that will be able to form the basis of such microcooling systems.

2. Construction of the microcooler

The manufacturing process for the microcooler employs silicon etching technology because it allows such a variety of different forms to be used. The required structure of the heat exchanger can be established by masking and anisotropic etching of a silicon wafer (100). The actual form in this case is shown in Fig. 2. Compared with the types of heat exchanger first introduced by Tuckerman and Pease [2] and Phillips [3] which had only single channels, our microcooler employs a pyramid structure to create a larger area for heat transfer. The truncated cone arrangement causes the coolant to constantly change its direction so that new turbulent boundary layers, which produce the best heat transfer coefficient, are always being formed.

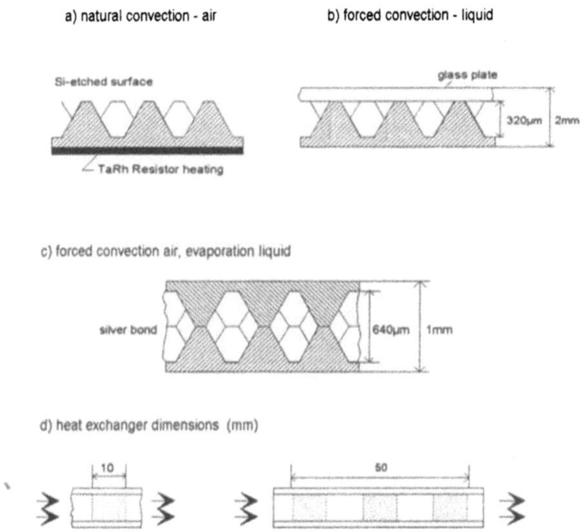

Fig. 3: Si-Micro Heat Exchanger Designs

A number of different structural designs have been developed depending on the particular type of application and the type of coolant being used:

- Natural convection, Fig. 3a
In this case the pyramid structure is not covered and only the greater area provided by the microfins is utilized for direct heat exchange with air, the heat exchanger being mounted vertically.

- Forced convection with water, Figs. 2 and 3b
Here the silicon structure is covered by a bonded glass plate. Ceramic connectors secured to each end of the heat exchanger with adhesive provide for the inlet and outlet of water. The effective connecting area is 1 cm^2.

- Forced convection with air, evaporating water, Fig. 3c
Larger flow passages are necessary in order to reduce the resistance to the flow of air and to provide space for the vapour bubbles which form. For this purpose, two deep-etched (320 μm) silicon surfaces have been bonded together with conductive silver. This doubles the cross-section available for flow and the silver solder ensures high thermal conductivity between the two halves.

- Enlarged heat-exchanger connecting area, Fig. 3d
In order to increase the connecting area available for heat exchange without having to make a new mask, several separate heat exchangers have been combined into one on the wafer in this design. To examine forced convection with air, three heat-exchanger units were

connected in series; the construction being as shown in Fig. 3c - stacked and bonded together. The effective area for heat exchange is increased to 5 cm² in this manner.
In order to integrate the heat source directly into the heat exchanger in each of the designs, a layer of TaRho was vapour-deposited for use as a heating resistance (Fig. 3a). It enables contact thermal resistance to the heat source to be avoided when necessary.
The microcoolers were manufactured in our Silicon Process Technology Department (Stephani, Bartsch).

3. Performance characteristics of various microcoolers

3.1 Natural and forced convection with air

In order to study the dissipation of heat by natural convection, the heat source was mounted flush in a vertical board as shown in Fig. 4a. The heat source was simulated by a copper block with a rectangular connecting area of 1 cm². The power losses could be adjusted by

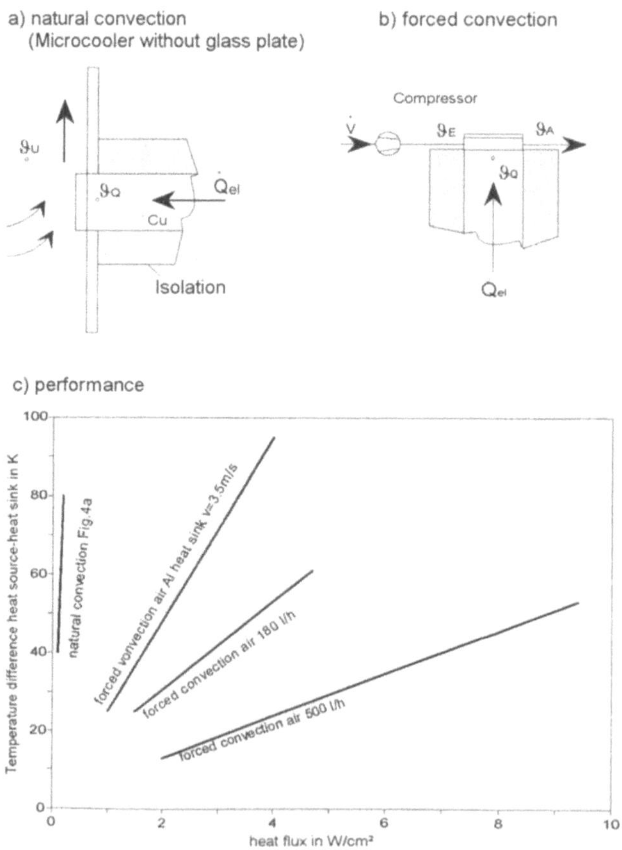

Fig.4: Microcooler - natural and forced convection with air

means of a heating element in the copper block which was totally insulated so that heat dissipation could only take place through the connecting face. The temperature in the copper block was measured by a Pt100 thermometer, which was mounted within a drill hole a few millimeters below the connecting face. The temperature measurement was corrected by the temperature difference which is required for the heat conduction within the copper block from the point of the temperature measurement to the connecting face of the heat exchanger. At this face, the silicon heat exchanger was fitted with microfins as shown in Fig. 3a secured with heat conducting paste. The microfins almost double the surface area for heat transfer. The results of this arrangement are shown in Fig. 4. With a temperature difference of 60 K, it is possible to dissipate 0.2 W/cm², which corresponds to a heat transfer coefficient of 33 W/m²K. Therefore, increasing the surface area with the microfins doubles the heat dissipation compared with a smooth surface while adding only minimal extra physical volume for the heat exchanger.
In the case of forced convection with air, the design of the silicon heat exchanger is as shown in Fig. 3c and it has a physical volume of 0.2 cm³. The test rig with a heated copper block is shown in Fig. 4b. As expected, the performance was much better with this arrangement. A small electrically-powered air compressor was used to produce an air flow of 180 l/h through the heat exchanger with a pressure drop of 0.1 bar. With a temperature difference of 60 K it was found possible to dissipate 10 W of heat in this way (Fig. 4c). This corresponds to a heat transfer coefficient of 1600 W/m²K, which is an extremely high value indeed for air cooling. By using high-pressure compressed air it was possible to achieve even higher values of flow rate of up to 500 l/h with a pressure drop of 0.25 bar. The performance of the silicon heat exchanger using air was then almost doubled (Fig. 4c).

3.2 Forced convection with water

The microcooler illustrated in Fig. 3b was used for the experiments with water. When used as a heat sink, the test rig was as shown in Fig. 4b and, when used for integrated cooling, as

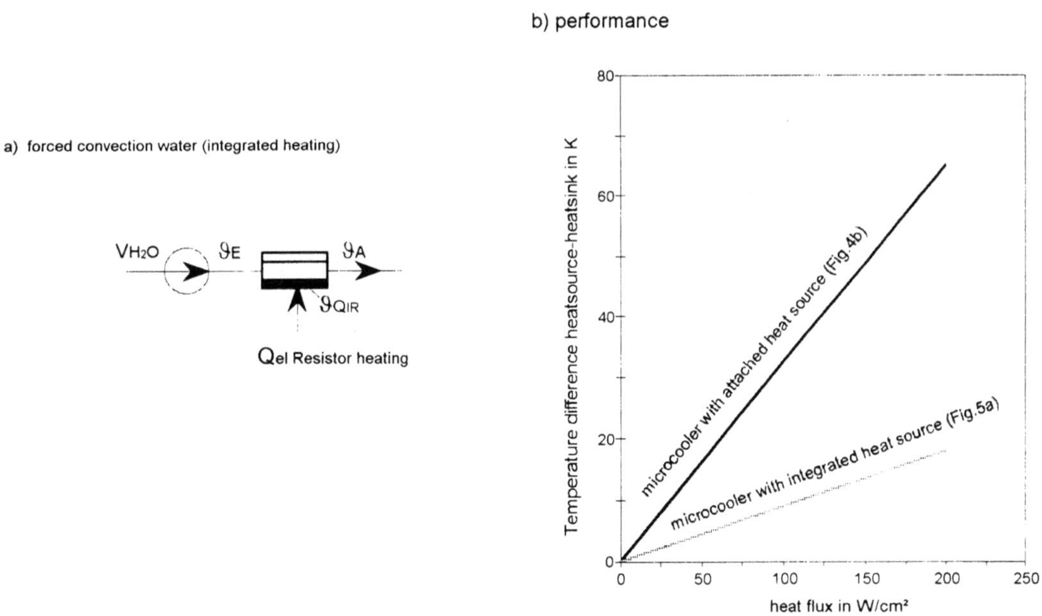

Fig. 5: Microcooler - forced convection water

shown in Fig. 5a. The heat source was provided by the layer of TaRho deposited on the back and used as a electric heating resistance. The flow rate was 12 l/h with a pressure drop of 0.1 bar. By dint of having the lowest thermal resistance, integrated cooling produced the best performance. With a temperature difference of 20 K it was found possible to dissipate 200 W of heat, which corresponds to a heat transfer coefficient of 100,000 W/m²K. With heat-sink cooling the values differ according to the type of thermal interface employed. The lowest values of thermal resistance of all were achieved with heat conducting paste, indium foil and liquid metal.

3.3 Thermosyphon circuit

The advantage of a closed thermosyphon circuit is that it has no moving parts and is self-regulating. Also, the heat dissipation to the ambient air can be moved to a place where there is sufficient space. Therefore, the thermosyphon principle is suitable for higher

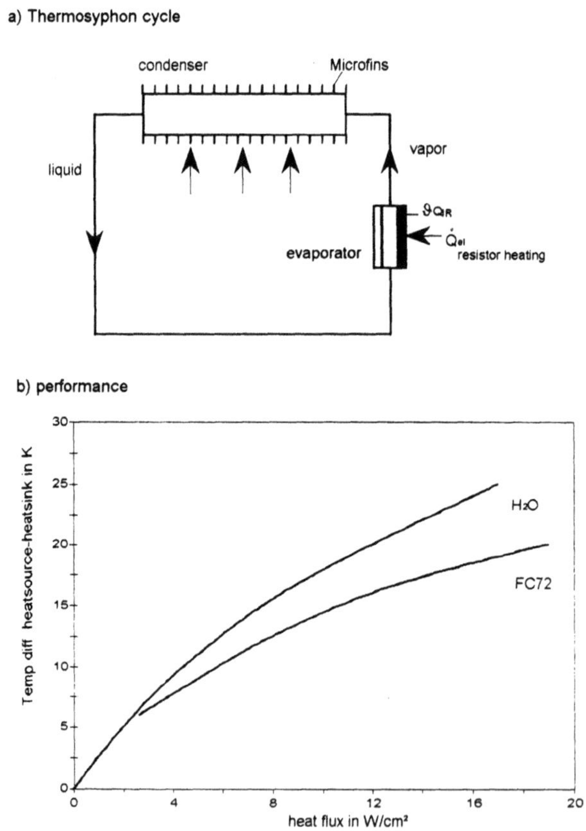

Fig.6: Thermosyphon cycle with Microcooler as evaporator and condensor

values power loss. Due to the high rate of heat transfer, the space required for the evaporator section locally at the heat source can be kept very small. The layout of the thermosyphon circuit is shown in Fig. 6a. The evaporator is a micro heat exchanger of the type shown in Fig. 3c. The evaporator can either be connected to the heat source through a contact interface or is integrated into the heat source itself - here in the test rig it is simulated by the vapour-deposited heating resistor. Ceramic connectors are bonded to the inlet and outlet of the evaporator. A micro heat exchanger of the type shown in Fig. 3d is used as the condenser. To improve the transfer of heat to the cooling air there are three heat-exchange elements as shown in Fig. 3a bonded to the top and bottom; this doubles the heat-exchange surface area (microfins). Flexible plastic hoses of 3 mm diameter secured to the appropriate connectors with adhesive are used to link the evaporator and condenser together. This allows greater flexibility in the physical positioning of the condenser. The only important point to remember is that the condenser must be above the evaporator so that gravity can assist the circulation of the coolant. The circuit was evacuated and filled with approximately 1 ml of coolant (water and 3M FC-72). Within the range of temperature under consideration (< 100°C) the vapour pressures of both coolants are below the ambient pressure so no demands are made on the pressure integrity of the circuit.

The results achieved with the thermosyphon circuit are shown in Fig. 6b. Despite the small size of the heat exchanger and connecting tubes, there were no problems with the thermosyphon action. With both coolants it is possible to dissipate large amounts of heat at only small temperature differences, achieving heat transfer coefficients of 5000 to 8000 W/m^2K. The values for FC-72 are slightly better, probably due to its substantially higher vapour density compared to that of water. Consequently, the proportion of surface area available for evaporating the liquid coolant is greater.

4. Typical application to a Type 486 microprocessor

The Intel 486DX microprocessor was chosen to provide a typical example of a microcooler application. Basically, this particular processor does not have any serious thermal problems, provided there is sufficient space available for mounting a heat sink. The standard heat sink is a finned unit with 9 mm fins and a physical volume of 13.5 cm^3. It is clamped to the ceramic pin grid array (PGA) package of the processor. In some cases, however, due to a continuing increase in the density of packaging, there is no longer enough space to mount the heat sink, which necessarily forces the designer to operate the processor at a lower clocking frequency in order to reduce the power losses and so keep the temperatures within the prescribed limits. Naturally, the performance capability of the processor is reduced as a result. However, this disadvantage can be avoided by using a suitable microcooling system. Even now it is obvious that there will be thermal problems with the next generation of 586 microprocessors because their power losses will be substantially greater than the present 4 W at approximately 15 W. The same applies to the development of multi-chip modules.

4.1 Thermal analysis of the microprocessor

The first step was to carry out a thermal analysis of the processor in order to obtain baseline data for the projected use of a microcooling system. The processor, together with its base, was mounted on a vertical printed-circuit board as shown in Fig. 7a. The power losses were simulated by means of a film heating element bonded to the processor on the back of the package. The precise figure of power loss could be adjusted with a controller. The whole rig was placed in a wind tunnel and exposed to a parallel flow of air, the velocity of which could be adjusted within a range from 0 to 3.5 m/s. The surface temperatures of the package (40 x 40 mm^2) were measured at the centre and edges by means of an infrared camera through a sapphire window in the side of the wind tunnel. The temperature of the processor itself was measured by means of a sensor on the back. The performance characteristic of the processor without a heat sink is illustrated in Fig. 7b. With the cooling air turned off, the temperatures increase sharply with the power losses, showing approximately identical differences between the processor, centre and edge readings. When the processor is being cooled with air, there is a sharp drop in temperature as the velocity rises, even when there is no heat sink. However, any further increase in velocity above 3.5 m/s only produces a slight further decrease in temperature. From the thermal analyses in Fig. 7 it is clear that the centre surface of the package is the best point of attachment for a microcooling system since direct cooling of the processor at the rear is impossible for technical reasons.

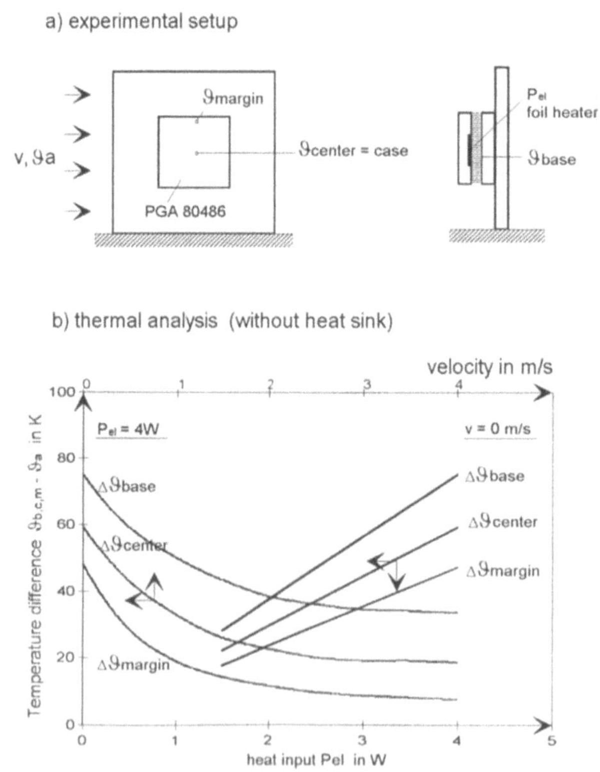

Fig. 7: Thermal analysis of microprocessor Intel 486 without heat sink

4.2 Performance comparison with microcooling systems

The PGA package was tested in the wind tunnel without heat sink and with several different types of heat sink. The standard unit with 9 mm fins and an oversized unit with 40 mm fins were studied at an air velocity of 3.2 m/s. For the purpose of comparison, the type of microcooler shown in Fig. 4b with compressed air and the type of Fig. 3a with forced convection air and the type in Fig. 5 with forced convection water and the type of Fig.6 with thermosyphon cycle were used.

The cooling effects of each can be compared from the reductions in case temperature shown in Fig. 8. The temperature difference achieved with the micro cooler and compressed air (10 l/min) is about 5 K lower than without heat sink , although the connecting area is only 4 cm². The microfins with a height of only 0,5 mm mounted over the whole processor case (16 cm²) show a temperature reduction which is slightly lower.
The cooling effect of both is at low heat inputs (2 W) comparable to the standard 9 mm finned heat sink. Due to the increasing heat spread in the processor case the cooling effect of the micro coolers decreases with higher heat inputs.
The large finned heat sink (40 mm), nearly 5 times the physical volume at 70 cm³ provides only an improvement of 2,5 K compared with the smaller short finned unit of 15 cm³. The same performance is shown by the micro thermosyphon when the physical volume is only 0,5 cm³ (1/30 of the long finned heat sink). The connecting area is only 6 cm² (1/8 of the processor surface). The best heat reduction is achieved with the micro cooler using forced convection water. The temperature rise is 1 K at 2 W and 5 K at 5 W heat input. With an different masking process it will be possible in the near future to manufacture micro heat exchangers with an connecting area that matches the surface of the processor. With these heat exchangers a far better heat reduction can be expected even at higher heat input

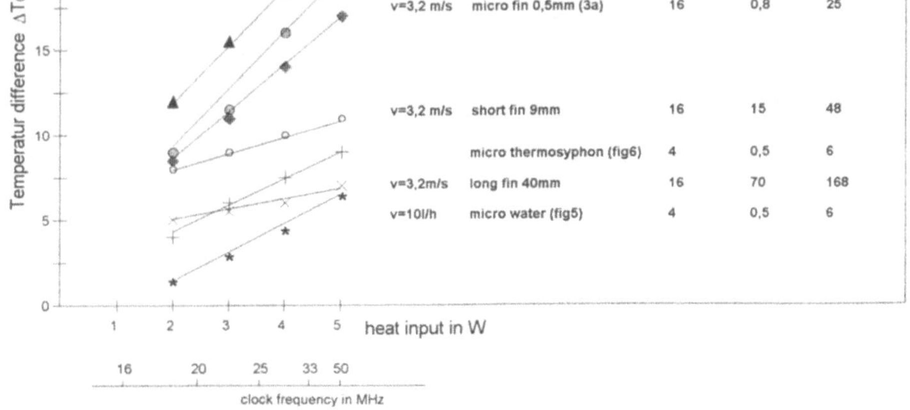

Fig. 8: Case temperature of 486 PGA-Package with various heat sinks (contact with thermal grease)

Thus, the use of microcooling systems is opening up a whole new range of possibilities in the field of electronics as the dissipation of power losses no longer represents a barrier to even higher density packaging.

5. Outlook

Although cooling techniques employing silicon micro mechanics are still in their infancy, they already show great promise for other areas of technology, such as sensors, power electronics and medical equipment. However, there are a number of questions to be answered before mass production is taken up. The main subjects of concern are how the heat exchangers will be integrated into the design, the arrangement of the coolant inlets and outlets, reliability and cost. From the point of view of thermal performance, therefore, the increasing density of device packaging in electronics presents a welcome challenge which will give rise to entirely new concepts and methods of manufacture for cooling systems.

References

[1] B.Gromoll, Trends in cooling technology for high-density packaging, COMP EURO 92, Proc., pp. 303-309, The Hague, Netherlands, May 4-8, 1992

[2] D.B. Tuckerman, R.F. Pease, High performance heat sinking for VLSI, IEEE Electr. Dev. Lett. EDL-2, 126, 1981

[3] R.J. Phillips, Microchannel heat sink, Lincoln Lab. J.,vol 1, pp. 31-48, 1988

Temperature controlled measurements of the critical heat flux on microelectronic heat sources in natural convection and jet impingement cooling

K. Köberle[1], H. Auracher[2],
[1] Institut für Technische Thermodynamik und Thermische Verfahrenstechnik, Universität Stuttgart, Pfaffenwaldring 9, 70569 Stuttgart, Germany
[2] Institut für Energietechnik, Technische Universität Berlin, Marchstr. 18, 10587 Berlin, Germany

Abstract. Experiments with a temperature controlled system were performed to study the critical heat flux (CHF) on a small circular heater surface (1cm²) in natural and jet impingement cooling. As test fluid the Fluorinert FC72 was used. In pool boiling experiments the effects of saturation conditions, the orientation of the heater surface and of heater sidewalls on the CHF-point were studied. The jet impingement experiments were performed with free jets of different temperatures, nozzle exit velocities and heater / nozzle distances for a nozzle with 1mm diameter. In both cases boiling curves were obtained showing presicely the heat flux and the heating surface temperature at CHF.

1. Introduction

The fast development of computer technology requires an improvement in the cooling methods of microelectronic devices. Chip level heat fluxes of 100 W/cm² and more are predicted for the coming years. Heat transfer with phase change provides cooling rates of this level. The upper limit under these conditions is the critical heat flux (CHF). Hence, there is an urgent need for reliable CHF-data and CHF-prediction methods to determine the safety margin between operating conditions and CHF of a microelectronic heat source.

To the authors´ best knowledge, the critical heat flux on microelectronic chips has always been measured with heat flux controlled systems only. This implies the danger of burnout if CHF is reached. Therefore, the authors developed a temperature controlled method. It enables a steady-state measurement of entire boiling curves without the danger of burnout and thus a precise determination of CHF and, of particular interest, the corresponding chip surface temperature.

In the present report this method is applied in experiments with natural convection and jet impingement cooling using the same test heater (simulated chip). This enables a reliable comparison of the upper limit of cooling rates in both systems, and a study of some basic effects on CHF and especially on its corresponding temperature which cannot be precisely determined by heat flux controlled systems.

C. J. Hoogendoorn et al. (eds.), Thermal Management of Electronic Systems, 233–242.

CHF in pool boiling has been studied extensively by experimental and analytical methods. Surveys have been presented by Dhir [1] and Auracher [2]. Katto [3] and Bergles [4] recently discussed the physical mechanism of CHF in pool boiling. Measurement techniques for steady-state experiments have been reviewed by Auracher [5].

One of the most commonly used models for the prediction of CHF on infinite horizontal surfaces was proposed by Zuber [6]. Some authors [7;8;9] extended Zuber's theory to different geometries. In general it was found that small surfaces show elevated CHF-values, while the CHF on large surfaces is independent of the heater dimension and corresponds closely to the Zuber prediction. Lienhard and coworkers [10] reported for finite horizontal plates an effect of heater size on CHF only in the case where the heating surface was smaller than three times Taylor's most dangerous wavelength. Carvalho and Bergles [11] studied the effect of thermal properties and thickness of the heater on CHF. They found a negligible influence of these parameters at thicknesses above a minimum value δ_{min}, e.g. δ_{min}=0.29mm for nickel and 0.64mm for stainless steel. In the present report a heater thickness far above δ_{min} is used.

The effect of system pressure follows directly from the change of the fluid properties in Zuber's equation. At lower reduced pressures, $p/p_{crit} \leq 0.3$, an increase of p results in higher values of CHF, primarily due to the increase in vapor density.

An induced convection effect upon CHF in pool boiling was qualitatively studied by Lienhard [12] and Haramura [13].These studies confirm that a secondary flow induced by rising bubbles increases CHF. This effect was quite small [13] but the corresponding surface temperature was considerably increased.

No results on the effect of heating surface orientation on CHF are available in the literature. Studies [14,15] about this effect on heat transfer in fully developed nucleate boiling show no significant influence.

A very effective alternative to pool boiling is jet impingement cooling of microelectronic heat sources. Studies in the literature present three different modes: free jets [16-20], submerged jets [21-24], and a mixed form of submerged jets and channel flow [25]. In all these studies the heat flux controlled technique was used. The general conclusion is that in fully developed nucleate boiling, heat transfer is independent of jet or test-section parameters like the nozzle diameter, d, the heater / nozzle distance, x, the jet velocity, u, and the diameter, D, of the heater surface. It was found that the boiling characteristics for various velocities merge into a common asymptote of fully developed nucleate boiling and that the curves for different subcoolings, ΔT_{sub}, are displaced slightly. On the other hand CHF depends on d ,x ,u, D and ΔT_{sub}. For example it increases with increasing jet velocity. In some experimental studies a power-law according to $\dot{q}_{CHF} \sim u^n$ was found. The exponent n is determined by the cooling configuration [25]. For a circular impinging jet n was found to be in the range: 0.16<n<0.44. Katto and Kunihiro [21] studied the influence of heater / nozzle distance x with submerged jets and found that CHF decreases with increasing x, apparently due to dissipation in the jet before it reachs the surface. Only limited data is available showing the effect of subcooling on CHF in jet impingement cooling. Monde and Katto [16] found in their free jet experiments with R113 that CHF is proportional to the square of subcooling ΔT_{sub}. In contrast, Ma and Bergles [23] measured CHF-data for R113 in the order of 50% higher than Monde and Katto though they used a submerged jet. In addition they observed only a weak dependence on ΔT_{sub}. The discrepancy between the data could be due to differences in the experimental procedure.

This paper presents experimental results for the test fluid C_6F_{14} (FC72, 3M-Company), boiling on a circular heater with a diameter of 11.3mm. The pool boiling experiments were carried out for different saturation temperatures, each test with horizontally and vertically arranged heating surface. Furthermore, in the case of horizontal heater surface the influence of induced sideflows was studied. The jet impingement experiments were performed with free jets of different temperatures, nozzle exit velocities and heater / nozzle distances x. Most of the experiments were carried out with a nozzle diameter of 1mm. In both cooling modes boiling curves were obtained showing precisely the heat flux and the heating surface temperature at CHF.

2. Experimental

2.1 Pool Boiling Facilities and Control System

The entire system used for the pool boiling experiments has been described elsewhere [27]. Its main part is a stainless steel test chamber with the test heater mounted either horizontally or vertically, Fig. 1.

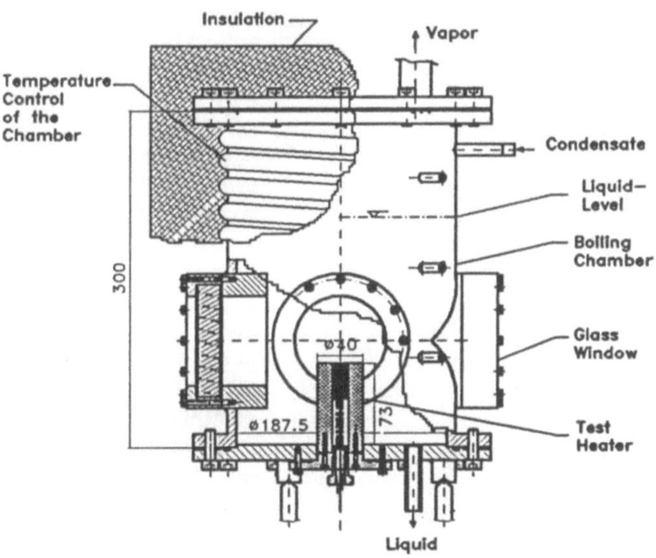

Fig.1: *Boiling chamber with horizontal heating surface of the test heater. For tests with a vertical surface the heater is mounted through one of the windows.*

The vapor removed from the vessel is liquified in a reflux condenser. In this way a precise system pressure control is achieved. The test heater consists of a cylindrical copper block, DC-heated by a sheathed resistance wire on the copper block wall. The maximum possible surface heat flux is 4.99 MW/m^2. In the top region the heater diameter is reduced to 11.3mm and a nickel plate of 0.6 mm thickness is soldered onto the copper block. Three thermocouples are located at the interface between nickel and copper. All thermocouples (NiCr-Ni) are metal sheathed with 0.25 mm outside diameter and 0.05 mm wire thickness. One of the thermocouples beneath the nickel plate serves as a sensor for the control system to give the deviation between set-point and actual temperature. The heater is insulated and sealed by a polyimide block (λ = 0.22 W/mK). Calibration experiments and a theoretical estimate by a numerical solution of the heat conduction equation were carried out to account for the heat losses through the polyimide block.

The temperature-controlled system used in this study enables steady-state experiments along the entire boiling curve. The method operates with electrical heating and feedback control of the heating surface temperature. Details are presented in [5] and [27]. It is important that the control system does not affect the boiling mechanism. This was theoretically proved elsewhere [25] and will be experimentally verified in the present report.

Boiling curves for saturation temperatures of $T_{sat} = 60$, 70 and 80°C were measured between the onset of nucleate boiling via CHF to the minimum heat flux of film boiling and vice versa with stepwise changes of the temperature difference ΔT_{sat}. With a horizontal heating surface additional tests were carried out to study the influence of induced sideflows on CHF. In these experiments a glass tube of an inside diameter equal to the heating surface diameter and a height of 80mm was fixed on the polyimide surface of the test heater. Before each run, the fluid was thoroughly degassed by vigorous boiling of the fluid in the chamber allowing the noncondensible gases to vent through the condenser.

2.2 Jet Impingement Facilities

The experiments with jet impingement cooling were performed in a closed loop consisting of the boiling vessel with the test heater, the nozzle unit (Fig. 2) and the reflux condenser, an encapsulated magnetic driven pump and units for filtering, metering and preheating or cooling of the flow. A circular jet was used impinging vertically on the center of the heated surface of vertical orientation, Fig. 2. It is, however, possible to move the nozzle in each direction by means of a

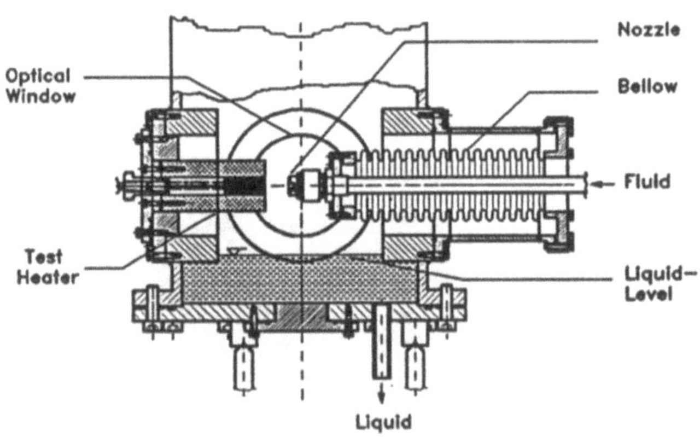

Fig. 2: Configuration of nozzle unit and test heater for jet impingement cooling experiments.

stainless steel bellow and a miniaturized xyz-micrometer motion table which enables the precise positioning (±0.2mm) of the jet. The first experiments were carried out with a fixed nozzle diameter of d=1mm (except for the results in Fig. 6) varying the following parameters: distance heater / nozzle: x = 2.5; 5; 10; 15mm; jet velocity: u = 5; 10; 15m/s; subcooling: ΔT_{sub} = 15; 25K.

3. Results

3.1 Pool Boiling

A representative boiling curve is shown in Fig. 3. It demonstrates that the control action does not affect the boiling mechanism. The test was first carried out under temperature controlled conditions by means of feedback control and then under heat flux controlled conditions. Clearly, the power input to the heater had to be

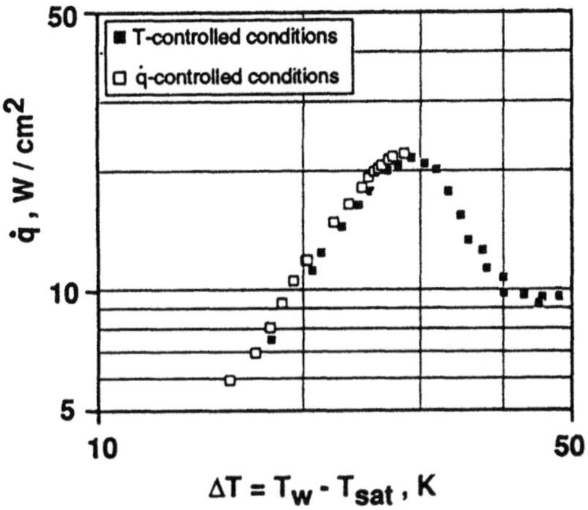

Fig. 3: Comparison of boiling curves measured under temperature and heat flux controlled conditions. FC 72, T_{sat}=60°C, heater surface horizontal.

switched off after reaching CHF in the latter experiments. The results of both runs are in good agreement (the small deviation is not typical, see also Fig. 6).

Fig. 3 may lead to the conclusion that heat flux controlled experiments are able to provide reliable data for the CHF-point. Actually $\dot{q}_{CHF}$ is never completely constant but oscillates more or less due to the boiling mechanism at this point. Therefore, especially if the often used foil heaters are employed, the power has to be switched off before the "time averaged $\dot{q}_{CHF}$ " is reached to prevent burnout. The situation is less dangerous in the present case due to the considerable thermal inertia of the heater. Of course the data obtained from heat flux controlled experiments represent the "safe side of CHF", but in each case the real surface temperature at CHF cannot be found by heat flux controlled studies.

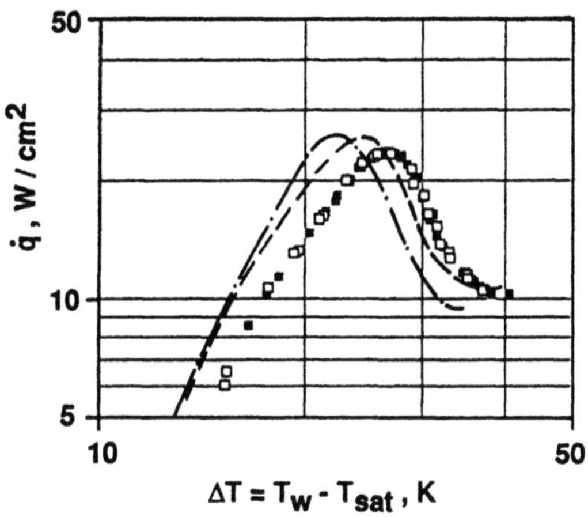

Fig. 4: Effect of system pressure on boiling curves of FC 72. Heater surface horizontal.
T_{sat} = 60°C: ■ ΔT - decreasing
□ ΔT - increasing
T_{sat} = 70°C: — — — — —
T_{sat} = 80°C: —·—·—·—

The effect of saturation pressure on boiling curves is shown in Fig. 4. CHF increases in agreement with Zuber's theory but more significantly the temperature at CHF shifts to lower values with increasing pressure. All curves in Fig. 4 were measured with stepwise increasing and stepwise decreasing ΔT. It is demonstrated in one of the curves (Tsat=60°C) that no hysteresis occurs between fully developed nucleate boiling and film boiling.

Fig. 5 presents a summary of the pool boiling experiments. All CHF-data are higher than those predicted from Zuber's correlation. This is in agreement with Lienhard and Dhir's [7] results that infinite flat plates exhibit CHF-values of about 14% higher than Zuber's original values. Jomard [26] confirmed these results qualitatively in his experiments with a large heater (50 x 50mm). In addition for horizontal plates with a size smaller than about $3\lambda_d$, with λ_d as the most dangerous wavelength [7] for Taylor instability, Lienhard et al. [10] observed a further increase of CHF. This effect is also included in the present results since λ_d for FC72 is between 7.76mm and 7.04mm in the range $60°C \leq T_{sat} \leq 80°C$.

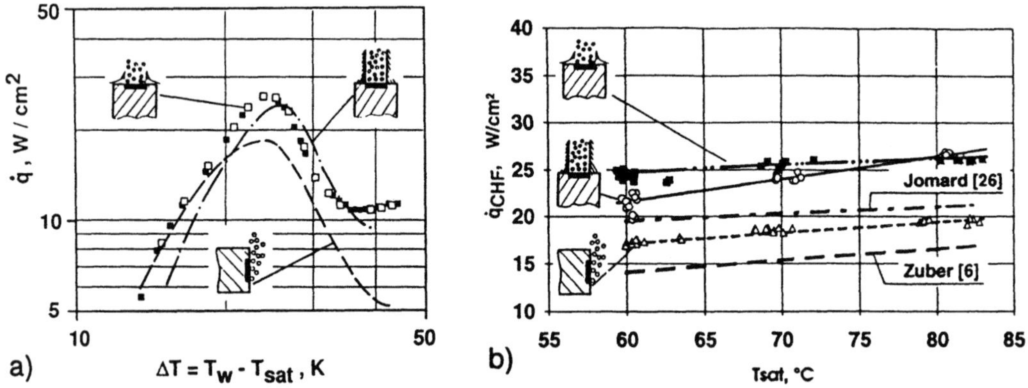

Fig 5: *Effect of saturation condition, orientation of the heater surface and induced sideflow to, a) the boiling curves at T_{sat}=70°C, and b) CHF, for FC72.*

If no sidewalls at the heating surface exist CHF is increased (Fig. 5) most likely due to induced secondary flows, in agreement with Lienhard's [12] and Haramura's [13] results. In addition ΔT_{CHF} decreases (Fig. 5a) indicating a better heat transfer coefficient at the CHF-point. However, this effect could only be confirmed for lower saturation temperatures (Fig. 5b). It will be subject of further studies to find a reliable physical explanation for this effect.

As shown in Fig. 5 the surface orientation strongly influences the CHF. The CHF-values of vertical orientation are more than 30% smaller than the corresponding values of horizontal orientation of the heater surface without sidewalls. Fig. 5a indicates that in fully developed nucleate boiling no significant influence of surface orientation exists in the case without sidewalls which is in agreement with the results of Nishikawa [14] and Jung et al. [15]. Furthermore, the boiling curves for this case show that ΔT_{CHF} is independent of the surface orientation. This holds for all system conditions studied in the present report.

3.2 Jet Impingement

A typical boiling curve for jet impingement cooling is presented in Fig. 6. These experiments were carried out with a nozzle diameter of d=2mm, all the others with d=1mm. Data are presented for temperature controlled and also for heat flux controlled conditions. Similar to the pool boiling experiments (Fig. 3) the boiling curves of both techniques are in good agreement and demonstrate again that control action has no influence on the heat transfer mechanism. It is furthermore obvious that also in jet i
mpingement cooling no hysteresis exists in the CHF region.

The effect of jet velocity u on heat transfer is shown in Fig. 7a for a constant heater / nozzle distance x=5mm. A considerable influence of jet velocity on the boiling curves is observed. With increasing u heat transfer in nucleate boiling is significantly improved and the CHF-point moves to higher values $\dot{q}$ and ΔT, respectively.

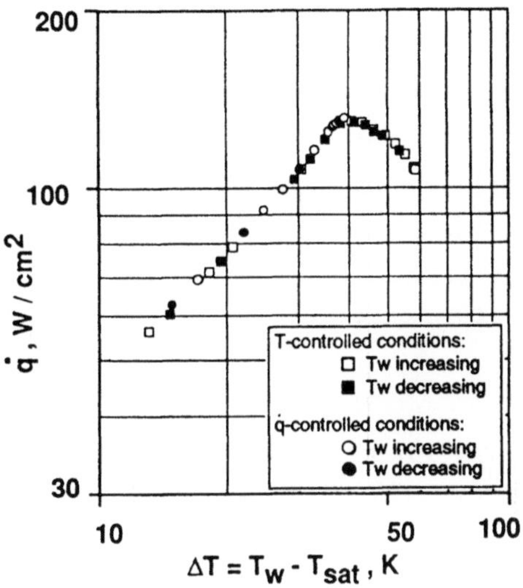

***Fig. 6:** Typical boiling curve for subcooled jet impingement cooling measured under temperature controlled and heat flux controlled conditions.*
FC72; T_{sat}=56°C; d=2mm; ΔT_{sub}=13K; u=10m/s; x=15mm.

The influence of jet velocity on CHF for different values of the distance x is presented in Fig. 7b. As already found in the literature a power law according to $\dot{q}_{CHF} \sim u^n$ exits (see chapter 1). Obviously the exponent n depends strongly on the heater nozzle distance. A variation in the range 0.11<n<0.34 was found for 15mm>x>2.5mm.

The effect of the heater / nozzle distance x on the boiling curves is shown in Fig. 8a. Heat transfer in nucleate boiling significantly decreases if the nozzle distance from the surface is increased. The same is true for CHF. The correspon-ding superheat ΔT_{CHF} shifts to slightly smaller values. Fig. 8b presents a summary of the results for $\dot{q}_{CHF}$ versus the distance x.

The effect of subcooling ΔT_{sub} on CHF is shown in Fig 8b and 9. For a constant jet velocity of u=15m/s and a jet orifice of d=1mm the dependence of CHF on heater / nozzle distance is plotted for two values of subcooling ΔT_{sub}=15 and 25K (Fig. 8b).

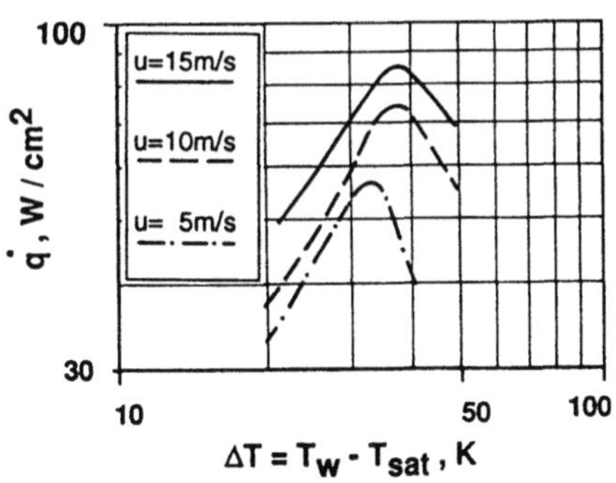

a)

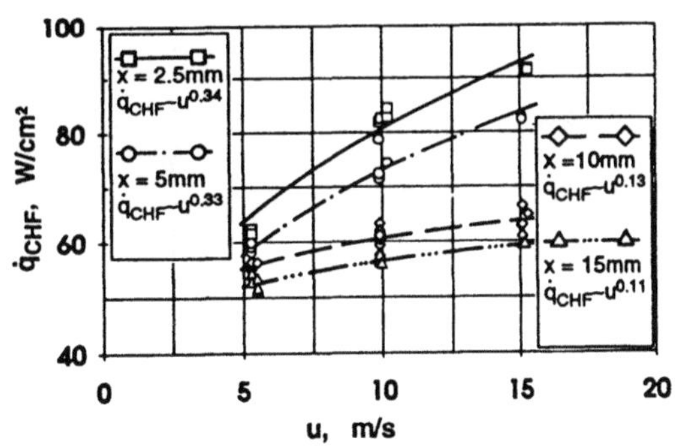

b)

***Fig. 7:** Effect of velocity u (FC72, T_{sat}=56°C, ΔT_{sub}=25K, d=1mm);*
a) on boiling curves, x=5mm; b) on CHF for different values of heater / nozzle distance.

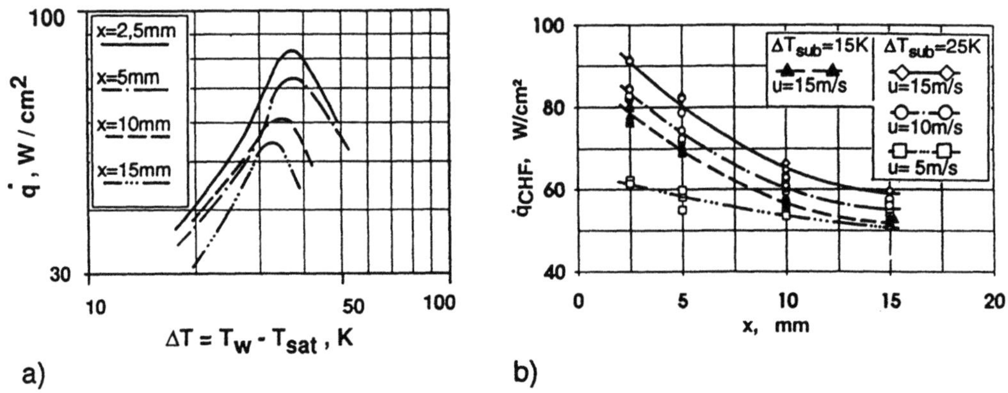

***Fig. 8:** Effect of heater / nozzle distance x, (FC72; T_{sat}=56°C; ΔT_{sub}=15K; 25K; d=1mm); a) on the boiling curves for jet velocity u=10m/s; b) on CHF for different jet velocities.*

Fig.9 presents boiling curves for different subcoolings. The nucleate boiling curves coincide if $\dot{q}$ is plotted versus T_W-T_{sat}, but CHF increases significantly as well as the corresponding superheat ΔT_{CHF}. The most widely accepted explanation of the subcooling effect on CHF is that the vapor leaving the heating surface condenses in the subcooled liquid and thus delays the onset of Helmholtz-instability.

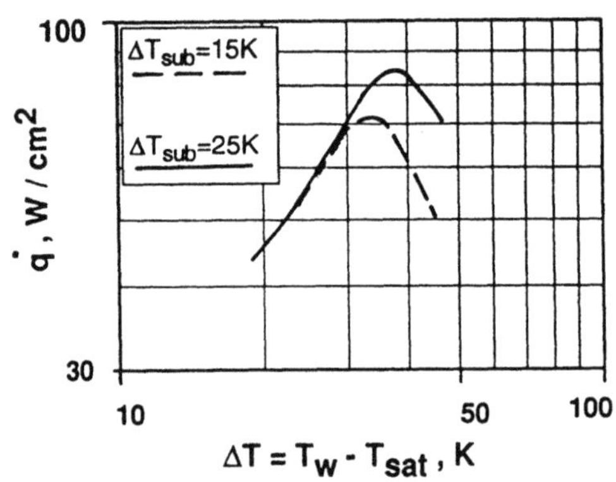

***Fig. 9:** Effect of subcooling on the boiling curves.*
FC72, T_{sat}=56°C, d=1mm, u=15m/s, x=5mm.

4. Conclusions

1. The control system operating with electrical heating and feedback control of the wall temperature enables steady-state experiments around CHF in pool boiling and also in jet impingment cooling. The boiling curves measured under heat flux and under temperature controlled conditions demonstrate that control action does not affect the boiling mechanism. This control system is a suitable technique to determine precisely $\dot{q}_{CHF}$ and the corresponding temperature difference ΔT_{CHF} without the danger of physical destruction of the heater.

2. Pool boiling experiments were carried out with the fluorcarbon FC72 under different saturation conditions and for horizontal and vertical orientation of the heater surface. The boiling curves show no hysteresis in the region around CHF. Horizontal orientation yields higher CHF-values than vertical orientation. A secondary flow induced by rising bubbles improves the heat transfer in nucleate boiling and at CHF. However, this effect vanishes at higher saturation pressure.

3. The jet impingment experiments with free jets of a nozzle diameter of 1mm yield heat fluxes about three times higher (up to about 90W/cm²) than in pool boiling. Larger jet diameters further improve $\dot{q}_{CHF}$ significantly to values far above 100W/cm². Increasing jet velocity and subcooling and decreasing heater / nozzle distance give a considerable enhancement of CHF. It will a be subject of further studies to optimize the jet impingment configuration by a systematic variation of all relevant parameters applying the temperature controlled technique.

Acknowledgements

The financial support of this work from the "Deutsche Forschungsgemeinschaft (DFG)" is gratefully acknowledged. The authors also thank 3M Deutschland GmbH for providing us with test fluid FC72.

Nomenclature

d	jet diameter, m	T_W	wall temperature, °C
D	heater surface diameter, m	u	jet velocity, m/s
p/p_{crit}	reduced pressure, -	x	heater / nozzle distance, m
$\dot{q}$	heat flux , W/cm²		
$\dot{q}_{CHF}$	critical heat flux, W/cm²	δ_{min}	minimum heater thickness, m
ΔT	T_W-T_{sat}, wall superheat, K		
T_{sat}	saturation temperature, °C	λ	thermal conductivity, W/mK
T_f	liquid temperature, °C	λ_d	most dangerous Taylor wavelength, m
ΔT_{sub}	T_{sat}-T_f, subcooling, K		

References

1. **Dhir, V.K.**: Nucleate and Transition Boiling Heat Transfer under Pool and External Flow Boiling. Proc. 9th Int. Heat Transfer Conf., Vol. 1, pp. 129-155 (1990).
2. **Auracher, H.; Köberle, K.**: Critical Heat Flux in Microelectronic Heat Sources. Int. Seminar on Evaporative Cooling, Novosibirsk, USSR (1991).
3. **Katto, Y.**: Critical Heat Flux in Pool Boiling. Proc. Engineering Foundation Conference, Pool and External Flow Boiling, Santa Barbara, CA, USA,1992, pp. 151-164 (1992).
4. **Bergles, A.E.**: What is the Real Mechanism of CHF in Pool Boiling? Proc. Engineering Foundation Conference, Pool and External Flow Boiling, Santa Barbara, CA, USA, 1992, pp. 165-170 (1992).
5. **Auracher, H.**: Transition Boiling in Natural Convection Systems. Proc. Engineering Foundation Conference, Pool and External Flow Boiling, Santa Barbara, CA, USA, 1992, pp. 219-236 (1992).
6. **Zuber, N.**: On the Stability of Boiling Heat Transfer. Transactions of the ASME, Vol. 80, pp. 711-720 (1958).

7. **Lienhard, J.H.; Dhir, V.K.**: The Extended Hydrodynamic Theory of the Peak and Minimum Pool Boiling Heat-Fluxes. NASA CR 2270 (1973).
8. **Park, K.A.; Bergles, A.E.**: Effects of Size of Simulated Microelectronic Chips on Boiling and Critical Heat Flux. J. Heat Transfer, Vol. 110, pp. 728-734 (1988).
9. **Bar-Cohen, A.; McNeil, A.**: Parametric Effects on Pool Boiling Critical Heat Flux in Dielectric Liquids. Proc. Engineering Foundation Conference, Pool and External Flow Boiling, Santa Barbara, CA, USA,1992, pp. 171-175, (1992).
10. **Lienhard, J.H.; Dhir, V.K.; Riherd, D.M.**: Peak Boiling Heat Flux Measurements on Finite Horizontal Flat Plates.J. Heat Transfer, Vol. 95, Series C, pp. 477-482 (1973).
11. **Carvalho, R.D.M.; Bergles, A.E.**: The Effects of Heater Thermal Conductance / Capacitance on the Pool Boiling Critical Heat Flux. Proc. Engineering Foundation Conference, Pool and External Flow Boiling, Santa Barbara, CA, USA, 1992, pp. 203-211 (1992).
12. **Lienhard, J.H.; Keeling, K.B.**: An Induced-Convection Effect Upon the Peak-Boiling Heat Flux. J. Heat Transfer, Vol. 92, pp. 1-5 (1970).
13. **Haramura, Y.**: Effect of Temperature Changing Rate and Wettability on Pseudo-Steady Transition Boiling of Saturated Water. Presented on : 3rd Int. Symposium on Heat Transfer - 1992, Beijing, China.
14. **Nishikawa, K; Fujita, Y; Uchida, S.; Ohta, H.**: Effect of Heating Surface Orientation on Nucleate Boiling Heat Transfer. Proc. ASME-JSME Thermal Eng. Joint Conference, Honolulu, March 1983, ASME, New York, Vol. 1, pp. 129-136 (1983).
15. **Jung, D.S.; Venart, J.E.S.; Sousa, A.C.M.**: Effects of Enhanced Surfaces and Surface Orientation on Nucleate and Film Boiling Heat Transfer in R-11. Int. J. Heat Mass Transfer, Vol. 30, pp. 2627-2639 (1987)
16. **Monde, M.; Katto, Y.**: Burnout of a High Heat -Flux Boiling System with an Impinging Jet. Int. J. Heat Mass Transfer, Vol. 21, pp. 295-305 (1978).
17. **Katto, Y.; Shimizu, M.**: Upper Limit of CHF in the Forced Convection Boiling on a Heated Disk with a Small Impinging Jet. J. Heat Transfer, Vol. 101, pp. 265-269 (1979).
18. **Monde, M.; Okuma, Y.**: Critical Heat Flux in Saturated Forced Convective Boiling on a Heated Disk with an Impinging Jet - CHF in L-Regime. Int. J. Heat Mass Transfer, Vol. 28, pp. 547-552 (1985).
19. **Monde, M.**: Critical Heat Flux in Saturated Forced Convection Boiling on Heated Disk with an Impinging Jet. J. Heat Transfer, Vol. 109, pp. 991-996 (1987).
20. **Vader, D.T.; Incropera, F.P.; Viskanta, R.**: Convective Nucleate Boiling on a Heated Surface Cooled by an Impinging , Planar Jet of Water. J. Heat Transfer, Vol. 114, pp. 152-160 (1992).
21. **Katto, Y.; Kunihiro, M.**: Study of the Mechanism of Burn-Out in Boiling System of High Burn-Out Heat Flux. Bulletin of the JSME, Vol. 16, pp. 1357-1366 (1973).
22. **Miyasaka, Y.; Inada, S.; Owasa, Y**: Critical Heat Flux and Subcooled Nucleate Boiling in Transition between a Two-dimensional Water Jet and a Heated Surface. Journal of Chemical Engineering of Japan, Vol. 13, pp. 29-35 (1980).
23. **Ma, C.F.; Bergles, A.E.**: Boiling Jet Impingment Cooling of Simulated Microelectronic Chips. Heat Transfer in Electronic Equipment-1983, ASME, New York, HTD-Vol. 28, pp. 5-12 (1983).
24. **Ma, C.F.; Bergles, A.E.**: Jet Impingment Nucleate Boiling. Int. J. Heat Mass Transfer, Vol. 29, pp. 1095-1101 (1986).
25. **Mudawar, I.; Wadsworth, D.C.**: Critical Heat Flux in Subcooled Flow Boiling of Fluorcarbon Liquid on a Simulated Elctronic Chip in a Vertical Rectangular Channel. Int. J. Heat Mass Transfer, Vol. 32, pp. 379-394 (1991).
26. **Jomard, T.; Lallemand, M.**: Correlations de Chaleur en Ebullition Libre de Fluides de Replacement du R113. Proc. of the Meeting of Commission B1, I.I.F. - I.I.R., Herzlia, Israel, pp. 115-121 (1990).
27. **Auracher, H; Köberle, K.; Marquardt, W.**: A Method for Temperature Controlled Measurements of the Critical Heat Flux on Microelectronic Heat Sources, Proc. 3rd UK Nat. Conf. and 1st Europ. Conf. on Thermal Sciences, Birmingham, UK, 1992, Vol.1, pp. 189-195 (1992).

Pool boiling heat transfer of dielectric fluids for immersion electronic cooling: effects of pressure

G. GUGLIELMINI, M. MISALE & C. SCHENONE
Energy Engineering Department, University of Genoa, Via All'Opera Pia 15/a (I) 16145 Genova - Italy

Abstract - Experiments on nucleate saturated pool boiling were performed on dielectric fluids suitable for direct-immersion cooling of electronic devices. Four different fluids of three families were investigated: perfluoropolyetheres HT-55 and HT-70, perfluorocarbon FC-72, and chlorofluorocarbon R-113. Experiments were carried out on a plane horizontal surface made of copper. The effects of pressure on boiling curves and heat transfer coefficients were studied. Increasing pressure enhanced boiling performance and affected the nucleate boiling hysteresis for increasing and decreasing heat fluxes.

1. Introduction

In electronic systems, the continuous trend towards increasing the power and reducing the size of the devices, has highlighted the necessity for special cooling techniques to remove the high power which is dissipated.

Direct-immersion cooling by boiling dielectric fluids has been recognized as a very effective means of thermal control for electronic devices [1, 2] owing to its efficient heat dissipation mechanism. The efficiency of this cooling technique is related to the thermophysical properties of the fluids employed. In the past freons were used, but recently environmental problems have prompted their gradual replacement and fluorochemical refrigerants are beginning to be used as substitutes. A major difficulty with dielectric fluids concerns nucleation anomalies, incipience superheat excursions, hysteresis phenomena and reduced critical heat fluxes [3-5]. Although literature on boiling heat transfer is extensive in general, heat transfer data regarding highly wetting fluids are scarce.

Our study focused on the thermal performance of four dielectric fluids of different families (HT-55, HT-70, FC-72, R-113), and involved the influence of pressure on boiling process, heat transfer coefficients and boiling curves for increasing and decreasing heat fluxes.

C. J. Hoogendoorn et al. (eds.), Thermal Management of Electronic Systems, 243–253.

2. Experimental apparatus and procedure

The experimental apparatus (Fig. 1a) consists of an hermetic stainless steel vessel (380 mm I.D. and 450 mm high) containing the test section assembly immersed in a pool of dielectric fluid. An auxiliary heater is wrapped around the vessel to compensate for the heat loss. Saturation conditions of the liquid bulk are maintained by a water-cooled condenser placed over the test section.

The heat source module (Fig. 1b) consists of a cylindrical copper calorimeter heated on the lower side by a flat electric heater. The boiling area is the flat upper surface of the cylinder (1256 mm^2). In order to realize a sufficiently uniform and reproducible surface condition it was used a sandblasting technique with controlled grade particles. By this procedure the surface roughness obtained was: Ra=0.76 μm. The thermal heat flux through the copper test section and the surface temperature were measured by nine calibrated K thermocouples located at three planes spaced 26 mm, lodged inside the copper cylinder.

Liquid and vapor temperatures were measured by a set of calibrated thermocouples. The pressure inside the vessel was varied from 0.5 bar to 2.0 bar with an uncertainty of ± 0.01 bar.

The data obtained from the thermocouples were recorded by means of an high precision system. The chain of the instruments used for the detection of the signal given by thermocouples produces an accuracy of ± 0.1°C on the absolute temperature values and ±0.02°C on differential values. Further details about the apparatus are reported in [6].

Before testing each fluid a degassing operation was carried out to obtain the correspondence between pressure and saturation temperature. In all tests the following procedure was employed: the heating surface was left immersed in the saturated liquid overnight, with nil heat flux; the run was carried out by supplying the test section electric heater with gradually increasing power.

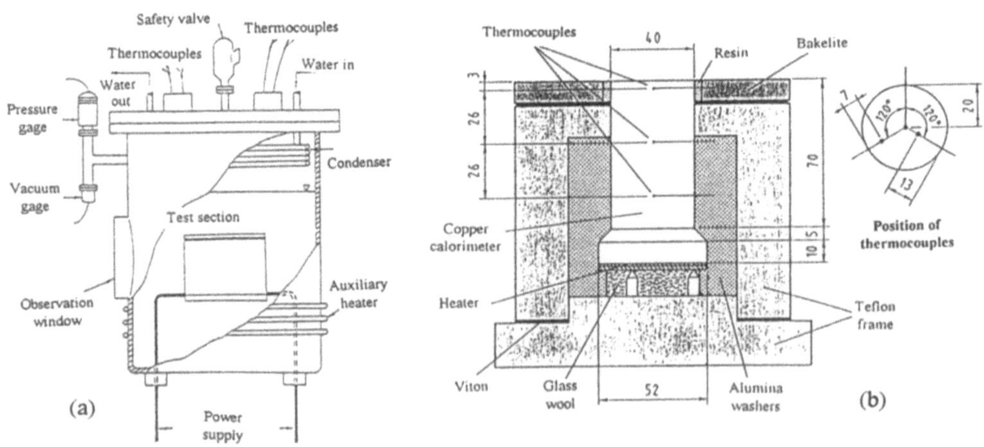

Fig. 1 - Experimental apparatus (a), and test section (b).

3. Results and discussion

3.1 Characteristic of boiling curves

Several test runs were performed on a plane copper surface in saturated dielectric fluids. Although the boiling curves differed according to fluid and saturation pressure, they showed a common behavior typical of highly-wetting liquids [7, 8]. A typical boiling curve is shown in Fig.2, which displays results for HT-70 at the pressure of 1.0 bar. At low heat fluxes, the mode of heat transfer was single-phase convection; raising the heat flux resulted in an increase in the wall superheat along the natural convection curve. At the point of boiling incipience, a drop in surface temperature, which is called temperature overshoot, was sometimes observed. Visual observation of the boiling surface indicated that incipience often occurred over a small patch. As heat flux increased, the boiling front expanded until the entire surface was undergoing nucleate boiling. This behavior, which has already been observed by other authors [8], could explain the steep slope of the boiling curve in the nucleate boiling region: when the heat flux rises, the boiling patch extends and encompasses a greater surface area, while the mean wall superheat remains approximately constant. The smaller the temperature overshoot, the larger the regime of patch boiling. Whenever superheat is very high at incipience, boiling tends to appear suddenly over the whole surface. Moreover, at decreasing flux, when boiling occurs over the entire surface, the wall temperature is slightly lower than at increasing flux, when only a portion of the surface is boiling. With further increases in heat flux, boiling spreads to the whole surface and the curve

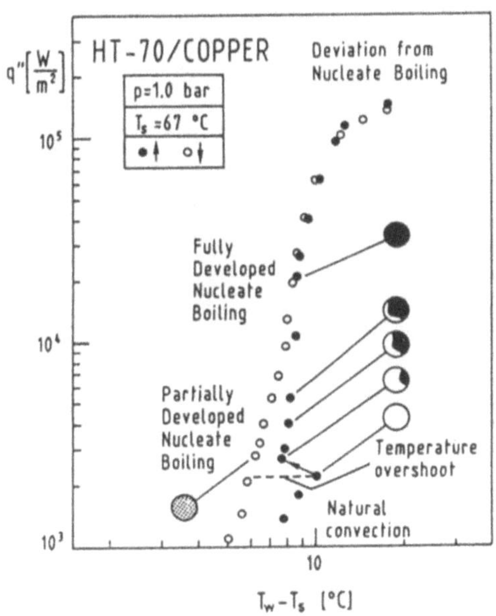

Fig. 2 - Typical boiling curve with boiling patch incipience mode.

assumes an approximately linear behavior in log-log coordinates. This mode is called developed or fully developed nucleate boiling and lasts until the slope decreases owing to the bubble crowding. From the point at which the temperature begins to increase nonlinearly on the log-log plot there is a deviation from nucleate boiling (DNB). After that point, the boiling regime is characterized by the formation of vapor mushrooms, as shown by camera shots. When successive increases were imposed on the heat flux, the experimental boiling curve reached a condition in which a small increase in the dissipated heat flux brought about a wall temperature variation that tended to became inordinate. The highest heat flux value that preceded this condition was assumed as the maximum heat flux (q''_{max}). Proceeding in this way, so as not to jeopardize the integrity of the test section, q''_{max} did not reach the critical heat flux (CHF). This may however be ascertained by default. At decreasing fluxes, below the DNB region, the same fully developed nucleate boiling regime was encountered. Further reduction caused a progressive collapse of nucleation sites, with a mixture of single-phase convection and nucleate boiling on the heating surface. This regime is called partially developed nucleate boiling. On continuous reduction of the wall heat flux and nucleation sites, the boiling curve will eventually merge smoothly into the single-phase convection curve.

3.2 Experimental data

HT-55

Figure 3 shows the data obtained for HT-55 at pressure values of 0.5 bar, 1.0 bar and 2.0 bar. For each pressure, three different regions can be observed: at

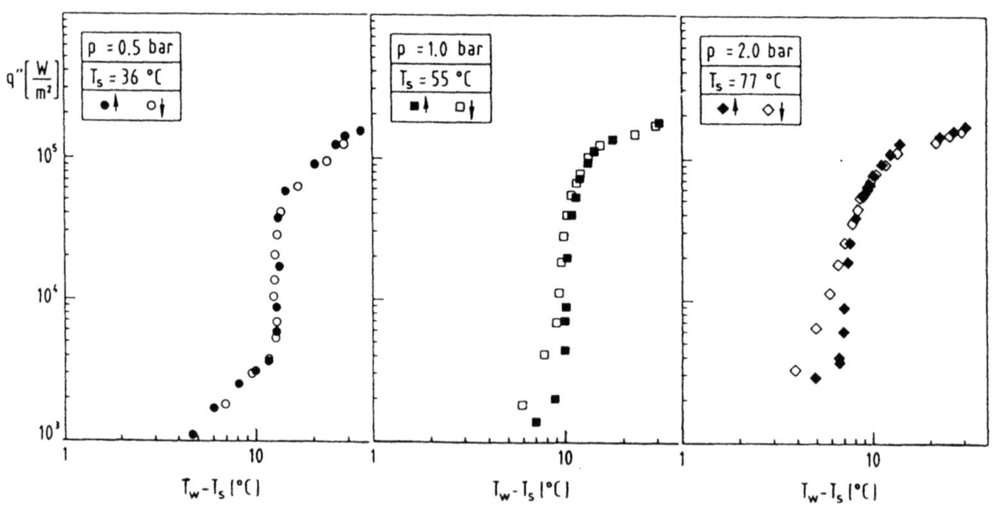

Fig.3 - Boiling curves for HT-55 on copper surface for different pressures.

low heat fluxes (lower than about $0.4\cdot10^4$ W/m^2) heat transfer was by natural convection; as the power to the test section was increased, the boiling incipience was observed, but the HT-55 data indicated no temperature overshoot for any of the three pressures; at higher heat fluxes, close to 10^5 W/m^2, the slope of the curves decreased and deviation from nucleate boiling occurred. For each pressure, the highest heat flux was determined, in the way previously described. The experimental values of q''_{max} for HT-55 are reported in Tab. 1 (see *3.3*), together with those obtained for the other fluids tested, and compared with the values of q''_{CHF} calculated by applying the Zuber-Lienhard correlation [9]:

$$q''_{CHF} = 1.14\ \pi/24\ \rho_v^{0.5}\ h_{fg}\ [g\sigma(\rho_l-\rho_v)]^{0.25} \qquad (1)$$

The boiling curve for 2.0 bar shows a sharp hysteresis effect: the increasing power points differ from the decreasing power points and these latter appear shifted to the left in relation to the former. It was observed [10] that subcooling and dissolved gas content effect nucleate boiling hysteresis and influence the different boiling regimes that accompany the increasing and decreasing heat flux portions of the curves. In this case the hysteresis seems to be pressure dependent, since this effect is not visible at 0.5 bar, negligible at 1.0 bar, but considerable at a pressure of 2.0 bar.

HT-70

Figure 4 shows the results for HT-70 for the pressures of 1.0, 1.5 and 2.0 bar.

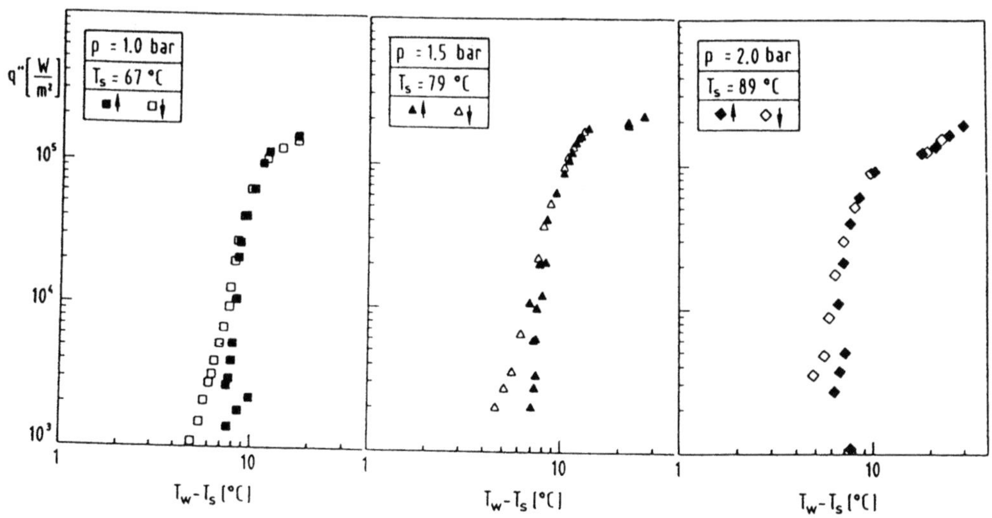

Fig.4 - Boiling curves for HT-70 on copper surface for different pressures.

The curves for 1.0 and 2.0 bar showed temperature overshoots at boiling incipience which were not visible at the intermediate pressure. Wall superheat hysteresis between increasing and decreasing heat fluxes was observed for all three pressures. In the nucleate boiling region, the curves followed the expected plot, and the heat transfer efficiency rose with pressure, though less than that of FC-72. In the DNB region, at a heat flux of more than about $10 \cdot 10^4$ W/m^2, there was a large shift of the boiling curves toward the right. The highest heat fluxes tested, q''_{max}, are reported in Tab. 1 (see *3.3*).

FC-72

The experiments for FC-72 (Fig. 5) appear to conform with the results previously obtained for HT-55 and HT-70. Again three pressure levels were chosen (0.5, 1.0 and 2.0 bar) and for each one a test was run at increasing and decreasing power. At the lowest pressure a sharp temperature overshoot was observed, and incipience occurred at the same time over the entire sample surface, with a direct shift from natural convection to nucleate boiling curve. The data show almost the same path for the three curves, but at high heat fluxes the slope of the 0.5 bar curve decreases more rapidly. The heat transfer coefficients and q''_{max} increase together with the pressure.

R-113

Boiling tests for R-113 showed a slightly different boiling behavior from that obtained for the other dielectric fluids (Fig. 6): while the established nucleate boiling regime showed a trend similar to HT and FC fluids, the slope of the boiling curve beyond the DNB was fairly similar to that in the fully developed nucleate boiling region. No temperature overshoot was visible at any pressure, while the hysteresis effect was very sharp for the 2.0 bar pressure. The values of q''_{max} were higher for R-113 than for other fluids at all pressures; yet, at the same pressure, the boiling curve was shifted towards the right as compared to those described in the previous paragraphs.

3.3 Pressure effects

Figures 3-6 clearly indicate that, before the nucleation sites become active, the data points, for the different pressures, fall approximately along the same straight line; when boiling incipience occurs, three different plots become distinguishable, one for each pressure. These plots remain distinct in the whole fully developed nucleate boiling region. Near the critical heat flux the plots seem to gather, but q''_{max} appears to be strongly influenced by the pressure. The effect of pressure on boiling behavior was similar for the four dielectric fluids tested: at intermediate fluxes, as pressure is increased, nucleate boiling heat transfer is enhanced and a sharp shift of the boiling curves towards the left is observed. A less than linear increase of the heat transfer coefficient with saturation pressure was found: every time pressure doubles, an increase of 1.3-1.5 times the heat transfer coefficient was measured. This result agrees with what has been reported in the literature for ordinary fluids in the pressure range near the atmospheric pressure and with what is predicted by correlations based on the reduced pressure [11, 12].

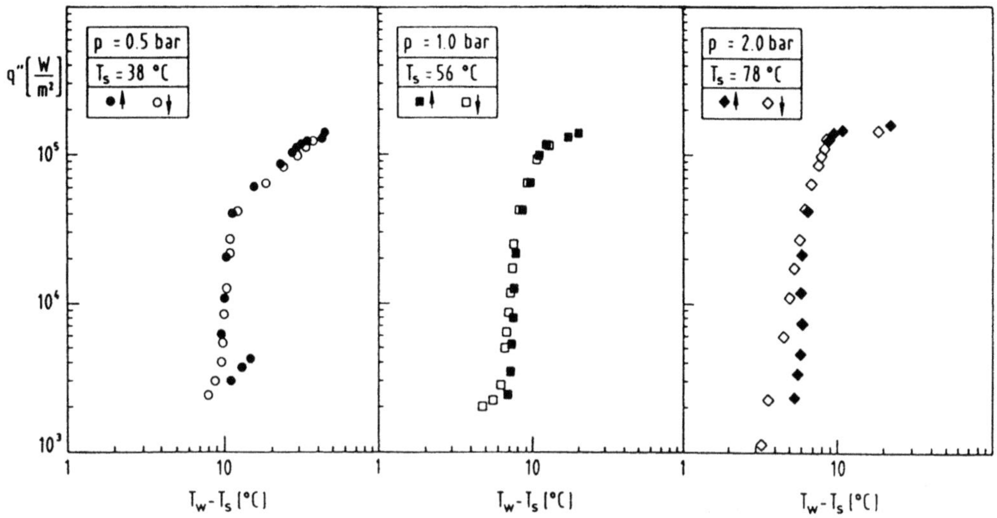

Fig.5 - Boiling curves for FC-72 on copper surface for different pressures.

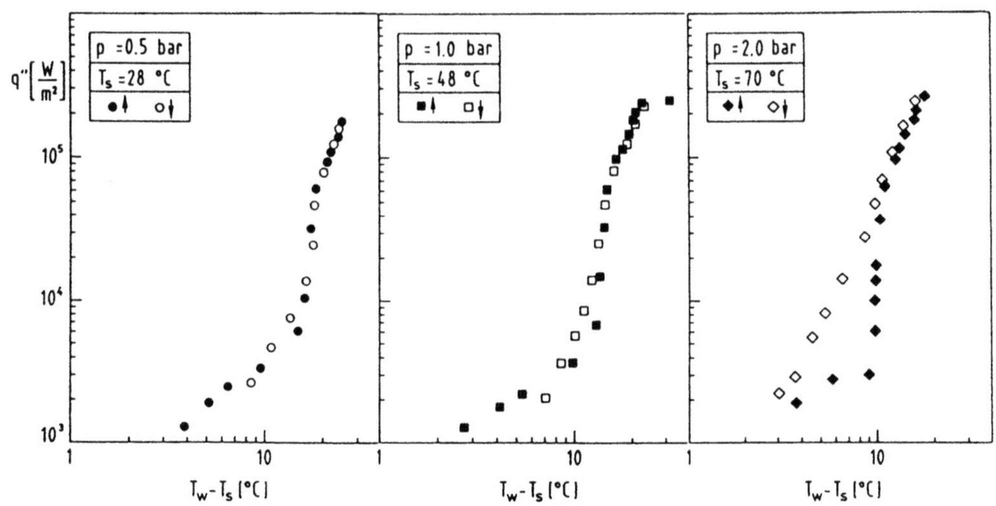

Fig.6 - Boiling curves for R-113 on copper surface for different pressures.

In order to analyze the pressure effects on the nucleate boiling the well-known Rohsenow correlation was utilized [13]. The experimental data for the different fluid/copper combinations were correlated by a least squares technique to evaluate the values of C_{sf} and exponent *a* to use in the Rohsenow relationship:

$$\frac{C_p \cdot (T_w - T_s)}{h_{fg}} = C_{sf} \left[\frac{q''}{\mu_l \cdot h_{fg}} \sqrt{\frac{\sigma}{g(\rho_l - \rho_v)}} \right]^a Pr^b \qquad (2)$$

The constants C_{sf}, *a*, and *b* are reported below:

	C_{sf}	*a*	*b*
HT-55/Copper	0.0042	0.267	1.7
HT-70/Copper	0.0032	0.200	1.7
FC-72/Copper	0.0027	0.161	1.7
R-113/Copper	0.0041	0.323	1.7

Figure 7 and 8 show the Rohsenow diagrams relating to the runs conducted with R-113 and FC-72 fluids respectively. The experimental data utilized for the evaluation of C_{sf} and *a* were obtained during nucleate boiling working with decreasing fluxes.

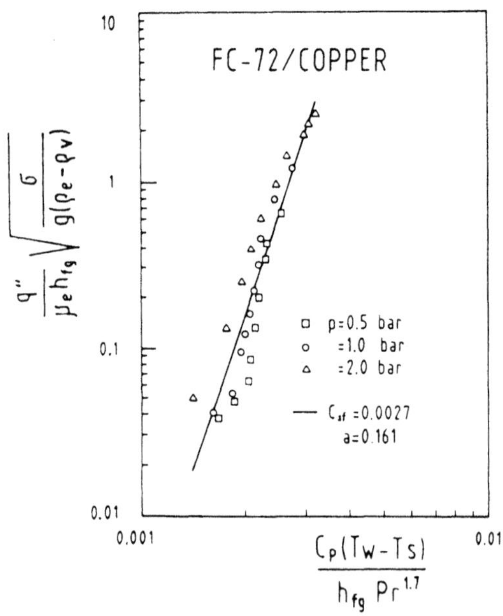

Fig.7 - Plot of Rohsenow correlation for FC-72.

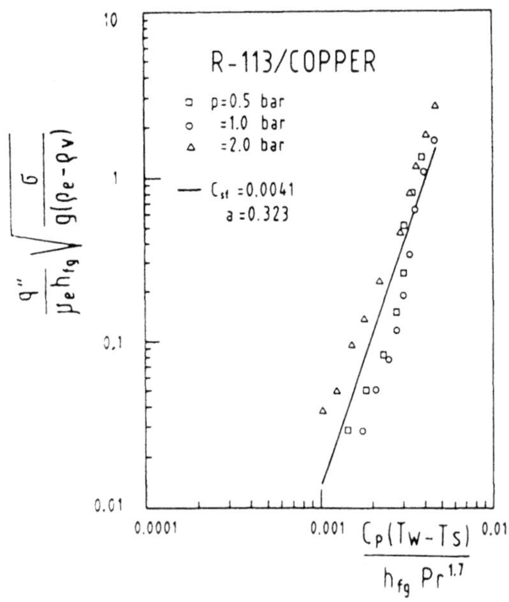

Fig.8 - Plot of Rohsenow correlation for R-113.

Table 1.

	HT-55	HT-70	FC-72	R-113
0.5bar	15.0 / 12.4	-	14.0 / 12.5	17.9 / 17.4
1.0bar	15.9 / 15.8	14.5 / 16.0	14.7 / 15.5	24.4 / 22.3
1.5bar	-	20.1 / 17.6	-	-
2.0bar	16.8 / 19.5	19.1 / 18.2	15.7 / 18.8	25.7 / 27.8

q''_{max} / q''_{CHF} 10^{-4} [W/m^2]

Tab. 1 shows the values of q''_{max} at different pressures for all the liquids tested in comparison with the q''_{CHF} predicted by the Zuber-Lienhard correlation. q''_{max} tends to increase with the pressure, but less than what correlation predicts. At the pressure of 1 bar there is a very good agreement between the calculated and the measured values. All the experimental data differ from those calculated of ± 17% and therefore Zuber-Lienhard correlation can be used with a reasonable approximation. R-113, which has the lowest heat transfer coefficients, shows the highest q''_{max} values, up to $25.7 \cdot 10^4$ W/m^2 at 2 bar.

Boiling incipience tended to occur at lower wall superheat values for higher pressures. A certain effect of pressure on the hysteresis of the boiling curves was observed, but it seemed to change from fluid to fluid. In the regime of partially developed nucleate boiling the differences in the wall superheat values between increasing and decreasing heat fluxes appear more evident when the pressure is increased. The incipient temperature overshoot was more frequently observed for the lowest vapor pressures.

4. Conclusions

Saturated pool boiling heat transfer of four commercial dielectric fluids (HT-55, HT-70, FC-72, R-113) at different pressures was described. The results obtained suggest the following comments:

1. Visual observation indicated that boiling incipience occurred over a small patch, with the gradual extension of the boiling area over the test surface as heat flux increased.
2. The incipience temperature overshoot was rarely observed and this was always less then 8 °C.
3. In the regime of partially developed boiling the portion of the boiling curves obtained increasing and decreasing heat fluxes showed a nucleate boiling hysteresis. This effect appeared to rise with pressure for all the fluids tested.
4. The pressure increase produced an enhancement of nucleate boiling heat transfer, of the same order for all the dielectric fluids tested. The increment of the heat transfer coefficients with saturation pressure is well represented by Rohsenow correlation.

5. The experimental q''_{max} values agreed with q''_{CHF} values calculated by Zuber-Lienhard correlation with an approximation of ±17% for all the pressures. q''_{max} tended to increase with pressure but less than what correlation predicts.
6. From the experimental data the constants of Rohsenow correlation were calculated for all the fluid/copper combinations.

Nomenclature

a, *b*, C_{sf} boiling constants, eq. (2)
C_p specific heat, J/kgK
g acceleration of gravity, m/s²
h_{fg} heat of vaporization, J/kg
p pressure, bar
Pr Prandtl number
q" heat flux, W/m²
Ra surface roughness, μm
T temperature, °C
μ dynamic viscosity, kg/ms
ρ density, kg/m³
σ surface tension, N/m

Subscripts
CHF critical
l liquid
max maximum
s saturation
v vapor
w wall

References

1. Incropera, F.P., 1990, Liquid Immersion Cooling of Electronic Components, Heat Transfer in Electronic and Microelectronic Equipment, Elsevier Publ. Comp., Washington, D.C., pp.407-444.
2. Bergles, A., 1987, Liquid Cooling for Electronic Equipment, Paper Presented at International Symposium on Cooling Technology for Electronic Equipment, Honolulu, Hawaii.
3. Bar-Cohen, A., and Simon, T.W., 1988, Wall Superheat Excursion in the Boiling Incipience of Dielectric Fluids, *Heat Transfer Engineering*, Vol. 9, pp.19-31.
4. Bergles, A., 1988, High Flux Boiling Applied to Microelectronics Thermal Control, *Int. Comm. in Heat and Mass Transfer*, Vol.15, pp.509-531.
5. Danielson, R.D., Tousignant, L., and Bar-Cohen, A., 1987, Saturated Pool Boiling Characteristics of Commercially Available Perfluorinated Inert Liquids, Proc. 1987 ASME/JSME Thermal Engng. Conf., Hawaii, Vol.3, pp.419-430.
6. Ferrando, F., Guglielmini, G., Misale, M., and Schenone, C., 1991, An Experimental Apparatus for Studying Dielectric Fluids Boiling, DINE Report EGR/16, Genova.

7. Mudawar, I., and Anderson, T.M., 1990, Parametric Investigation Into the Effects of Pressure, Subcooling, Surface Augmentation and Choice of Coolant on Pool Boiling in the Design of Cooling Systems for High-Power-Density Electronic Chips, *J. of Electronic Packaging*, Vol.112, pp. 375-382.
8. Heindel, T.J., Ramadhyani, S., and Incropera, F.P., 1992, Liquid Immersion Cooling of a Longitudinal Array of Discrete Heat Sources in Protruding Substrates: II - Forced Convection Boiling, *J. of Electronic Packaging*, Vol.114, pp.63-70.
9. Lienhard, J.H., Dhir, V.K., and Riherd, D.M., 1973, Peak Pool Boiling Heat-Flux Measurements on Finite Horizontal Flat Plates, *J. of Heat Transfer*, Vol.95, n.4, pp.477-482.
10. You, S.M., Simon, T.W., and Bar-Cohen, A., 1990, Experiments on Nucleate Boiling Heat Transfer with a Highly-Wetting Dielectric Fluid: Effects of Pressure, Subcooling and Dissolved Gas Content, Cryogenic and Immersion Cooling of Optics and Electronic Equipment, edited by Simon, T.W., and Oktay, S., ASME HTD-Vol.131, pp.45-52.
11. Mostinski, I.L., 1963, Application of the Rule of Corresponding States for the Calculation of Heat Transfer and Critical Heat Flux, *Teploenergetica*, Vol.4; English abstract, 1963, *Br.Chem. Eng.*, Vol.8, p.580.
12. Cooper, M.G., 1984, Heat Flow Rates in Saturated Nucleate Pool Boiling - A wide-ranging examination using reduced properties, *Advances in Heat Transfer*, Vol.16, pp.157-239.
13. Rohsenow, W.M., 1952, A Method for Correlating Heat Transfer Data for Surface Boiling of Liquids, *Trans.ASME*, Vol.74, pp.969-976.

7. THERMAL CHARACTERIZATION OF ELECTRONIC PARTS

THERMAL DESIGN OF A 419 PINS CERAMIC PIN GRID ARRAY PACKAGE

Thierry FROMONT
BULL S.A
BP 53 78340 Les Clayes sous Bois FRANCE

ABSTRACT

Thermal Finite Element Analysis of a Pin Grid Array construction is performed at the design stage of the package. Thermomechanical analysis of assembly processes in close relationship with the thermal behaviour of the package are also performed.

INTRODUCTION

Due to the influence of the junction temperature on electrical performance, thermal aspects are important parameters for the design of a package encapsulating complex custom VLSI chips with up to 5 W power dissipation.

Despite this fairly low power dissipation range, thermal problems were carefully examined because of the low temperature rise allowed to the chip above air inlet temperature.

During the design stage of the package, thermal Finite Element Analysis using the ANSYS software [1] was performed. This allowed us to respect the short development time frame by avoiding numerous experiments on prototypes. Since the thermal design of the package is in close relationship with thermomechanical and mechanical behaviour, these two other topics were also addressed.

Finite Element Analysis results are presented and experimental results are compared with design objectives.

PACKAGE DESCIPTION

The package described in figure 1 is composed of a Pin Grid Array ceramic base (cavity down) receiving a custom VLSI chip. A copper alloy thermal plate is epoxy bonded on the backside of the ceramic body.

C. J. Hoogendoorn et al. (eds.), Thermal Management of Electronic Systems, 257–265.

This thermal plate allows a good heat transfer from the package to the removable heat sink.

The ceramic base is 49.8 mm X 49.8 mm and 2.78 mm thick with 12 conductive layers. The package has 419 pins arranged in two staggered 2.54mm square grids. The 15 mm x 15 mm chip, with 428 pads is wire bonded to a double deck cavity before the package is hermetically sealed by a kovar lid.

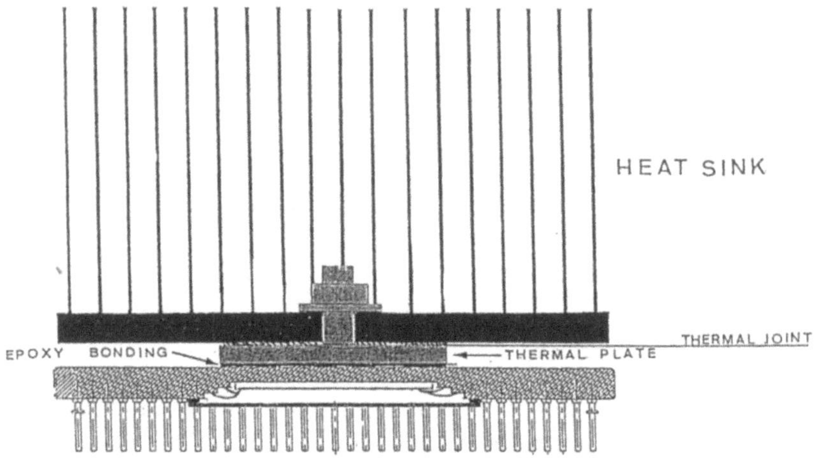

Figure 1: Package cross section

THERMAL DESIGN

To meet the system performance requirements, the maximum junction temperature shall be less than 70°C . To allow typical low acoustic noise, the cabinet environment provides a 1m/s air speed between the printed circuit boards with a maximum air inlet temperature of 35°C. One of the configurations has 4 packages in the same air column with 5W dissipation per package.

Taking into account the air temperature rise, the thermal resistance junction to ambient shall be less than 3.5°C/W.

To reach this value, Finite Element Analysis of the junction to heat sink resistance and the heat sink to air resistance was performed for several designs.

The first analysis led to the choice of the die bonding material and the second led to the final design of the heat sink dimensions and realisation process, taking into account space allocation in the cabinet and cost considerations.

Before performing the Finite Element Analysis, the thermal conductivities of the candidate products were measured.

Because of interface contact resistance, probably due to high temperature hermetic sealing, the measurement of the bulk material thermal conductivities is not relevant. In order to use more realistic values during the analysis, the thermal conductivities of the products were measured including interface effects after being submited to all the manufacturing flow chart processes.

The equivalent thermal conductivities used in the models, including bond interface effects, are shown in table 1.

PRODUCT	EQUIVALENT THERMAL CONDUCTIVITY
DIE ATTACH SILVER GLASS	7 W/mK
DIE ATTACH EPOXY	0.3 W/mK
THERMAL PLATE ADHESIVE	1.5 W/mK
THERMAL JOINT	0.21°C/W for 22 mm sq sample

Table 1: EQUIVALENT THERMAL CONDUCTIVITIES

Heat transfer coefficients from literature taking into account the Langhaar velocity distribution were used for the heat sink modelling [2].

FINITE ELEMENT THERMAL ANALYSIS AND RESULTS

JUNCTION TO HEAT SINK-THERMAL RESISTANCE

The finite element model post processing results of the junction to heat-sink resistance are shown in figure 2. Due to the symmetry of the structure, only one quarter of the package has been analyzed. It consists of 3D isoparametric thermal elements. This model includes a 15 mm square and 0.5 mm thick silicon chip, bonded on a 1.27 mm thick ceramic base with 200 µm thick silver glass or 100 µm of epoxy adhesive. A 22 mm square and 4 mm thick copper alloy thermal plate bonded with 127 µm thick adhesive layer, and covered by a 150 µm thick thermal joint finalizes the model. A 1 W power load is applied at the chip level and the thermal joint surface in contact with the heat sink is maintained at 0°C (This will ensure coherency with the heat-sink model). Therefore, the maximum junction temperature value is the value of the thermal resistance from chip to heat-sink.

Figure 2 shows the temperature profile from junction to heat sink. Thermal resistance is 1.2°C/W for a chip bonded with silver glass. For a chip bonded with epoxy adhesive the resistance value is 2.5°C/W.

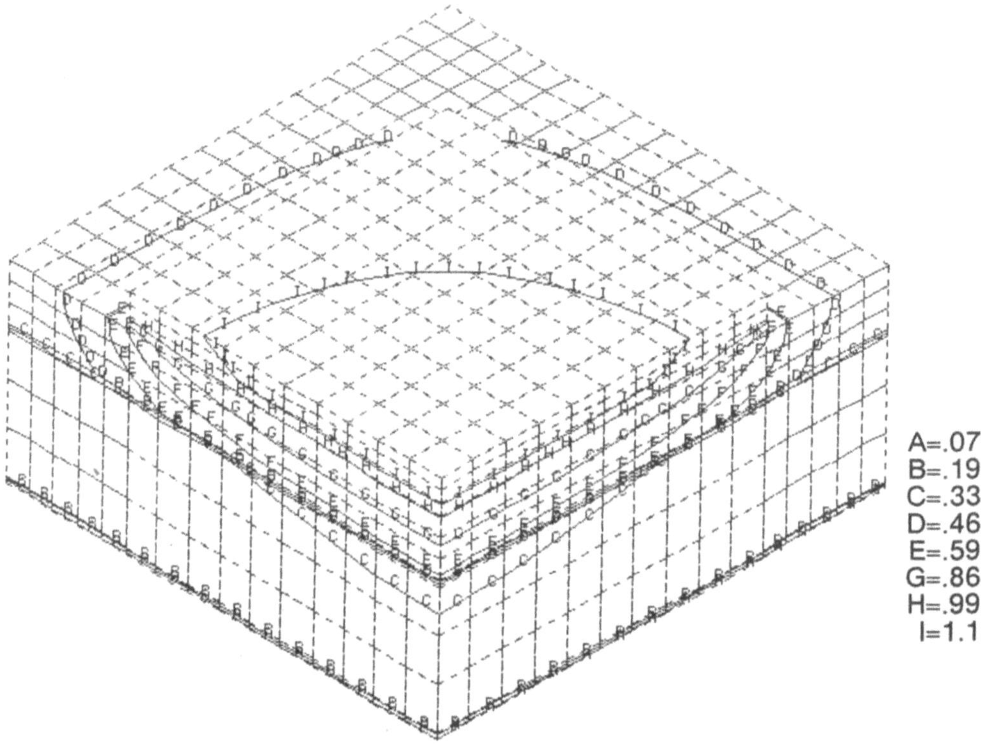

Figure 2: Junction to heat sink temperature profile (°C)

HEAT-SINK THERMAL RESISTANCE

The finite element model post processing results of the heat-sink with a 1 m/s air speed beween fins are shown in figure 3. Since air temperature rise is not taken into account at this step of the analysis, the symmetry of the structure allows to model only one quarter of the heat-sink.
This model includes a 48 mm square and 4 mm thickness base on which are 15 fins of 0.5 mm thick and 24 mm height at 3.3 mm pitch. For coherency with the previous model, a 1 W power load is applied at the contact area between heat-sink and thermal joint and the forced convection heat transfer coefficient is applied on the fins and the part of the base subjected to air flow. Air temperature rise is not taken into account at this analysis level.

Figure 3 shows the temperature profile of the heat-sink located at the entrance of the air column. It has a resistance of 1.4°C/W. Because of the variation of the heat transfer coefficient along the heat sink column, the heat-sink located at the exit of the air column has a resistance of 1.9°C/W.

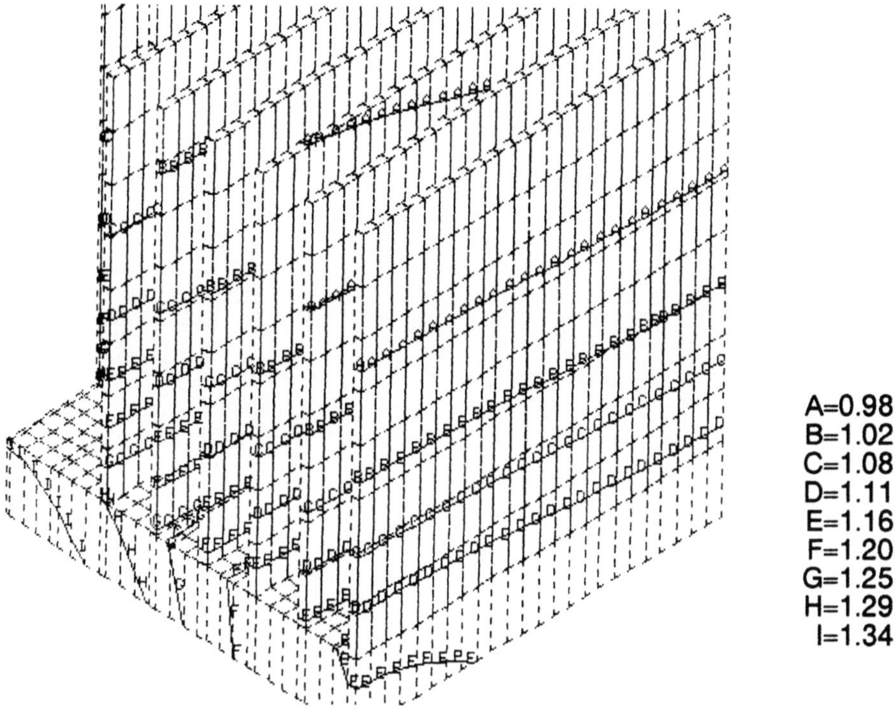

Figure 3: Heat-sink temperature profile (°C)

JUNCTION TEMPERATURE IN THE CABINET

Due to the heat sink design, silver glass die bonding is mandatory in order to match the maximum junction temperature requirement of 70°C. Taking into account the air temperature rise in the cabinet, junction temperature of the VLSIs in one component's column were computed to check that they were below 70°C for a 35°C air inlet and 5 W dissipation per package.

THERMOMECHANICAL AND MECHANICAL DESIGN

Due to the thermal mismatch between the ceramic package and the copper alloy thermal plate, the epoxy bonding assembly process induces stress in the whole structure when cooling from curing temperature to minimum storage temperature. The ceramic shall not be subjected to tensile stress greater than 150 MPa and shear stress in the adhesive layer shall be in accordance with data provided by the adhesive supplier, based on lap shear test.

After the package is inserted in the socket the screwing of the heat sink on the thermal plate is a key parameter for the thermal control. It shall ensure a mean pressure of 1.2 to 1.5 MPa in the thermal joint. The screwing torque, equilibrated by the reactions on the pins in the socket, induces stress in the epoxy adhesive and the ceramic layer.

There shall also be no plastic deformation in the bolt.

The material characteristics used in the models are given in table 2 and figure 4.

MATERIAL	YOUNG MODULUS GPa	THERMAL COEFFICIENT OF EXPANSION 10^{-6}/°C
CERAMIC	255	6.5
ADHESIVE	see figure 4	60
COPPER ALLOY	103	17
ALUMINUM ALLOY	70	21

TABLE 2: MATERIAL PROPERTIES

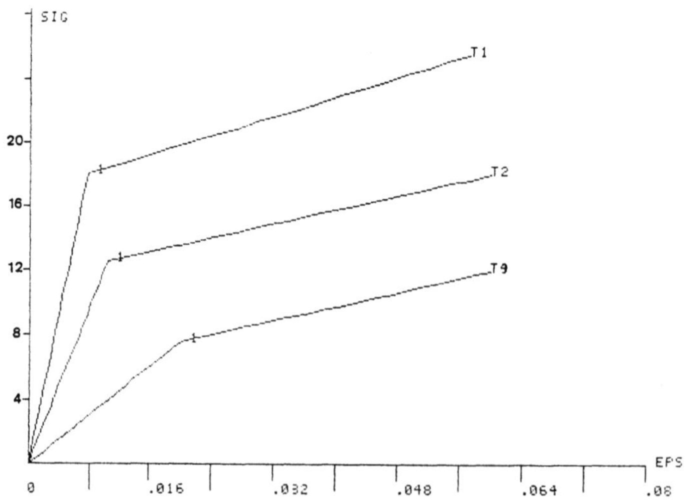

Figure 4: Stress-strain curve for the thermal plate adhesive

THERMOMECHANICAL AND MECHANICAL FINITE ELEMENT ANALYSIS RESULTS

THERMAL PLATE ASSEMBLY

A finite element model of the thermal plate bonding process has been built. Due to its symmetry properties only one quarter of the structure is analyzed. It consists of 3D isoparametric elements. The model includes the central area of the ceramic package, 127 μm of adhesive and the thermal plate. For CPU time limitations, the model is coarse, but results are accurate enough at this stage of the design. It has been assumed that the structure is stress free at 150°C and is cooled at -40°C. Plastic behaviour of the adhesive (neglecting creep) is taken into account.

The first principal stress in the ceramic is less than 75 MPa and shear stress at the adhesive interface ranges from 1.8 MPa at the center of the package, to 25 MPa in the corners of the adhesive layer. The maximum plastic deformation is about 6% in the corner, with a mean value less than 3%; ensuring therefore good fatigue properties of the assembly (Fig 5).

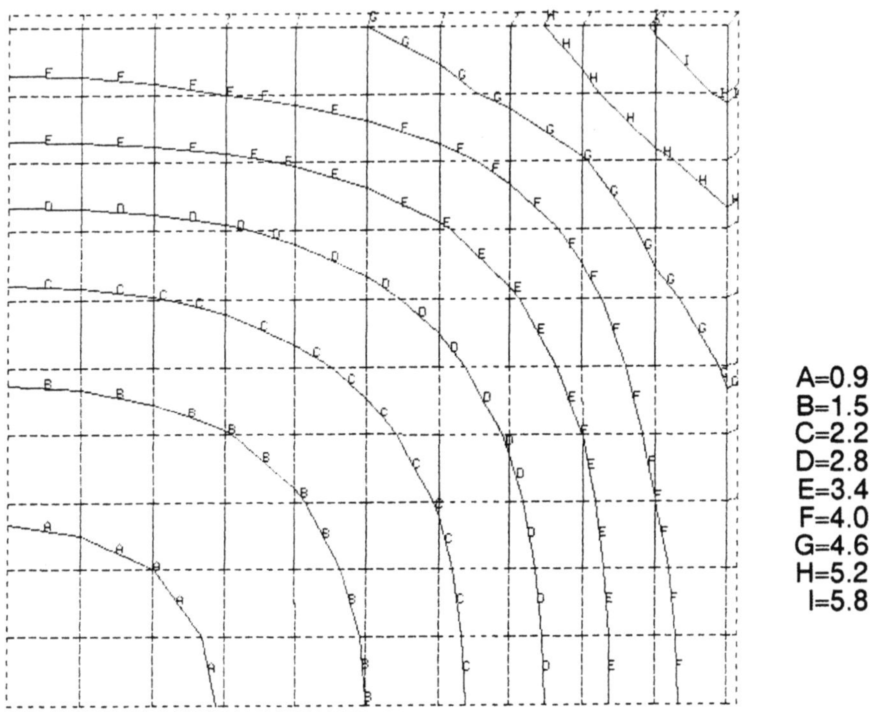

Figure 5: Plastic strain in the adhesive layer(%)

HEAT-SINK ASSEMBLY

In order to estimate the stress in the stucture during the screwing of the heat sink on the thermal plate a three dimensional analysis was performed. The model includes the ceramic substrate, the adhesive layer, the thermal plate using three dimensional isoparametric elements and the PGA pins using 3D spar elements. The pin's tips are blocked in displacement and a mechanical load simulating the 1.4Nm screwing torque is applied at the top of the thermal plate bolt.

The maximum tensile stress in the ceramic is less than 1.4 MPa. The maximum shear stress in the adhesive layer is less than 1 MPa. Finally, during this process the maximum horizontal reaction on the pins is about 0.3N. All these stresses are in the acceptable range of values (Fig 6).

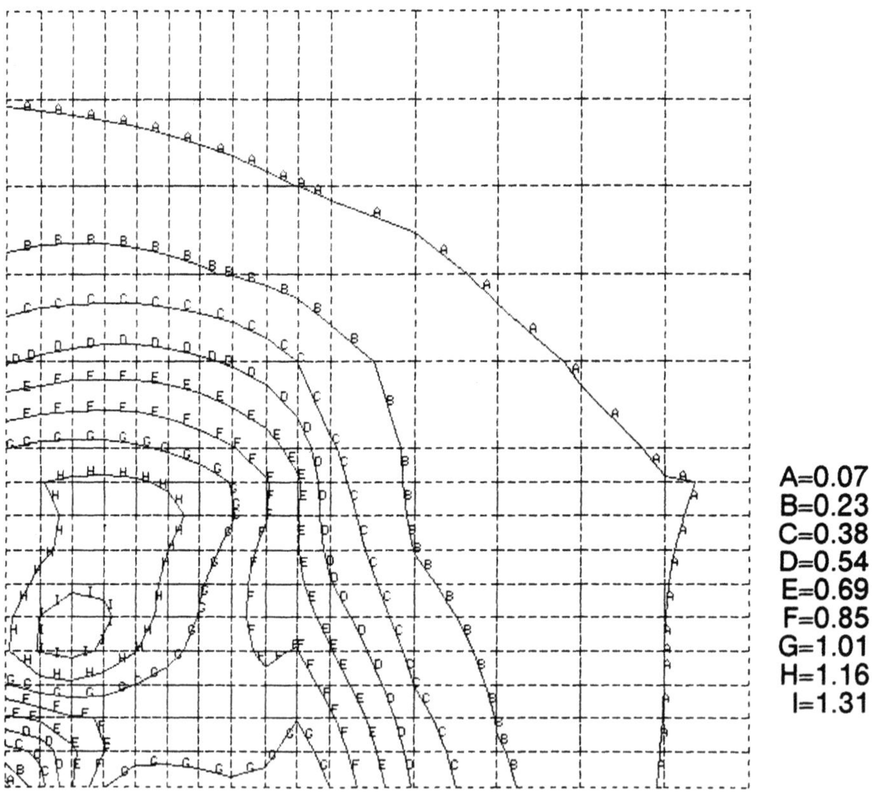

Figure 6: Tensile stress in ceramic during heat sink screwing (in MPa)
One quarter of the model

THERMAL JOINT COMPRESSION

The design point for the pressure in the thermal joint is 1.2 MPa to ensure a good heat transfert behaviour. A non-linear analysis has been undertaken to verify that the thermal plate and heat sink base thicknesses ensure a uniform pressure within the thermal plate area.
A two dimensional axisymmetric model has been defined. Gap elements are used between the thermal joint and the two metal pieces to allow any loss of contact (due to plates bending) to be detected. The load is applied at the top of the bolt while preventing from any dispacement the area of the heat sink base which is in contact with the nut.

The stress in the thermal plate is less that the elastic limit of the copper alloy (fig 7) and the pressure distribution in the thermal joint is greater than 1.2 MPa (figure 8).

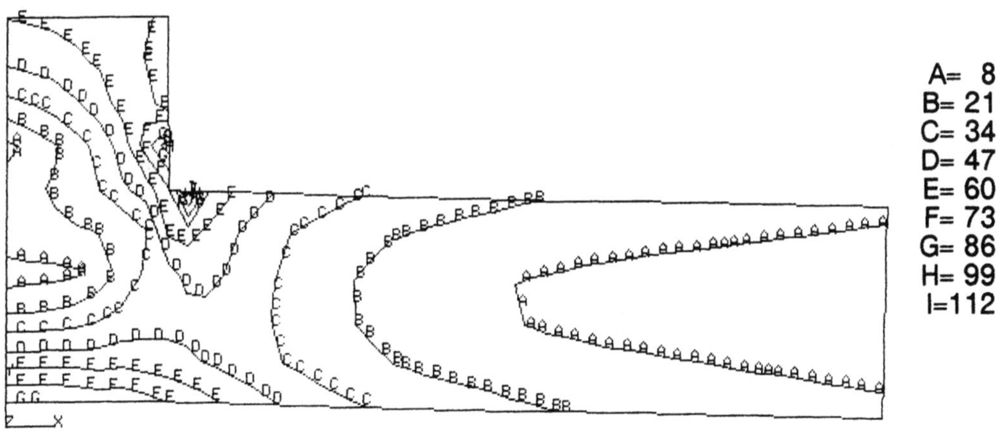

Figure 7: Stress in thermal plate (MPa)

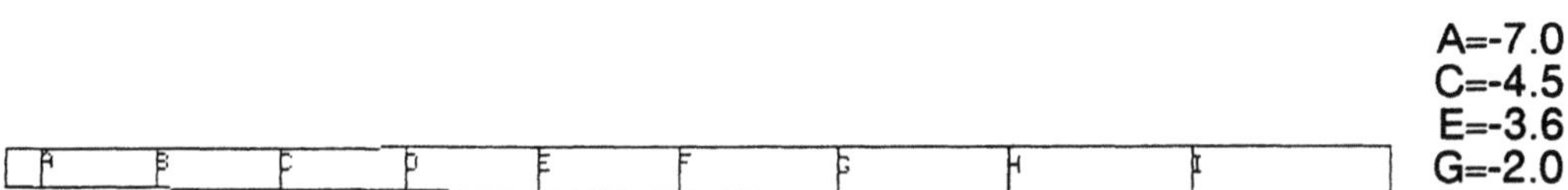

Figure 8: Pressure in the thermal joint (MPa)

EXPERIMENTS

To correlate with simulation results, we performed experiments on specific test vehicles close from the designed structure using an NMRC 12mm square thermal test chip. The internal diodes were calibrated to get the forward voltage drop correlation with temperature. Thermal plate temperature was measured using thermocouples. Thermal resistance from junction to thermal plate were in the range of 1.4°C/W for the 12X12mm chip, therefore ensuring a thermal resistance junction to heat sink in the range of .9 to 1.1°C/W for a 15mm square chip.

Measurements on heat-sink in the final cabinet environment using resistors as heat source led to thermal resistance from 1.1°C/W to 1.6°C/W in an air column.

One batch of packages equiped with thermal plates were submitted to 100 thermal cycles -40°C to 125°C. No degradation was observed and no evolution of the pull strength has been noticed. Heat-sink were screwed on package's thermal plates without any problem.

CONCLUSION

The final parts designed by simulation gave experimental results in accordance with the objectives. By this way the development was performed within the time schedule, limiting intermediate mock-up and experiments. This points out the advantages of using Finite Element Analysis for design.

ACKNOWLEDGMENTS

The work presented here has been sponsored in part by research projects from the European community (ESPRIT project APACHIP). Thanks are due to G.DEHAINE, X.SAINT-MARTIN and S.BOUDOUX for their fruitful discussions.

REFERENCES

[1] SWANSON ANALYSYS SYSTEM Inc "Ansys User's Manual".
[2] General Electric: Data book Heat Transfert and Fluid Flow section 503.2 January 1971.

Design of an optimal heat-sink geometry for forced convection air cooling of Multi-Chip Modules

B. BOESMANS[1], F. CHRISTIAENS[2], J. BERGHMANS[1], E. BEYNE[2]

[1]*Department of Mechanical Engineering, Katholieke Universiteit Leuven, Celestijnenlaan 300A, B-3001 Heverlee, BELGIUM*
[2]*Interuniversity Micro Electronics Center (IMEC), Kapeldreef 75, B-3001 Heverlee, BELGIUM*

Abstract. An assessment is made of the limits to the cooling capability of forced convection air cooling of multi-chip modules. Therefore heat-sinks of different geometries are analysed, using a new design criterion which involves both the effects of heat transfer performance and pressure drop. It is shown that, when pressure drop is taken into account, plate fin heat-sinks have a better cooling performance than offset-strip fin or pin fin heat-sinks. The dimensions of a plate-fin heat-sink are optimized using an analytical model. The optimal plate-fin heat-sink has a cooling performance comparable to what is normally achieved by integral water cooling. To confirm the thermal performance of the optimal plate fin heat-sink, measurements are carried out in a low speed windtunnel.

Nomenclature

A	channel cross-section [m^2]
B	PCB width [m]
B_s	module width [m]
c	heat capacity of air [J/kgK]
d,l,s,x	fin structure dimensions [m]
D	distance between adjacent PCB's [m]
D_h	hydraulic diameter of the channel = 2D [m]
f	Darcy-Weissbach friction factor
h	convective heat transfer coefficient [W/m^{2}K]
h_s	convective heat transfer coefficient based on fin base surface [W/m^{2}K]
HTA	heat transfer augmentation
j	Colburn-factor
k	thermal conductivity of air [W/mK]
k_{fin}	thermal conductivity of the fin [W/mK]
L	channel length [m]
L_s	module length [m]
N	number of fins per heat-sink
Nu	Nusselt-number
P	channel perimeter [m]
P_h	heated part of channel perimeter = B [m]
Pr	Prandtl-number
Δp	pressure drop over the channel [Pa]
Δp_s	pressure drop over one module [Pa]
$\dot{q}$	heat flux [W/m^2]
Q_s	power dissipated by one module [W]
Re	Reynolds-number
$(k/s)_i$	internal heat transfer coefficient [W/m^{2}K]
$T(z)$	(mixing-cup) air temperature in channel [K]
$T_w(z)$	temperature at channel wall [K]
$T_j(z)$	junction temperature [K]
u_{max}	highest air velocity in fin structure [m/s]
$\dot{V}$	volumetric air flow rate [m^3/s]
X_{HTA}	HTA performance figure defined in text

Greek symbols

η_{fin}	fin efficiency
θ	dimensionless temperature defined in text
ρ	density of air [kg/m^3]
ξ_h	channel aspect-ratio = $P_hL/A = L/D$

Subscripts

h	hydraulic or heated
s	module
i	internal
*	hypothetical configuration
0	reference configuration

C. J. Hoogendoorn et al. (eds.), Thermal Management of Electronic Systems, 267–276.

1. Introduction

Multi-Chip Modules (MCMs) have recently won increasing popularity and importance because of their ability to handle high-speed signals and their high packaging density. On the other hand, the MCM-approach puts high demands on the heat removal from the IC's to the ambient, with heat fluxes attaining several Watts/cm^2. Cooling by forced convection of air will remain for some time the major and preferred cooling scheme for many applications of electronics cooling, e.g. in the telecommunications industries. The advantages of air cooling are well known: besides being the least expensive cooling scheme, air cooling provides an unequaled degree of reliability with minimum maintenance requirements. However as power dissipation per unit surface area increases, forced convection air cooling is only possible if heat transfer augmentation techniques (HTA) are used. Possible heat transfer augmentation techniques for forced convection air cooling are extended surfaces (heat-sinks), turbulence promotors, thermal mixing, etc. These techniques improve on the convective heat transfer, but inevitably at the cost of an increased pressure drop. Therefore both heat transfer and pressure drop should be considered -simultaneously- for the design and the selection of heat transfer augmentation techniques.

The aim of this study is to assess the practical limits to the cooling capability of forced circulation air cooling, applied to multi-chip modules with extended surfaces. Obviously some criterion, involving both heat transfer and pressure drop, is needed to quantify cooling capability, and to compare different possible solutions to the cooling problem.

In the paper, an original criterion (and a figure-of-merit) is developed for the optimal design and the selection of heat transfer augmentation techniques for forced circulation air cooling of constant power input devices. It is essentially based on a minimization of the highest junction temperature in the system at a constant power input. Contrary to other existing criteria however, this minimization is performed under 'fixed fan'-conditions. From a practical point of view it is indeed meaningful to compare different HTA cooling schemes while operating with the same fan, or to optimize the dimensions of a particular HTA for a given fan. Furthermore it is shown that, under some -clearly defined- conditions, the effect of increased pressure drop plays only a minor role on the HTA performance. For these conditions, a simple design based on a constant flow rate assumption, is justified.

The design criterion is then applied to the design of (dense) fin structures, intended for the cooling of multi-chip modules. Different heat sink geometries (pin fin array, plate fins and staggered strip fins) are considered. In a first stage, the design is based on an analytical model of the heat transfer process. Extensive use is made of semi-empirical heat transfer correlations. These models allow a geometrical optimization and a comparison of the different fin structures. Without considering pressure drop, pin fins have the best cooling performance. With the new design criterion however, it is found that simple plate fin heat-sinks are much better.

In order to verify the results of the optimization study, measurements are carried out on the optimal plate fin heat-sink. The measured thermal resistance and pressure drop compare well with the results of the analytical model.

2. Design criterion

The optimal design and the selection of heat transfer augmentation techniques requires their cooling performance to be quantified. The mathematical expression of cooling performance should involve both heat transfer and pressure drop considerations.

In any case, cooling performance can be judged from the value of the junction temperature at the worst possible location in the system, at a given power dissipation rate. The optimal HTA design is thus equivalent to the minimization of the highest junction temperature in the system. This can easily be done if an additional restriction of constant pressure drop or constant flow rate is used. Neither of these restrictions however is a priori acceptable. Any HTA increases both the convective heat transfer coefficient and the friction factor at a fixed flow rate. In a real system however, flow rate is not fixed. With increasing friction, a real fan-unit produces lower flow rates. Therefore cooling performance deteriorates, partly because the convective heat transfer coefficient itself decreases, and partly because at lower flow rates the average air temperature in the channel increases.

Here it is proposed to minimize the highest junction temperature for a given fan (='fixed fan'-conditions), and for given overall dimensions of the flow channel in which the electronic devices are installed. Thus different HTA solutions are compared, all of them using the same fan.

In order to develop a mathematical expression for the cooling performance, a simple model is now set up. The real cooling problem is modeled as a heated flow channel with a fixed and steady heat flux $\dot{q}$ (or several of these channels in parallel). Part of the channel wall contains packages with heat generating electronic devices (see fig. 3).

For convenience, three different configurations (see fig. 1) are distinguished :

1. a reference configuration (identified with index 0) which can be freely chosen. Here plain convectively cooled packages without any HTA and with a fan-unit operating at its standard working point.($\dot{V}_0$, Δp_0) are taken as the reference.
2. the actual configuration (no index) with some form of HTA implemented. Due to the increased friction, both pressure drop Δp and flow rate $\dot{V}$ will differ from their reference values.
3. a hypothetical configuration (index $*$), identical to the actual configuration with HTA, but operating at the reference flow rate $\dot{V}_0$ instead of the actual flow rate.

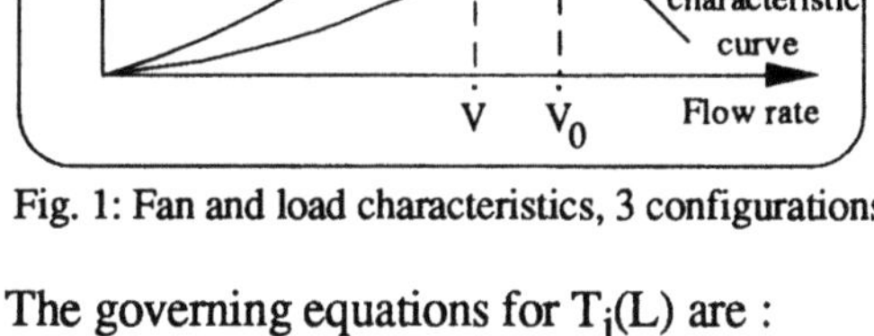

Fig. 1: Fan and load characteristics, 3 configurations.

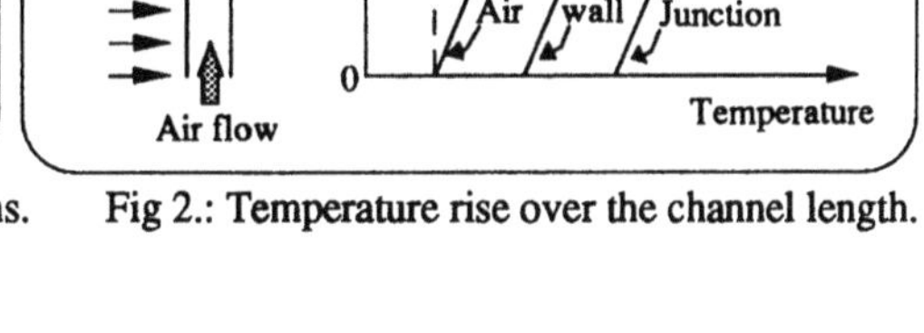

Fig 2.: Temperature rise over the channel length.

The governing equations for $T_j(L)$ are :

$$\dot{q} = (k/s)_i\,(T_j(L) - T_w(L)) \qquad (1)$$

$$\dot{q} = h\,(T_w(L) - T(L)) \qquad (2)$$

$$\dot{q} = \frac{\rho c \dot{V}}{P_h L}\,(T(L) - T(0)) \qquad (3)$$

$$\Delta p = f\,\frac{LP}{8A^3}\,\rho\,\dot{V}^2 \qquad (4)$$

$$\Delta p = F(\dot{V}) \qquad (5)$$

The overall thermal resistance between junction and air is split in a fixed internal (conductive) part $(s/k)_i$ and an external (convective) part $1/h$, as represented in the first and second equation. This external part incorporates the effects of HTA, such as extended

surfaces, and obviously depends on the flow rate in the channel. The third equation is the energy balance applied to the flow channel. The last two equations represent the channel load curve and the fan characteristic curve. The heat transfer coefficient h is based on the heated wall of the channel surface area. The friction factor f is based on the total channel area. Bear in mind that both h and f depend on $\dot{V}$, which is not known a priori.
From the first three equations:

$$T_j(L) = T(0) + \dot{q}\left\{\left(\frac{s}{k}\right)_i + \frac{1}{h} + \frac{P_h L}{\rho c \dot{V}}\right\} \tag{6}$$

The following dimensionless numbers are defined:

$$\theta = \frac{k}{D_h}\,\frac{T_j(L) - T(0)}{\dot{q}} \tag{7}$$

$$j = \frac{Nu}{Re\, Pr^{1/3}} \quad \text{(Colburn factor)} \tag{8}$$

The inverse of the dimensionless temperature θ indicates how difficult it is to attain the desired heat transfer performance, as specified through $T_j(L)$ and $\dot{q}$. The load curve (eq. 4) and the fan characteristic yield under fairly general conditions a result close to

$$\frac{\dot{V}}{\dot{V}_0} = \left(\frac{f_*}{f_0}\right)^{-n} \tag{9}$$

where the exponent n depends on the slope of the fan characteristic and the flow type (laminar or turbulent). Then it is possible to rewrite eq. 7 as follows (see ref. [6])

$$\theta = \left(\frac{s}{k}\right)_i \frac{k}{D_h} + \frac{1}{X_{HTA}\, Nu_0} \tag{10}$$

with X_{HTA} defined as

$$X_{HTA} = \left(\frac{j_*}{j_0}\right)\left(\frac{f_*}{f_0}\right)^{-n(1-b_2)}\left[\frac{1}{1 + j_*\, \xi_h\, Pr^{-2/3}\left(\frac{f_*}{f_0}\right)^{nb_2}}\right] \tag{11}$$

and $b_2 =$ 0.2 for turbulent flow
0.33 for laminar flow

In this expression all parameters are to be evaluated for the hypothetical (*) and the reference (0) configuration. The unknown flow rate $\dot{V}$ is eliminated from the problem.

The goal of the HTA development is the minimization of θ. Both Nu_0 and the first term of the RHS of eq.(10) have fixed values. Therefore X_{HTA} is a direct performance measure for the HTA technique. The design and evaluation of HTA techniques can be performed by maximizing X_{HTA}. But no matter how large X_{HTA} is made, it is of course impossible to lower θ below the 'internal resistance only'-limit :

$$\theta = \left(\frac{s}{k}\right)_i \frac{k}{D_h}$$

For a constant flow rate situation, $\dot{V} = \dot{V}_0$, whatever the pressure drop. Therefore n=0 and X_{HTA} reduces to

$$X_{HTA} = \left(\frac{j_*}{j_0}\right)\left[\frac{1}{1 + j_*\xi_h Pr^{-2/3}}\right]$$

The second factor represents the influence of the air temperature T(L) on the cooling performance. The closer $T_j(L)$ approaches T(L), the harder it gets to decrease $T_j(L)$ further. At this point it becomes possible to draw some important conclusions. For typical propeller-type fans, it is found that $n \approx 0.3$. Assuming turbulent flow in a short channel ($\xi_h \approx 0$), the HTA performance measure is

$$X_{HTA} = \left(\frac{j_*}{j_0}\right)\left(\frac{f_*}{f_0}\right)^{-0.24}$$

Even if both j_* and f_* are varying similarly, as can be expected from the Reynolds-Colburn analogy, a better cooling performance can be achieved by the application of the HTA technique. The negative effect of the increased pressure drop does not compensate for the positive effect of the better convective heat transfer. Furthermore, under the same conditions, a good approxi-mation of X_{HTA} is obtained by setting n=0, and by using simply

$$X_{HTA} = \left(\frac{j_*}{j_0}\right)$$

as an optimization criterion. This corresponds of course to a design performed at fixed flow rate conditions. It is important to note that this simplified approach is only valid for short channels ($\xi_h \rightarrow 0$). For many applications electronic packages are mounted on vertical PCB's in large cabinet-type racks (as shown in fig. 3). Then clearly the full formula for X_{HTA} should be used.

3. Analysis of extended surfaces

Using the HTA performance measure X_{HTA}, developed in the preceding paragraph, an assessment can be made of the limits to the cooling capability of forced convection air cooling for a specific application. This application consists of an assembly of several vertical flow channels, formed by parallel PCB's (see fig.3). The PCB's contain regularly distributed 2"x2" heat dissipating multi-chip modules. It is supposed that all modules are equal and that all of them have to be equipped with (the same) fin structures to enhance cooling performance. The flow channel has a 50 cm by 5 cm cross-section, and different channel lengths are considered. At the bottom of the flow channel a fan-unit supplies 5 m/s of air over the full cross-section in the reference configuration. It is always assumed

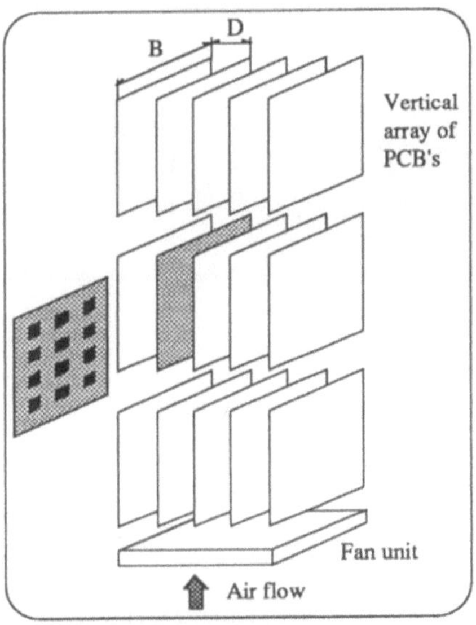

Fig. 3: Vertical flow channels formed by parallel PCB's.

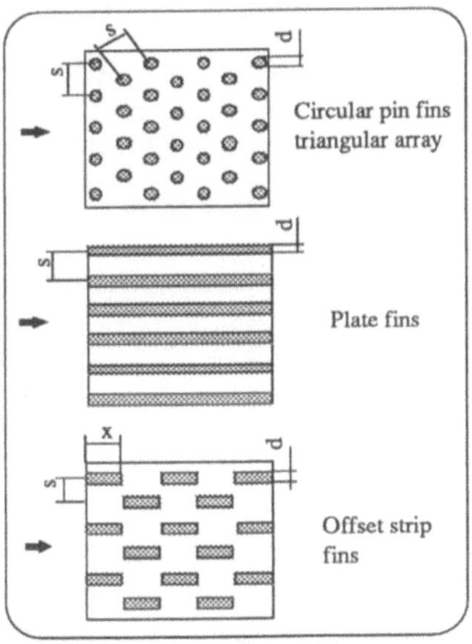

Fig. 4: Three different heat-sink geometries.

that whenever the channel cross-section is not fully used by the fin structures, an appropriate flow blockage is provided to ensure sufficient air flow along the fins.

As one possible kind of HTA implementation, different fin structures (=heat-sinks) are investigated. The fin structures are divided in the following categories: pin fin arrays, plate fins and staggered strip fins (see fig.4).

First of all, for every category the optimal dimensions are determined using an analytical model. The best fin structures in every category are then compared and a most appropriate category is chosen. This optimized structure represents the practical limit to what can be achieved in forced convection air cooling, by using current HTA (heat-sink) technology.

4. Analytical model

For every heat-sink category a fairly simple analytical model is developed for the calculation of X_{HTA} for given fin structure dimensions. The evaluation of X_{HTA} requires the evaluation of j_* and f_*, and these quantities are based on channel dimensions. Some careful but straightforward calculations show that they can be related to the corresponding parameters h_s and Δp_s, based on module dimensions.

A. Zukauskas (ref. [5]) **Pin fin staggered array** Jacob (ref. [1])	$Nu = 0.36\,Pr^{0.36}\,Re^{0.6}$ $h_{fin} = \frac{Nu\,k}{d}$ $\Delta p_s = \frac{2L_s}{\sqrt{3}s}\frac{\rho u_{max}^2}{2}\left(1 + \frac{0.472}{\left(\frac{s}{d} - 1\right)^{1.08}}\right)Re^{-0.16}$
Kakaç (ref. [2]) **Plate fin**	$x = \frac{L_s}{2\,s\,Re\,Pr}$ $Nu = 7.55 + \frac{0.024x^{-1.14}}{1 + 0.0358Pr^{0.17}x^{-0.64}}$ $h_{fin} = \frac{Nu\,k}{2s}$ $\Delta p_s = 4\,Pr\,x\frac{\rho u_{max}^2}{2}\left(\frac{3.44}{\sqrt{x\,Pr}} + \frac{24 + \frac{0.674}{4\,x\,Pr} - \frac{3.44}{\sqrt{x\,Pr}}}{1 + \frac{0.000029}{(x\,Pr)^2}}\right)$
Offset strip fins Manglik and Bergles (ref. [3])	$D_h = \frac{4(s-d)\,l\,x}{2(\,(s-d)\,x + l\,x + d\,l\,) + d(s-d)}$ $\alpha = \frac{s-d}{l} \quad \delta = \frac{d}{x} \quad \gamma = \frac{d}{s-d}$ $Nu = 0.6522\,Re^{0.4597}\,Pr^{0.33}\,\alpha^{-0.1541}\,\delta^{0.1499}\,\gamma^{-0.0678}\left(1 + 5.269\,10^{-5}\,Re^{1.340}\,\alpha^{0.504}\,\delta^{0.456}\,\gamma^{-1.055}\right)^{0.1}$ $h_{fin} = \frac{Nu\,k}{D_h}$ $f = 9.6243\,Re^{-0.7422}\,\alpha^{-0.1856}\,\delta^{0.3053}\,\gamma^{-0.2659}\,(1 + 7.669\,10^{-8}\,Re^{4.429}\,\alpha^{0.920}\,\delta^{3.767}\,\gamma^{0.236})^{0.1}$ $\Delta p_s = 4\,f\frac{L_s}{D_h}\frac{\rho u_{max}^2}{2}$

Table 1: Semi-empirical correlations used in the analytical models

Following common approximations are introduced to model the heat transfer from a fin structure :

- all fins have constant cross-section
- parallel fins have the same temperature ; i.e. side effects are neglected
- the heat transfer coefficient is uniform over the fin surface
- heat flow in the fins is assumed to be one-dimensional
- heat transfer at the fin tip is neglected
- natural convection and radiation are neglected
- the flow profile before the fin structure is assumed uniform

Any of these restrictions can be relaxed if required.

The convective heat transfer coefficient for an entire module is calculated as :

$$h_s = N \frac{A_{fin}}{B_s L_s} \eta_{fin} h_{fin}$$

with a fin-efficiency :

$$\eta_{fin} = \frac{\tanh(ml)}{ml} \qquad m^2 = \frac{h_{fin} P_{fin}}{k_{fin} S_{fin}}$$

Both h_{fin} and Δp_s are derived from semi-empirical correlations. Here only a brief overview of these correlations is given in table 1. An extensive discussion can be found in ref. [4]. Using the preceding equations it is possible to determine the combination of fin structure dimensions that maximizes X_{HTA}. This way, the 'best' fin structure design for every category, is obtained.

5. Results and discussion

The optimization of the fin dimensions is performed for two different channel lengths: a short channel of 0.3 m length, and a long channel of 1.8 m length, typically corresponding to a full-size rack. The calculations are performed with n=0.3, which is a typical value for propeller type fan-units, commonly used in the electronics industry. For purposes of comparison the optimization is also carried out without taking pressure drop into consideration (by setting n=0).

		pin fins	plate fins	strip fins
n = 0.3, L = 0.3 m	d	3 mm	1 mm	1 mm
	s/d	2	2.5	2.5
	l	50 mm	50 mm	50 mm
	x			20 mm
	X_{HTA}	5.4289	8.003	6.257
	Δp	343 Pa	140 Pa	233 Pa
n = 0.3, L = 1.8 m	d	4 mm	1 mm	1 mm
	s/d	3	7	7
	l	50 mm	50 mm	50 mm
	x			20 mm
	X_{HTA}	1.6218	2.606	1.9529
	Δp	924 Pa	309 Pa	437 Pa
n = 0, L = 0.3 m	d	1 mm	1 mm	1 mm
	s/d	1.2	1.2	1.2
	l	5 mm	50 mm	10 mm
	x			2 mm
	X_{HTA}	27.1286	22.42	25.85
	Δp	16638000 Pa	9057.8 Pa	9324400 Pa

Table 2: Optimization results.

The optimization results are presented in table 2. Without taking pressure drop into consideration (n=0), pin fins are found to have the best cooling performance. But an optimization only based on heat transfer invariably leads to extremely close fin packings and very high values of the friction factor f. From a practical point of view the pressure drop of these fin structures is unacceptable for ordinary fan-units.

The best heat transfer performance for plate fin structures is achieved by a dense structure, with d = 1 mm and s/d = 2.5. The somewhat higher pressure drop of a dense fin structure does not counterbalance its much better heat transfer. Obviously the use of dense fin structures requires special measures to ensure that the air flow passes through the fins. It is considered that this is quite simply possible.

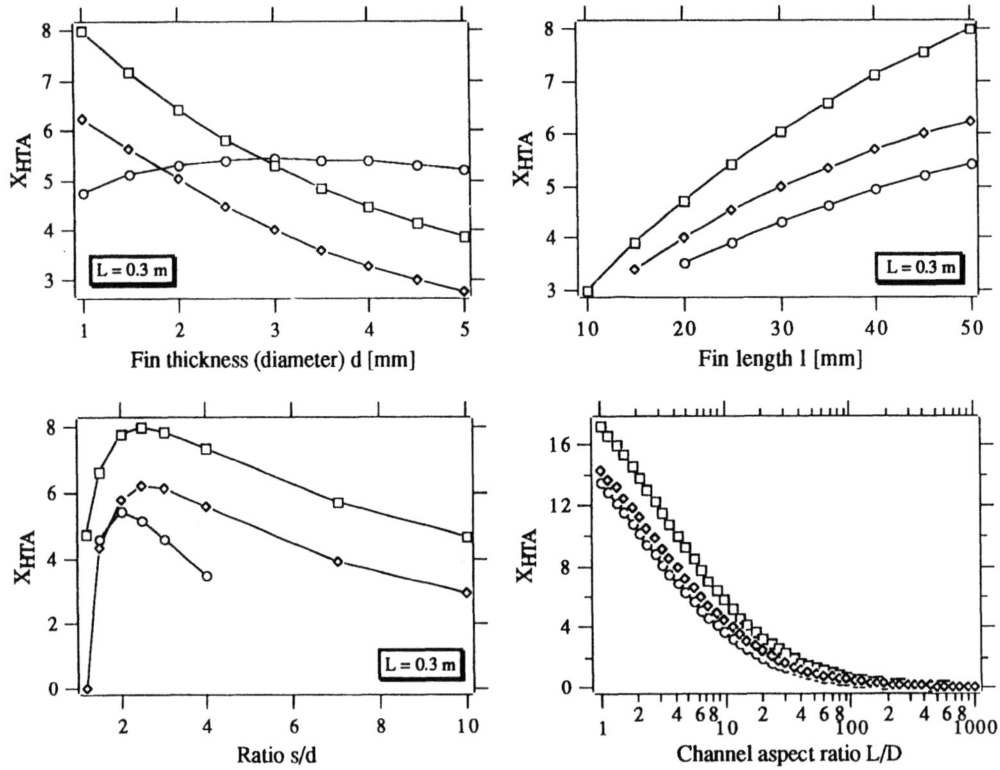

Fig. 5: Variation of X_{HTA} around the optimal geometry, for L=0.3.
❑ = Plate fins, O = Pin fins, ◊ = Strip fins

The optimal plate fin structure has a specific base-to-air thermal resistance of 6 °C.cm²/W in the specified configuration with L=0.3 m. This fin structure thus should allow a multi-chip module with an internal resistance rated for instance at 4 °C.cm²/W to dissipate up to 4 W/cm² (at a junction temperature of 80 °C and an ambient temperature of 40 °C). This performance is comparable to what can be achieved by integral liquid cooling techniques. Therefore it is concluded that forced convection air cooling using very dense plate fin structures is a viable cooling scheme for MCMs. The internal thermal design of the MCM-D technology should ideally yield an internal thermal resistance in the range of 2 to 4 °C.cm²/W. Higher internal resistances would deteriorate the cooling performance; a lower internal resistance would have very little effect when forced convection air cooling is used. It is judged that the quoted internal resistance is achievable, though it probably requires special design measures such as thermal vias.

For every fin structure category, the variation of its X_{HTA} around the optimal geometry has been investigated. The results are shown in fig. 5 for L=0.3 m. The results for L=1.8 m are similar. In all cases the fin length l is best taken equal to the available spacing between adjacent flow channel walls. For both the fin thickness d and the packing ratio s/d an optimum exists, although for some structures the optimal value of d may be less than the minimal practical dimension of $d \approx 1$ mm .

It is also very clear (see fig. 5) that X_{HTA} is much smaller in configurations with a long flow channel (high channel aspect ratio) than in short channels. This is due to the effect of the energy equation in the set of equations governing the highest junction temperature in the system. Finally it must be made clear that the optimal geometry of a fin structure depends on the configuration in which the fin structures are to be used. The results obtained in this paper represent some common configurations, but for any particular configuration the optimization procedure should preferably be repeated.

6. Experimental verification

In order to characterize the optimal heat-sink geometry in well defined circumstances, measurements were carried out in a low-speed windtunnel. As shown in fig. 6, an open blow-type windtunnel with test section of 35cm×28cm is used in the experiments. The windtunnel is equipped with a centrifugal fan which allows air velocities in the range of 0.8 to 5.5 m/s. The heat-sink (5cm×5cm×5cm) is placed in a suspended test channel. Around this test channel a wire netting is installed in order to get a flow resistance comparable to that of the heat-sink. At the base of the fin structure a flat resistance heater (50mm×50mm×2.5mm) is attached to dissipate the required power. The DC supply can be regulated by a voltage regulator to maintain a steady power supply of 30 W. A thermally conductive paste, introduced between the heater and the base, decreases the contact resistance. In order to prevent heat losses to the surroundings, thermal insulation (k=0.040 W/mK) is attached to the walls of the test channel and the backside of the heater.

The aim of the experiments is to characterize the thermal resistance and pressure drop across the heat-sink, as a function of the air velocity. The pressure drop is recorded using a differential pressure tube. Two static pressure taps are located at the upper surface of the test channel, one in front of the heat-sink and the other behind the heat-sink. A Data Acquisition Unit registers temperatures and total dissipated power. The thermal resistance is evaluated from the air inlet temperature, the average fin temperature, the total dissipated power and the heat losses to the surroundings. These are estimated using a linear conduction model and two temperature measurements. The flow velocity and the turbulence intensity is measured with a Laser Doppler Velocimeter.

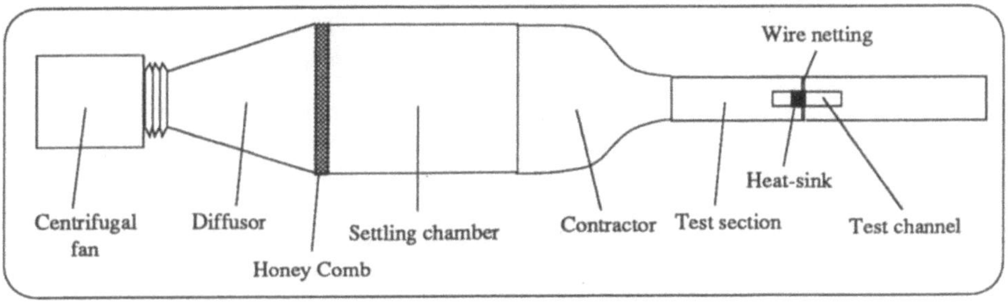

Fig. 6: Experimental measurement set-up.

Fig. 7 compares the measured thermal resistance and pressure drop with the analytically calculated values. As far as the flow resistance is concerned there is a good correspondence between measurements and theoretical predictions. For the thermal resistance, the theoretical predictions underestimate the measured values in the low velocity region (0-2m/s); at higher velocities there is a better agreement.

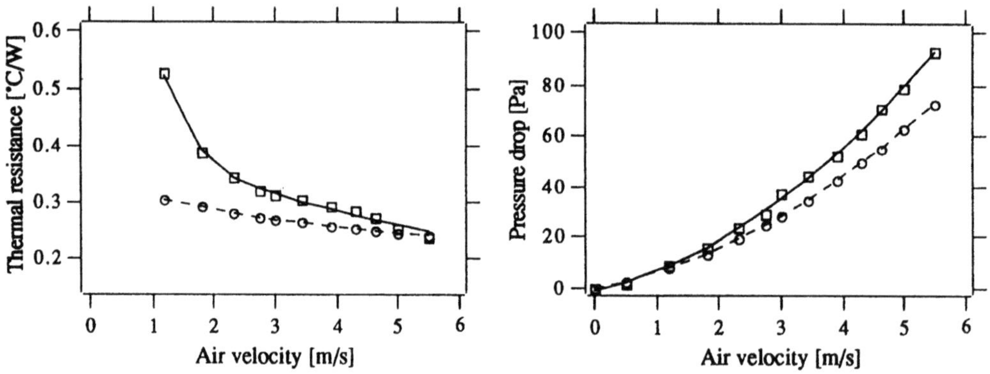

Fig. 7: Thermal resistance and pressure drop of the optimal plate fin heat-sink. (d=1 mm, s/d=2.5, l=50 mm, material = aluminum) ○ = analytical model (Kakaç), □ = measurement

7. Conclusion

A new design criterion for heat transfer augmentation in electronics cooling has been developed. It involves both the effects of heat transfer performance and pressure drop. Using this criterion different heat-sink geometries have been compared. It is found that the 'best' design of extended surfaces for forced circulation air cooling of electronic devices is a very dense plate fin structure.

This fin structure should allow a multi-chip module to dissipate up to 4 W/cm^2 (at a junction temperature of 80 °C and an ambient temperature of 40 °C). This cooling performance is comparable to what can be achieved by integral liquid cooling techniques. Therefore it is concluded that forced convection air cooling using very dense plate fin structures is a viable cooling scheme for MCMs.

References

[1] GUYER E.C., *Handbook of Applied Thermal Design*, Mc-Graw Hill, NY, 1989.

[2] KAKAC S., SHAH R.K., AUNG W., *Handbook of single-phase convective heat transfer*, Wiley, NY, 1987.

[3] MANGLIK R.M., BERGLES A.E., "The thermal-hydraulic design of the rectangular offset-strip-fin heat exchanger", in : Shah R.K., Kraus A.D., Metzger D. (eds.), *Compact Heat Exchangers*, Hemisphere, NY, 1990.

[4] CLAES D., VAN UYTVEN G., *Verbetering van de warmteoverdracht aan multichipmodules m.b.v. direkte luchtkoeling*, thesis K.U.Leuven, TMEC/LWR/1991-1992/EE-01, 1992.

[5] ZUKAUSKAS A., "Heat transfer from tubes in cross-flow", *Advances in Heat Transfer*, vol.8, Academic Press, NY, 1972.

[6] BOESMANS B., CHRISTIAENS F., *Development of an optimization criterion for heat transfer augmentation techniques*, Internal Report K.U. Leuven, 1992.

Thermal characterization and design for high performance multichip module type laminate

E. LAUWERS[1], A. ACKAERT[1], K. ALLAERT[1]
W. TEMMERMAN[2], W. NELEMANS[2], T. GOOSSENS[2]
[1]ALCATEL BELL Research Center, Antwerp, Belgium
[2]ALCATEL BELL Switching Hardware, Antwerp, Belgium

Abstract. Some of the main problems that arise during the design of high end electronic systems are related to transmission speed, interconnection density and power dissipation. Multi-chip modules offer a possible solution to reduce the density problems and they have advantages regarding electrical performance. The thermal management of MCM's however remains a problem area due to the increased packaging density.
In this paper, alternative solutions are presented to improve the thermal performance of the MCM-L technology, based on 1) internal heat spreaders with thermal via's for improved through-substrate conduction and 2) heat spreaders attached to the laminate with thermal wells to allow the connection of the dice directly to the heat spreader. The heat transfer to ambient is additionally improved by the use of heat sinks. The thermal performance of the MCM-L technology with attached heat spreader is further evaluated based on experimental data from measurements on different test modules and on the use of conjugate heat transfer simulations.

1 Introduction

For new applications within the telecommunication industry like B-ISDN, equipment designers are faced with different challenges. The transmission speed, even at board level, is increased upto 600 Mbit/s requiring accurate design of the electrical interconnections (controlled impedance, line termination, design rules for crosstalk). The complexity of the printed circuit board (PCB) is increased to more than 10 layers due to the component placement density and the interconnection requirements. The power dissipation per PCB is increased to more than 100 W, the dissipation per chip to more than 6 W. Standard solutions however, based on single chip packages, multilayer or more advanced PCB's and forced air cooling at rack level, are at the limits of their possibilities.
The need for more functionality per PCB and an ever increasing power dissipation per component, is a driving factor for the introduction of more advanced packaging methods like multichip modules. The multichip module concept has a number of advantages : the packaging density is increased by chip and wire ASIC assembly and reduction of the number of discrete components (for termination and decoupling) and the electrical performance is improved by the reduction in chip to chip distance and the minimization of the inductive and capacitive discontinuities between the chips. The rise in power dissipation density through the increase of the packaging density and the increase of dissipation per component however, must be compensated by an improvement of the

C. J. Hoogendoorn et al. (eds.), Thermal Management of Electronic Systems, 277–286.

thermal path to ambient. This paper considers the use of multichip module type laminate (MCM-L) technology as a solution for these thermal challenges.

2 Application environment

The application of multichip modules type laminate is investigated for a printed circuit board of the B-ISDN equipment [1]. The present PCB with single chip packages has a total power dissipation of about 100W (with a power density of 15W/dm²). The dissipation per component is maximum 6.5W. The power density per component is around 3W/cm², with worst case 26.5W/cm² (for a critical component with 6.5W for 5.0x4.9mm²). The functionality of the printed circuit board allows a redistribution of the different functions over a motherboard and two identical multichip modules. Each module contains five ASICs with associated memory chips and discrete components. The total power dissipation of each module is around 26W and the dimensions of the module are estimated at 130x60 mm².

The assembly of the motherboard and two modules (with a total dissipation of about 100W) is mounted in a rack configuration which is cooled by forced air cooling with a maximum air speed of 3 m/s. The normal operation junction temperature (Tj) of the ASIC's is 85°C for an ambient temperature (Ta) of 23°C.

3 MCM-L technology

3.1 General

The MCM-L technology is based on standard PCB technology. The multilayer structure of dielectric material and copper interconnection layers provides an acceptable interconnection density with line widths and spacings of 100 µm for the printed electrical connections. For the described application, two signal layers situated between two power planes (reference for impedance control) provide sufficient routing capability.

3.2 MCM-L technology for increased heat dissipation

For the considered MCM-L modules, the ASICs are mounted on one side of the multichip module using chip and wire technology. The heat transfer from the ASICs to ambient is mainly based on through-substrate conduction which is improved by the use of integrated heat spreaders with thermal via's (figure 1) or by the use of attached heat spreaders and thermal wells (figure 2).

The presence of thermal via's under the ASICs partly reduces the routing area, the direct connection of the silicon ASICs to the copper heat spreader requires dedicated die attach material to allow the considerable TCE mismatch. The chip and wire assembly is normally covered with glob top material for mechanical protection.

In the rest of the paper, the emphasis is mainly stressed on the multichip module construction with attached heat spreader.

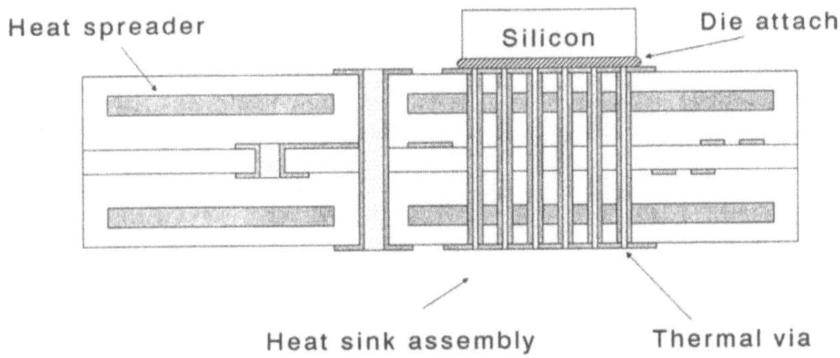

Figure 1 MCM-L technology with integrated heat spreader

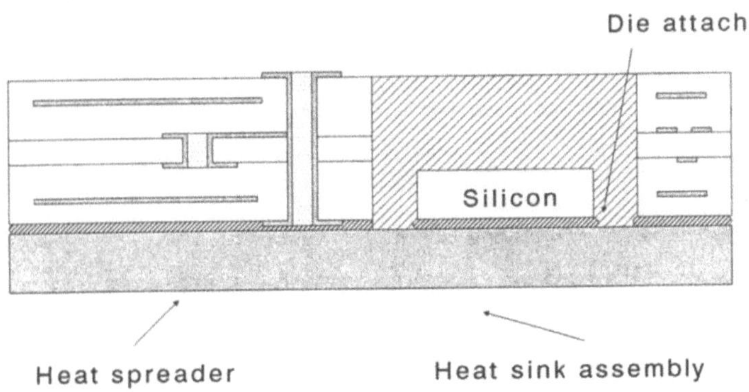

Figure 2 MCM-L technology with attached heat spreader

4 Thermal characterization of the MCM-L technology

4.1 Principle

The implemented method is based on the combined use of 1) experimental data obtained by measurements on thermal test modules in well defined test set-ups and of 2) thermal simulations based on thermal models which are developed for the different module constructions. The thermal models are tuned by comparison of the simulation results with the experimental data.

4.2 Thermal simulations

The thermal simulations are performed using the computational fluid dynamics (CFD)-tool FLOTHERM (registered trademark of Flomerics, UK) which is tailored for electronic systems. The thermal models consider all the specific characteristics of the MCM-L technology (e.g. influence of dissipating area, die attach, heat sink,..). The grid size is defined based on the size of the computational volume and on the details which are taken into account (required accuracy). For simulations at rack level, the thermal models for other elements of the equipment practice (e.g. subrack and rack environment, fan units) are the same as for simulations of classical PCB's [2].

4.3 Test module description

The thermal test module is constructed using an FR4-laminate structure attached to a copper heat spreader to improve the thermal behaviour. The thermal test chips with die size $7x8mm^2$ or $7x4mm^2$ are placed in thermal wells in order to make direct contact with the copper heat spreader (thermal well A & B). The test chip contains two type of diodes : one is used to generate heat dissipation and the other one for temperature measurement. The connection between the silicon dice and the copper is made with thermal conductive die attach material (AMICON 50262-5 or GRACE XE 80025) which compensates the TCE mismatch between the silicon and copper . The electrical connections are realised by

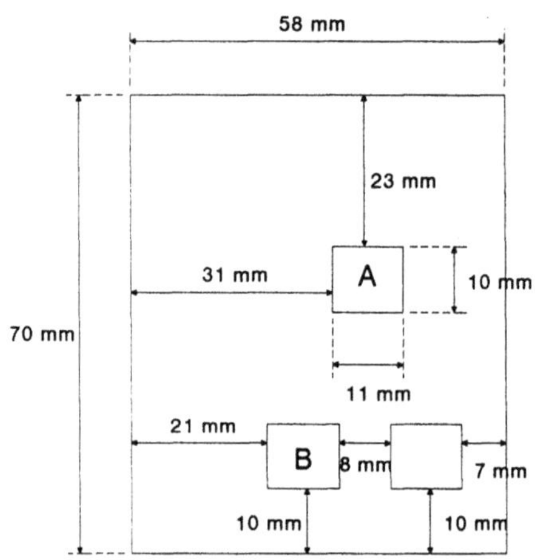

Figure 3 Physical dimensions of the thermal test module

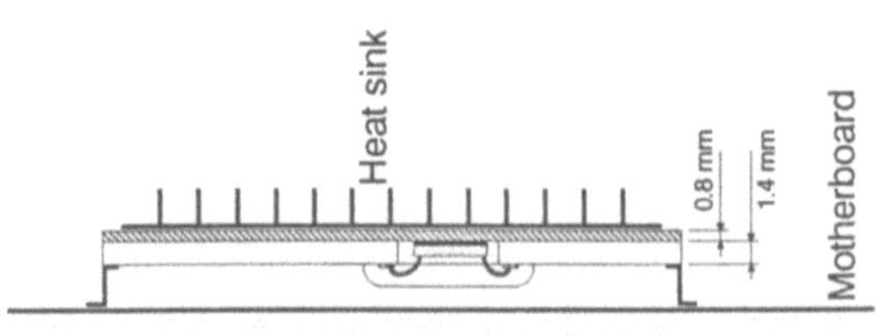

Figure 4 Assembly of the thermal test module to the motherboard

wire bonding (gold wire). The naked dice are not covered with glob top or any other coating, but they are protected with a plastic cover. The physical dimensions and the assembly of the test module to the motherboard are given in figure 3 and 4.

4.4 Test set-up description [3]

For measurements regarding the thermal resistance from junction to case (being the copper heat spreader for the described test module), the test set-up consists of a water cooled aluminum base plate to which the heat spreader of the test module is connected using a thermal conductive grease. This configuration is vertically mounted in a subrack above a flow chamber (test set-up Nr.1).
For measurements regarding the thermal resistance from junction to ambient, the test module with additional heat sink is mounted at a distance from the water cooled aluminum base plate using a plastic intermediate part. This configuration represents a thermal model for the actual PCB. The component side of the module is directed towards the base plate. This configuration (figure 2 for reference) is also vertically mounted in a subrack above a flow chamber (test set-up Nr.2). The air flow at the output of the flow chamber can be controlled with air speed ranging from 0 m/s to 3 m/s.

5 Measurements and simulations

For the MCM-L construction with attached heat spreader, the main characteristics which are important for the thermal behaviour of the module are evaluated with respect to thermal measurements and thermal simulations.

5.1 Material parameters

The thermal conductivity parameter of the different materials used in the MCM-L construction with attached heat sink are collected from literature [4] and datasheets. A good knowledge of these parameters is required in order to create a reliable model for thermal simulations. The thermal conductivity of the copper material, used for the construction of the test modules is measured using the method described in reference [5]. The material parameters, used for the thermal simulations are given in table 1.

Table 1 : Thermal conductivity [W/m.K]	
Silicon	153
Die attach	
AMICON 50262-5	0.9 - 1.0
GRACE XE 80025	0.9 - 1.0
Copper	350
Thermal grease	0.9

5.2 Silicon die

The influence of the silicon die on the thermal resistance to ambient is mainly defined by the power density. Parameters which are therefore important are die size, the dissipating area and the power dissipation per die.

5.3 Die attach

The die attach material has two main functions : the connection of the silicon die to the copper heat spreader with compensation of the TCE mismatch and the provision of a thermal path of the chip to the heat spreader. The thermal performance of the die attach is influenced by the thermal conductivity, the thickness, the presence of voids, delamination and the quality of the contact towards the silicon die and the copper heat spreader which are both influenced by the ability of the die attach to compensate the TCE mismatch.
Table 2 gives the experimental results for the thermal resistance from junction to cold plate based on measurements using test set-up No.1 for modules with thermal wells and modules with thermal via's (diameter 0.25 mm / pitch 0.5 mm / number 360). The modules with thermal via's directly under the dice, have the same dimensions as described in section 4.3 but with two internal heat spreaders of 0.2 mm The variation of the thermal resistance is explained by the difference in the quality of the die attach connection for the distinct test modules as can be remarked in figure 5 which represents a section at die attach level for test module A (die size 7x8mm^2), result obtained by acoustic microscopy.

Figure 5 Section of the die attach level for a thermal test module (die size 7x8mm2 / AMICON 50262-5) obtained by acoustic microscopy (light area : voids and/or delamination)

Obvious, good control of the die attach connection is important, especially for high power dissipation applications. A higher thermal conductivity for the die attach will reduce the influence of the negative effects.

Table 2 : Thermal resistance junction to cold plate [°C/W]						
Die size	Thermal well			Thermal via		
	max	**mean**	min	max	**mean**	min
7x8	1.73	**1.28**	0.95	1.89	**1.70**	1.45
7x4	3.14	**2.69**	2.37	3.40	**3.10**	2.87

5.4 Copper heat spreader

The 'heat spreading' effect of the attached copper plate is influenced by its thickness. The convection to ambient is influenced by the total area of the module. Using a thermal simulation model tuned for the thermal module with external heat spreader (die sizes 7x8mm^2) in test set-up No.2 without additional heat sink, thermal simulations have been performed with variation of the heat spreader thickness and variation of the module area. Figure 6 indicates the influence of the heat spreader thickness on the junction temperature (6W dissipation in each cavity A&B), figure 7 displays the influence of the heat spreader (and module) area on the junction temperature for the same configuration.
Consequently, a reduction of the heat spreader thickness to 0.4 mm will not have a large impact on the cooling performance and an increase of the module dimensions will allow cooling of the module even without additional heat sink.

5.5 Heat sink assembly

The thermal performance of the MCM-L assembly with heat sink is evaluated by experiments for two alternative heat sinks. The measurements are performed using a module mounted on the motherboard like in test set-up Nr.2. Given an input air temperature of 23 °C, an air flow of 1 or 3 m/s, a cold plate temperature of 23 °C and dissipation in both thermal wells A & B, the junction temperature are represented in figure 8 for a massive heat sink with large thermal mass type SK136 of Fisher Elektronik and a heat sink with low thermal mass and large cooling surface Lockhart 5-9- 500-14. For comparison, the junction temperatures for the module without heat sink were 101°C for 1 m/s air speed and 82°C for 3 m/s for 6W dissipation in both dice.
The use of a heat sink considerably improves the thermal performance of the module and as a result will allow the reduction of the dimensions of the module.

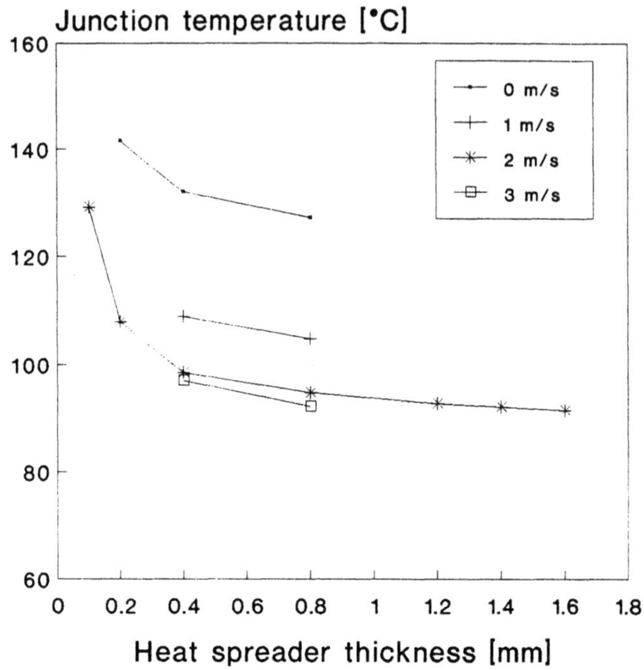

Figure 6 Junction temperature vs. heat spreader thickness based on thermal simulation (die size 7x8mm2)

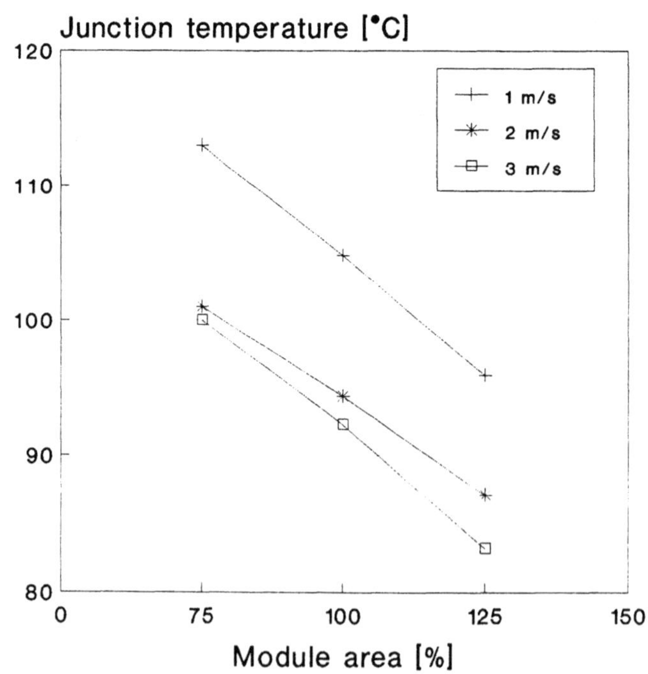

Figure 7 Junction temperature vs. module area (die size 7x8mm2)

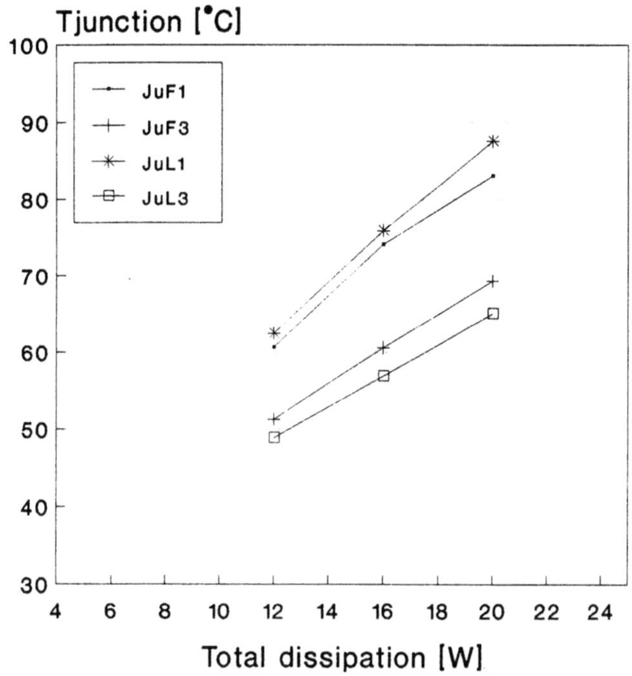

Figure 8 Junction temperature vs. total module dissipation for heat sink Fischer SK136 and Lockhart 5-9-500-14 for the thermal test module with extrenal heat spreader (die sizes 7x8mm2). The junction temperatures are indicated for an air speed of 1m/s and 3m/s.

6 Conclusion

The thermal characterization of the MCM-L module construction with attached heat spreader is performed using a combination of experiments and simulations.

Evaluation of the measurement and simulation results towards the application for B-ISDN, where a total of 26W at module level is specified for module dimensions of 130x60mm2, show that the solution with attached heat spreader is capable of handling the power dissipation with forced air cooling, even without an additional heat sink. When a reduction of the module dimensions is considered, the use of an additional heat sink become mandatory.

Important factors which have to be taken into account during the design of the MCM-L technology with attached heat spreader, are the die attach connection (thickness, thermal conductivity and quality of bonding), the heat spreader thickness, the module dimensions and the heat sink type.

Future work will be focused on the further evaluation of the MCM-L construction with integrated heat spreaders and thermal via's. The integration of the thermal models for the MCM-L module within thermal simulations at rack level remains another point of interest.

7 References

[1] A. Ackaert, L. Vandam, W. Delbare, K. Allaert, 'The use of MCM-L manufacturing and assembly technology for broadband applications',

[2] W. Temmerman, W. Nelemans, T. Goossens, 'Air flow modelling requirements nearby fans', Flotherm User Conference, May 1993

[3] W. Temmerman, W. Nelemans, R. Hendrickx, M. Ryckebusch, L. Rottiers, 'Application of a numerical tool for thermal evaluation of printed boards in communication systems', 1992 Intersociety Conference on Thermal Phenomena, p. 197-203

[4] A.D. Kraus, A. Bar-Cohen, 'Thermal analysis and control of electronic equipment', McGraw-Hill Book Company

[5] G. Demey, S. Demolder, 'A test substrate for thermal analysis of hybrid circuits', Hybrid Circuits, 1987 No.12, p 43-45

The thermal characteristics of a board-mounted 160 lead plastic quad flat pack.

M. DAVIES[1], J. LOHAN[1], J. PUNCH[1] & T. MOORE[2]
[1]*Department of Mechanical Engineering, University of Limerick, Limerick, Ireland.*
[2]*Analog Devices, Raheen Industrial Estate, Limerick, Ireland.*

Abstract. Thermal characterisation forms an integral part of the specification of all electronic components, and standard test methods have been developed to evaluate this feature. However, the applicability of these results is being questioned because a range of values can be obtained, depending on the interpretation of the test method and variation of test conditions. The experimental approach used in the study highlights the effects of some factors such as board copper content and orientation, lead contact with the board, power level, and local air temperature on the thermal characterisation of a board mounted 160 lead Plastic Quad Flat Pack (PQFP) in a natural convection environment. This information should also prove useful to electronic system designers.

1 Introduction

Heat generating electronic components mounted on printed circuit boards (PCBs) transfer heat to the surrounding cooling air by various different physical mechanisms and geometric paths. During manufacture and reliability testing the reverse occurs, where severe thermal environments are used to heat components during reflow soldering and reliability testing. To ensure that maximum operating temperatures are not exceeded during either process, the electronic system designer must know the components' thermal characteristics, the mechanisms of heat transfer to or from the components, and how components thermally interact. Thermal characterisation of electronic packages is governed by MIL and SEMI[1] test specifications, where the internal and external thermal behaviour is measured in terms of junction-to-case, θ_{jc}, and junction-to-ambient, θ_{ja}, thermal resistances respectively.

In recent years there has been growing concern expressed about the accuracy and applicability of both junction-to-case and junction-to-ambient thermal resistance data. While these tests form the basis of semiconductor device thermal characterisation, many authors, such as Dutta[2] and Shanker et al[3], agree that even by following standard methods considerable variation in these values can be obtained. Siegal[4] criticises the general perception that the junction-to-case and junction-to-ambient parameters are constant, and thus independent of power dissipation level or local air temperatures. It is well documented that these values should not be used to evaluate thermal designs over widely varying applications if accurate results are required. Bar Cohen et al[5] suggest that the real value of such thermal resistances should be as a figure of merit against which different components could be compared. This view is shared by the current authors.

C. J. Hoogendoorn et al. (eds.), Thermal Management of Electronic Systems, 287–296.

In order that thermal resistance data of different components can be compared directly, the test procedure should be rigidly defined so as to permit no variation in the test method. The test should bear some resemblance to practical operating conditions experienced by the component, however, these conditions are so varied that it is not possible to model them all with one test. This does not reduce the need for information which will allow the thermal behaviour of a component to be predicted in an application. In this respect, it is important to understand all the factors which effect thermal resistance and this paper was conceived with the intention of providing such information for a board mounted component.

This experimental investigation extended beyond an examination of power level and local air temperature variation effects, as described by previous authors, to include studies of the effect of board copper density, the board orientation in a gravitational field and the number of leads soldered to the PCB on the measured thermal resistance. Examination of these results serves to increase the existing awareness as to the variability of thermal resistance with test method. This fundamental information is not only useful for evaluating component behaviour in a particular application, but it also provides the foundation for the development of more effective solder reflow and temperature cycling thermal profiles, which will help achieve the desired process objectives without compromising component reliability.

2 Experimentation

Experimental thermal characterisation was carried out on a 160 lead plastic quad flat pack. The test methods for still and forced-air junction-to-ambient thermal resistance measurement of integrated circuit (IC) packages are specified by G38-87. This section provides details of the thermal test chip, test board and measurement system, and outlines the procedure used to record thermal resistance data.

2.1 Thermal Test Chip

The thermal test chip, conforming to SEMI G32-86, was contained in a 160 lead plastic quad flat package. Its dimensions are presented in figure 2.1.

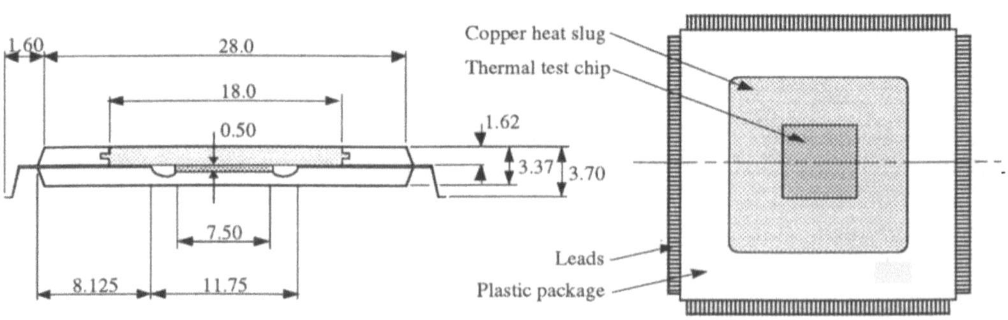

Fig. 2.1 Dimensions of (Square) 160 lead QFP in mm.

This test chip generates a known heat flux at a known junction temperature, permitting the measurement of junction-to-case and junction-to-ambient thermal resistances.

2.2 Thermal Test Board

To investigate the effects of board copper content and distribution, three thermal test boards, suitable for surface mount chip carriers and flat pack components, were used: one with copper on the component side only which conformed to SEMI G42-88, a second with copper on both sides of the board and a third with a minimum of copper beneath the component only. The single- and double-sided boards had the same distribution of copper, which covered 20% of the surface area of a side. This distribution is shown in figure 2.2.

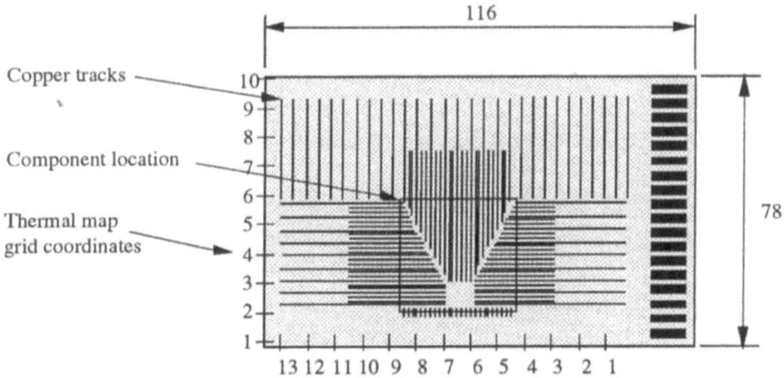

Fig. 2.2 Thermal test board.

2.3 Measurement System

A schematic of the measurement system is shown in figure 2.3 and is discussed in detail in section 2.5.

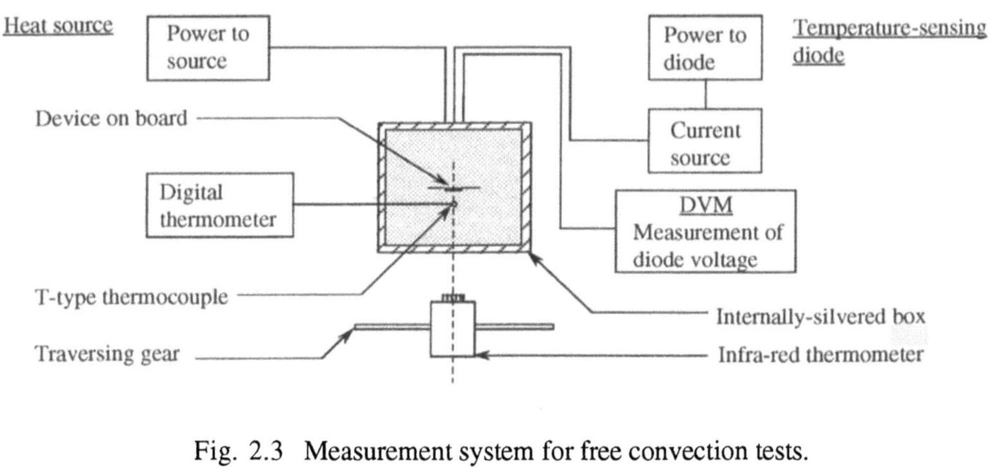

Fig. 2.3 Measurement system for free convection tests.

The board-mounted component was placed in a variable temperature wind tunnel, figure 2.4, to investigate its response to changing air temperature. Tests representing the thermal environments used during solder reflow assembly and reliability testing were carried out.

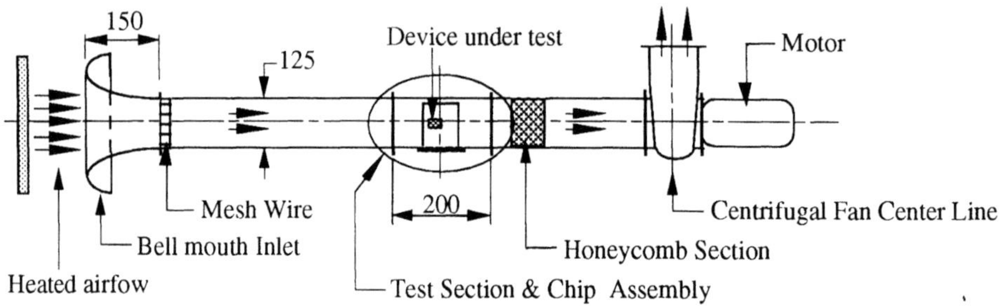

Fig.2.4 Variable air temperature wind tunnel.

2.4 Measurement of Junction-to-Ambient Thermal Resistance

Steady state still-air junction-to-ambient thermal resistance measurement was carried out, according to the procedure outlined in SEMI G38-87, to investigate the effects of board copper content and orientation in a gravitational field, lead contact with the board, power level and local air temperature variation.

a) Effect of board copper content and orientation in a gravitational field. During test each of the three thermal test boards was positioned centrally in an internally-silvered, square box of volume 0.02832 m^3. Five spatial orientations were used for each board: three with the plane of the board vertical, and two with the plane horizontal. These orientations are shown in table 2.1. In each case the heat generated by the component was held constant at 2.5W.

b) Effect of lead contact with the board. To assess the influence of lead contact, for the double-sided board in a vertical orientation, the number of leads soldcred to the board tracking was varied from a minimum of six to the maximum of one-hundred and sixty.

c) Effect of power level. Using the same board and orientation as in (b) above, the power input to the component was varied from 0.5 W to 3.0 W, in steps of 0.5 W, to investigate the effect of power level.

d) Effect of variable local air temperature. The test equipment shown in figure 2.4, was used to evaluate the effect of variable local air temperature, over the same board and orientation as in (b) above, in a forced convection airstream of velocity 4 m/s. In this case the local air temperature was varied between 30 and 105°C at a ramp rate of 15°C per minute. Using thin film thermocouples the time varying temperature response of the copper heat slug, plastic package, lead and board were measured. The effect of both internal and external heating, on the temperature distribution across the assembly, was assessed by combining a "power on" test with the outlined thermal environments by electrically powering the component with 2.5W during the heating cycle. This test replicates the thermal environments that devices are exposed to during reliability testing. A similar test was carried out to replicate solder reflow conditions where the local air temperature was increased to 200°C and the temperature response at the outlined locations again measured.

2.5 Measurement of Temperature Maps

Dissipating 2.5W in a natural convection environment, the surface temperature of each board, in each of the five orientations, was mapped using an infra-red thermometer mounted on traversing gear. To ensure uniform emissivity for the infra-red camera, the board and component were sprayed black. Spot measurements, of 2 mm diameter, were taken at 7.5 mm centres, covering all of the board using a 13 x 10 grid. A fine scan thermal map of the package was obtained using a 12 x 12 grid, covering the outline of the component at 3 mm centres. As part of the systematic study of the factors influencing the junction-to-ambient thermal resistance, temperature distributions on the facing surfaces of the component and the board were also measured using thin-film thermocouples for the same power level.

Table 2.1 Board orientations and views for temperature maps.

Orientation No.	Views for temperature maps	
	View No.	Description
1 / 2 	V_1	Source side
3 / 4 	V_2	Source side
5 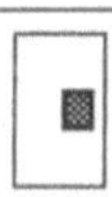	V_3 V_{fs}	Source side Fine scan of component

A square slot, of side 165 mm, was cut in one side of the internally-silvered box to permit measurement of the temperature maps. A silvered card was mounted around the lens of the infra-red camera to seal the box during measurement.

3 Results and Discussion

The results are presented and discussed in terms of the various parameters tested: board orientation and copper content, lead contact, power level and variable air temperature.

3.1 Board Orientation and Copper Content

The thermal maps of the board surfaces are presented as contour plots on figures 3.1 to 3.4. Two important effects are observed. Firstly, comparing figures 3.1 and 3.2, both measured in the horizontal orientation, the board is hottest in the regions of greatest copper density, indicating that heat spreads most effectively through the copper tracking. This is evident when figure 3.1 is compared with figure 2.2 as the isotherms follow the same three lobe pattern as the copper tracking. Secondly, in the vertical orientation, the board is heated by a thermal plume emanating vertically from the component, figures 3.3 and 3.4. A similar temperature wake was found by Moffat et al[6] for forced convection. Essentially, the board acts as a cooling fin, the plume and the copper tracking spreading heat through the fin. Comparing figure 3.4 with figure 3.2, in particular the 30 to 45°C isotherms, the contribution of the thermal plume is clear as a greater proportion of the board is heated.

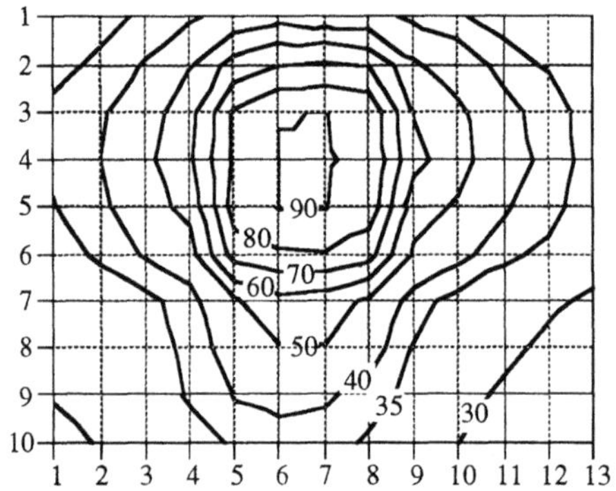

Fig. 3.1 Thermal map of board 1 (Double-sided) in orientation 1, view 1 (Source side).

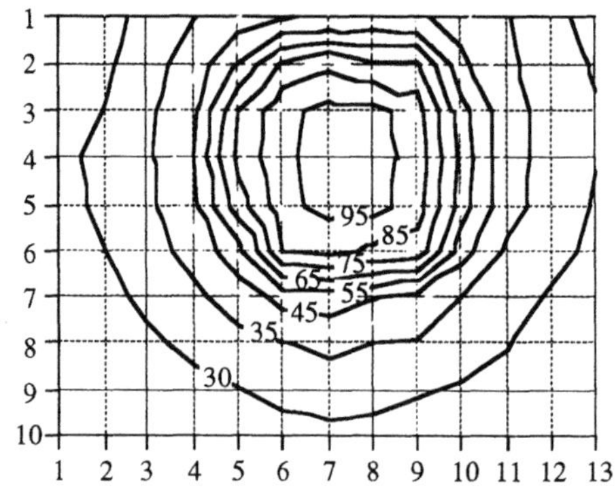

Fig. 3.2 Thermal map of board 3 (Least copper) in orientation 1, view 1 (Source side).

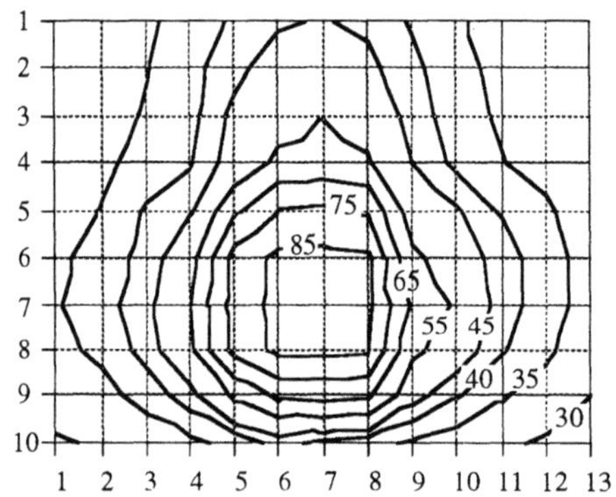

Fig. 3.3 Thermal map of board 1 (Double-sided) in orientation 3, view 2 (Source side).

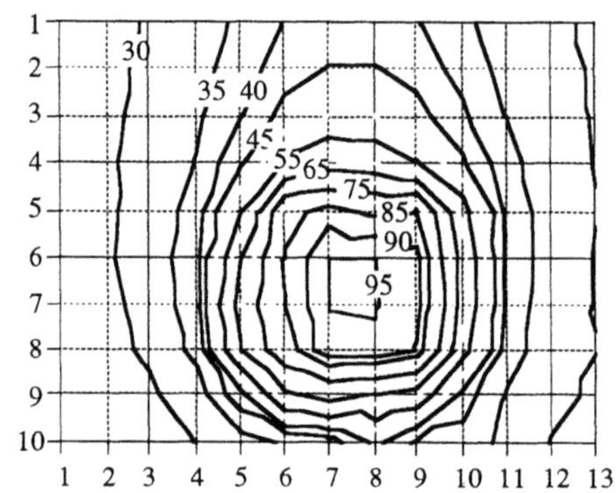

Fig. 3.4 Thermal map of board 3 (Least copper) in orientation 3, view 2 (Source side).

This increased heat dissipation over the horizontal position is reflected by lower package temperatures, with figures 3.1 and 3.3 in particular showing this feature. However, if other components are placed in the wake of this emanating thermal plume their heat dissipation may be restricted. Figures 3.3 and 3.4 show the combined effects of copper content and thermal plume while due to the horizontal orientation only the former effect is active in figures 3.1 and 3.2. Comparison of both sets highlights the contribution of each heat dissipating effect. For each board, little variation occurs in the junction-to-ambient thermal resistance for the three different vertical orientations, figure 3.5. This suggests that the convective heat transfer coefficient is a weak function of orientation and offers some justification for the use of a flat plate coefficient in the energy balance analysis, section 3.4. However, average junction-to-ambient thermal resistance increased by 8% when placed in the horizontal orientation. Overall, average thermal resistance values of the single-sided board are slightly greater than those of the double-sided board, but the minimum copper board yields significantly higher values, indicating that conduction through the copper tracking on the component side of the board is a powerful mechanism of heat dissipation.

Figure 3.6, obtained from the infra-red fine scan of the component surface, indicates how significant differences in the value of junction-to-case thermal resistance can be obtained by the location of the case temperature probe. A 19% difference was recorded across the heat slug alone.

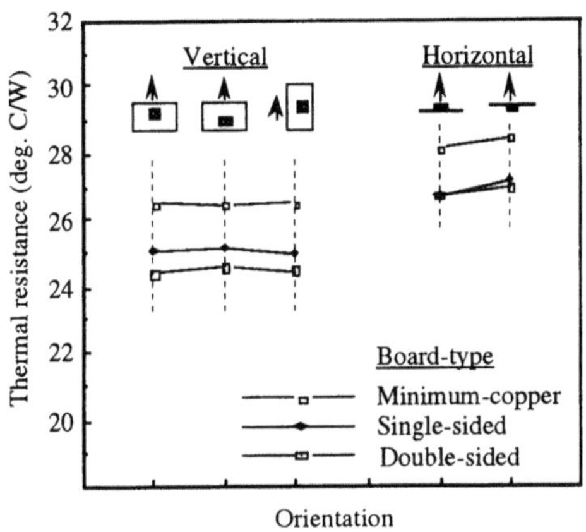

Fig. 3.5 Thermal resistance as a function of board type and board orientation.

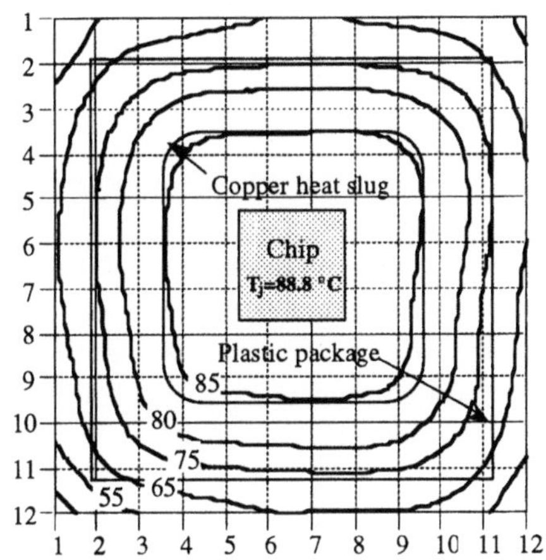

Fig. 3.6 Fine scan thermal map of test chip top surface in orientation 5.

3.2 Lead Solder Contact

The conductive thermal resistance along the leads is inversely proportional to the lead total cross sectional area, and hence to the number of leads soldered to the board. Measured values decrease from 24.2°C/W for 6 leads soldered to 20.6°C/W for all leads soldered, figure 3.7, revealing that conduction into the board through the leads is an important mode of heat dissipation.

3.3 Power Level

Figure 3.8 shows that thermal resistance is inversely proportional to power level, its value dropping by some 10% for the power range considered. The data represents a 29% variation in junction-to-ambient thermal resistance values due to combined power level and solder contact variation. It was also noted that the value of junction-to-case thermal resistance increased by 19% over the same range in power level and increased by another 12% by reducing the number of soldered leads from 120 to 8.

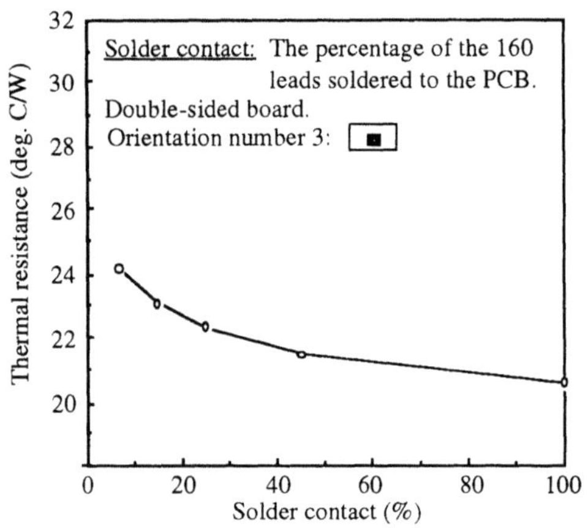

Fig. 3.7 Still-air thermal resistance as a function of solder contact.

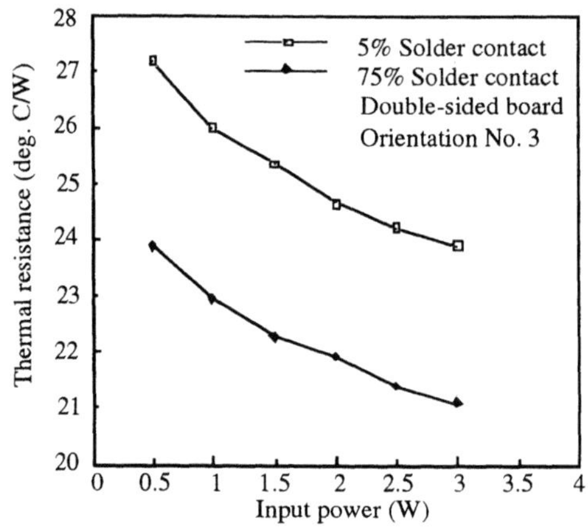

Fig. 3.8 Still-air thermal resistance as a function of input power.

3.4 Energy Balance on Component

This analysis was carried out to investigate the modes of heat dissipation from the component, and the paths through which these modes occur. The component, on the double-sided PCB, was positioned centrally in the insulated box. Temperature profiles were measured on the component rear surface and the adjacent surface of the board using thin-film thermocouples, and on the exposed surface of the component using infra-red thermometry. Variation of temperature over these surfaces was slight, however, and area-averaged values are shown in figure 3.9. An energy balance was carried out to indicate the quantity of heat lost by each mode: by convection from the exposed surface; by conduction and radiation across the air gap; and by conduction through the leads into the board. These values, figure 3.9, give an order of magnitude assessment of the heat paths.

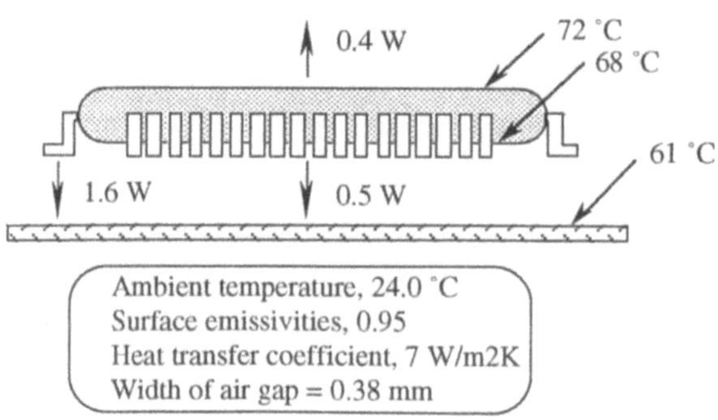

Energy balance		(W)
Exposed surface: Convection	0.4	
		0.4
Gap: Conduction	0.4	
Radiation	0.1	
		0.5
Lead conduction:	1.6	
		1.6
Total (Power level)		2.5

Fig. 3.9 Energy balance on the component

3.5 Variation of Local Air Temperature

Local air temperature variation not only occurs during operation in a confined space, but components are also exposed to severe thermal environments during thermal cycling reliability testing, figure 3.10, and solder reflow assembly, figure 3.11. It is important to predict the temperature response of component elements to such environments because reliability is a strong function of temperature. .Figure 3.10 shows the temperature response of the chip junction, component lead and the PCB beneath the component to a local air temperature, changing at the rate of 15°C per minute and moving at 4 m/s. These conditions are typical of the stress screening environments to which components are exposed.

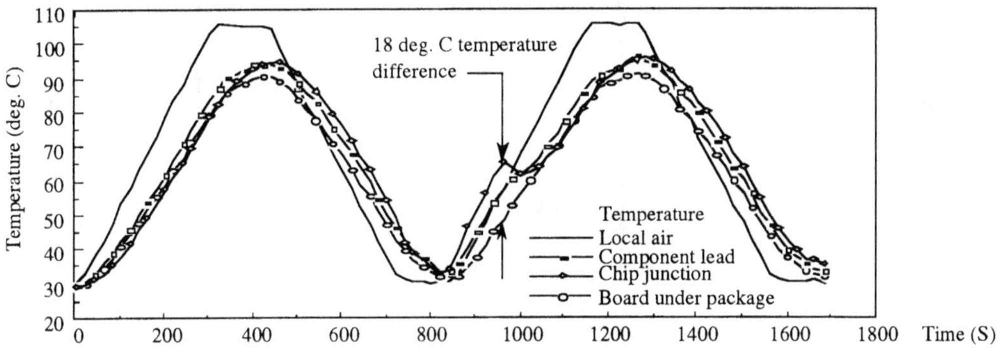

Fig. 3.10 Component response to temperature cycle.

The most notable feature of figure 3.10 is the large 18°C temperature difference induced across the assembly during the second thermal cycle. This shows an average increase in temperature difference of 11°C from the first temperature cycle and was caused by powering the device, with 2.5W, for the first two minutes of the cycle. This indicates the ability of "power on" temperature cycling to induce greater thermal stress on a component over "power off" temperature cycling and Engelmaier[7] recommends its use as a more effective method of assessing the quality and reliability of solder joints. Crook et al[8] attributes the detection of 20 to 25% more latent defects to power on temperature cycling.

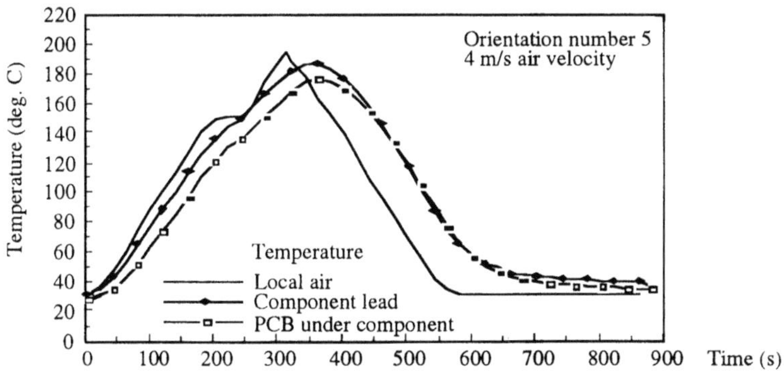

Fig. 3.11 Component response to typical solder reflow thermal profile.

A similar temperature response was obtained for the solder reflow temperature profile, with the low thermal capacity lead responding fastest to the high ramp rates of 40 and 60°C per minute. Operating with an air velocity of 4 m/s the lead attained a temperature of 180°C whereas the junction, and other locations measured, remained on average 13°C cooler. Clearly, exposure of the junction to such extreme temperatures for even seconds has an adverse affect on its subsequent reliability.

4 Conclusions

* The PCB acts as a cooling fin: heat spreads along the fin from the component by conduction through the copper tracking and leads, and a naturally convecting plume.
* Little variation occurs in the junction-to-ambient thermal resistance in the three different vertical orientations, but the horizontal orientations yield 8% higher values.
* The content and distribution of copper tracking on the component side of the board are very influential factors on the junction-to-ambient thermal resistance.
* Thermal resistance is inversely proportional to power level and to lead soldered area.
* The heat balance on the component showed that conduction through the leads accounted for approximately two thirds of the total heat dissipation.
* Power on temperature cycling induced the greatest level of thermal stress across the component indicating its ability to precipitate latent defects. The low thermal capacity lead responded fastest to the solder reflow thermal profile.

Acknowledgements

The authors gratefully acknowledge the support of Eolas through the Power Electronics Program in Advanced Technology and the co-operation of Amkor Electronics who supplied the plastic package.

References

1. SEMI International Standards (1992), Packaging Volume.
2. Dutta, V.B. (1988), *Junction-to-Case Thermal Resistance - Still a myth ?*, Proceedings of the 4th Annual IEEE SEMI-THERM Symposium, pp 8-11.
3. Shanker, B.J. and Lambertson, R.T. (1986), *Sources of error in package thermal resistance measurements*, Proceedings of the 36th Electronics Components Conference, pp 161-168, Seattle, WA.
4. Siegal, B.S. (1988), *Factors affecting semiconductor device thermal resistance measurements*, Proceedings of the 4th Annual IEEE SEMI-THERM Symposium, pp 12-18.
5. Bar-Cohen, A., Elperin, T. and Eliasi, R.(1989), *Θ_{JC} characterisation of chip packages -- justification, limitations, and future*, Proceedings of the 5th IEEE SEMI-THERM Symposium, pp 1-4.
6. Moffat, R.J., Ariviza, D.E., and Ortega, A. (1985), *Cooling electronic components: Forced convection experiments with an air cooled array*, 23rd AIChE/ASME Nat.'l Heat Transfer Conference, Denver, CO.
7. Engelmaier, W., AT&T Bell Laboratories (1985), *Functional cycles and surface mounting attachment reliability*, Circuit World, Vol II, No. 3.
8. Crook, K., Jawaid, S., Salazar, K. and Trujillo, J. (1991), *Stress screening on PMAX and PMIN modules and implementation results*, Digital Equipment Corporation, TR91-0084/SSEH (Revision B).

Investigation of Die Attach Integrity Using Transient Thermal Analysis Techniques

DAGUANG LIU*, JOHN BARRETT*, & SÉAN CIAN Ó MATHÚNA†
* *National Microelectronics Research Center, University College Cork, Ireland,*
† *Power Electronics Ireland, NMRC, University College Cork, Ireland*

Abstract The objective of this paper is to present an investigation of die attach integrity by transient thermal analysis techniques which include: Thermal Transient Analysis System (TTAS) using thermal test chip; Infrared Thermal Imaging System; and thermal modeling using both Finite Difference Method (FDM) and RC network. The measured results are compared among these techniques. The effects of die attach with voids at 25%, 50%, 75% is clearly detected using transient thermal test system and transient Infrared thermal. A 3-D finite difference model of the package structure (40 pin CERDIP) is developed and optimized. Using RC network, a π network model was found to provide the most accurate correlation to the measured results compared to L type and T type network.

I. Introduction

The requirement for methods to characterize thermal performance of electronic packages, die attach materials, and in-situ die attach layers is becoming more important as time to market decreases and need to reduce engineering costs increases. Accurate thermal models to predict the thermal performance of ICs are also becoming an integral part of the overall system design.

Normally the thermal performance of an IC package is described by steady state junction temperature, but this is of little help in pulsed applications and may lead to a redundant and expensive thermal design. This is because in pulse mode, the junction temperature of devices is described by transient junction temperatures with maximum temperatures defined by the average power and not the peak power. Transient junction temperature is not only decided by the power and length of power pulse, but also by the thermal and physical properties of packages, such as package dimensions, thermal conductivities, thermal capacities, and die attach integrity.

Thermal test chips have been used to investigate die attach integrity under steady and transient state [1, 20]. The steady state thermal performance of ICs under small randomly distributed voids and large contiguous voids has been examined by Malalingam, et al [12]. Mahalingam, et al [11] have also investigated transient thermal performance under high current power. Test systems have been developed to

C. J. Hoogendoorn et al. (eds.), Thermal Management of Electronic Systems, 297–306.

measure thermal impedance of IC packages by Pshaenich [21] and Cognetti, et al [7]. The Infrared technique has been used to measure the surface temperature of ICs while it is dissipating power [9, 15, 6]. Area with voids or delaminations can be detected by Infrared techniques [5]. Numerical methods are effective tools in predicting thermal performance of IC packages [4, 19]. Like RC network in electrical circuit analysis, electrical analogy of thermal RC network has been used to predict junction temperature and verify the measured results [18, 16, 23, 2].

The thermal test chip (CMOS2) described in this paper, which has four separate heat sources and eighteen temperature sensors, will be applied to the analysis of the overall die attach integrity because voids in die attach will result in the asymmetric temperature distribution. In this paper, the voids in die attach are intentionally designed for the purpose of studying the effects of voids in die attach. The Thermal Transient Analysis System (TTAS) described in this paper is based on the electrical test method. The function of the system is:

- To measure the transient junction temperature rise of thermal test chips.
- To detect the integrity of die attach, which is accomplished by measuring the transient junction temperature rise of devices.

TTAS will be combined with Infrared Thermal Imaging System, which will be extended to obtain transient thermal imaging. This combination has not been reported before. THERMIC [17], a 3-D finite difference program, will be used to study the effects of voids in die attach on transient thermal performance of IC packages. As the analysis of RC network, which is composed of lumped R and C, usually requires less computer time and memory size compared to the distributed RC line [10, 22], three types of RC network — π type, L type, and T type, will be presented and compared with the measured results of TTAS later in this paper. HSPICE [8], a circuit analysis program for the simulation of electrical circuits in steady state, and transient state, will be used to determine the transient thermal response of a package using an equivalent electrical circuit based on the physical and thermal properties of the package.

II. Thermal Test Chip and Thermal Test System

The thermal test chip CMOS2, which is used for thermal characterization of packages [15], is one of a family [3, 14, 13] of test chips which have been developed for monitoring IC package performance parameters such as thermal resistance, environmental reliability and thermal-mechanical stress.

As can be seen from Figure 1(a) , the CMOS2 with dimensions of 5×5 mm^2 consists of four large area polysilicon heat sources symmetric about the center of CMOS2 and eighteen temperature sensing diodes. Though eighteen diodes are available on CMOS2, only six of them are actually used in measuring junction temperature because of the limitations of bond pads in package and wire length and angle of wirebonding machine. The locations of the six diodes of CMOS2 are shown in Figure 1(b).

Figure 1: (a) The Diagram of CMOS2 thermal test chip (b) Locations of six diodes of CMOS2 and heat sources in real packaging

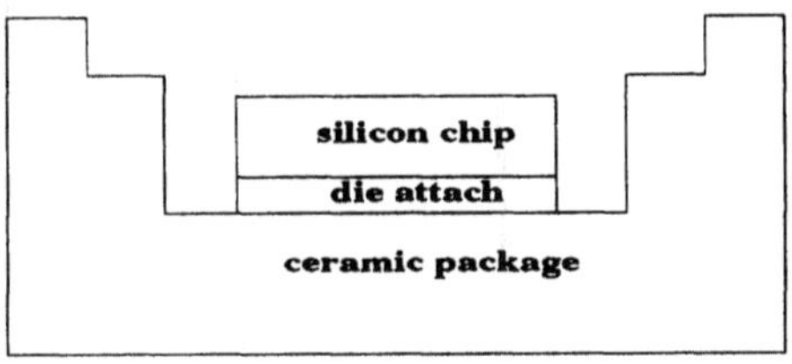

Figure 2: Lateral view of 40 pin CERDIP

Temperature coefficient of forward voltage drop of CMOS2 diode is -2.4 mv/^{0}C, which is obtained from the calibration of CMOS2 by TRES (Thermal Resistance Evaluation System). TRES is an in-house system developed to calibrate thermal test chip and measure steady state thermal resistance [15]. The packaged devices used in this study consist of the CMOS2 thermal test chip mounted onto a 40 pin CERDIP with silver filled Polyimide as die attach material (see Figure 2).

The Thermal Transient Analysis System (TTAS) (see Figure 3(a)) associated with thermal test chip CMOS2 is used to evaluate the effects of voids in die attach and measure transient junction temperature rise. The TTAS was designed with the following specifications:

- One reading per millisecond
- Source 10 μA current into diode
- Offer maximum power 503.0 watts
- Power pulse length from 0.02 μseconds to 9995 seconds

The capability of the system is extended to obtain transient thermal images due to the combination of the TTAS system with the Infrared Thermal Imaging System(see Figure 3(b)). The Infrared Thermal Imaging System described in this paper is a real-time thermal imaging system for temperature measurement and analysis of thermal images. Measurements can be taken at specified intervals to produce a complete

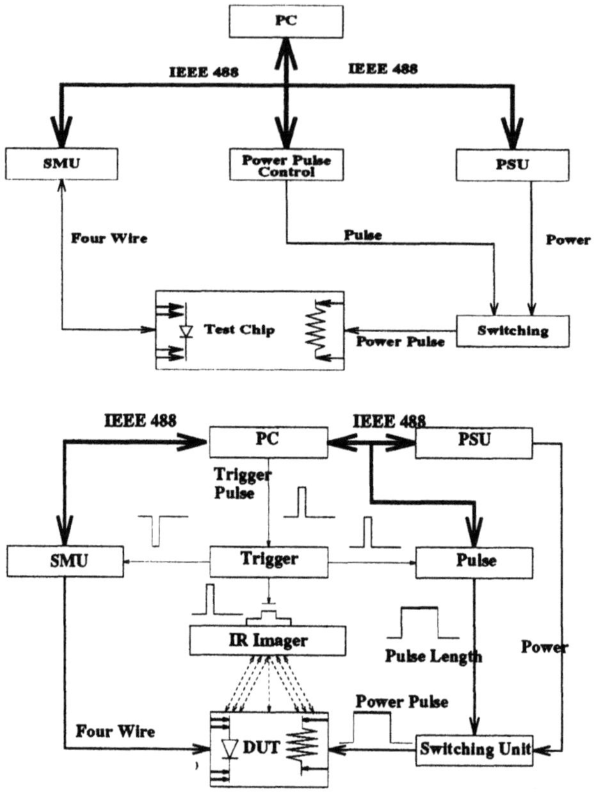

Figure 3: (a) Diagram of Thermal Transient Analysis System (b) The Combination of Thermal Transient Analysis System and Infrared Thermal Imaging System

picture of object performance under test conditions. The Infrared Thermal Imaging System has an ability to acquire 25 full images of the device per second (take one image at 40 milliseconds interval) with a temperature accuracy of $\pm$ 2%, and geometrical resolution of 192 μm (33 ^{o} lens) and temperature resolution of 0.7 ^{o}C. Another 100 times faster measurement of the Infrared Thermal Imaging System is available, which is 0.4 milliseconds per line compared to 40 milliseconds per image. This measurement could give one line thermal image of the device in time interval 0.4 milliseconds with $\pm$ 2% accuracy, temperature resolution 0.7 ^{o}C. The measured line is 27 mm long, and 192 μm width. The software incorporated in the system enables a thorough analysis of the results to be obtained.

III. Results and Discussions

Thermal Transient Measurement

Figure 4 shows the pattern of voids and the position of diodes of CMOS2 in desired structure.

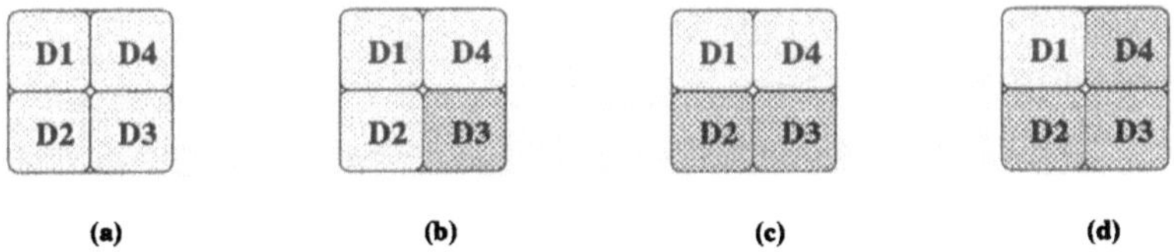

Figure 4: (a) Die attach with voids free (b) Die attach with 25% voids (c) Die attach with 50% voids (d) Die attach with 75% voids. Diagram of voids pattern in THERMIC model

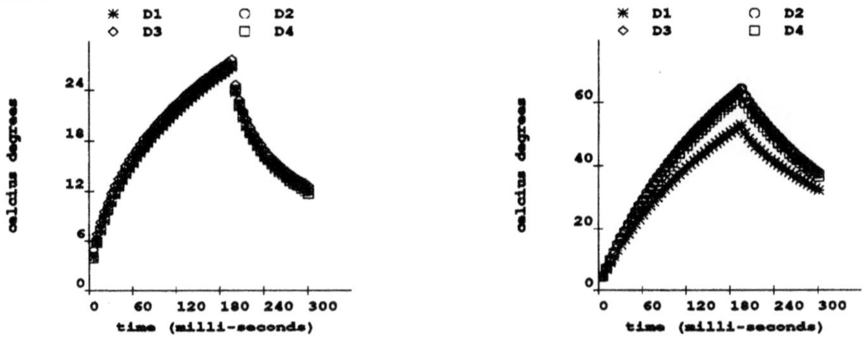

Figure 5: (a) 0% voids. Power 12 watts, length of power pulse 200 milliseconds. Temperature rise of diodes D1, D2, D3, D4. (b) 75% voids. Power 12 watts, length of power pulse 200 milliseconds. Temperature rise of diodes D1, D2, D3, D4. Only D1 is above the die attach.

Figure 5(a) shows junction temperature rise of CMOS2 with voids free die attach. D1, D2, D3, D4 share the same temperature changing during the power pulse. Figure 5(b) shows that presence of voids in die attach (75%) which cause non-uniform increase of junction temperature. Diodes D2, D3, D4 are localized above the voids, so junction temperatures of D2, D3, D4 are higher by 10°C than that of D1, which is above the die attach.

Thermal Transient Measurement vs. Infrared Measurement

Table 1 shows the measured results of TTAS and Infrared. The temperature above the voids clearly has a higher junction temperature (see **bold underline** number in Table 1). The measured results of TTAS and Infrared in Table 1 show excellent agreement. The source of the difference is due to the accuracy of the systems. The accuracy of the Infrared measurement is ± 2%, and TTAS has an accuracy of ± 4%. The measurement mode is 25 full images per second, time interval is 40 milliseconds per image.

Thermal Transient Measurement vs. THERMIC

The THERMIC program calculates temperatures in 3-D for various electronic micro structures by means of a Finite Difference Method (FDM). The model based on

% Voids	0%			25%			50%			75%		
Diode No.	Junction Temperature °C											
	TTAS	IR	Model	TTAS	IR	Model	TTAS	IR	Model	TTAS	IR	Model
1	66.7	64.7	71.5	71.3	71.7	74.6	86.1	85.7	83.5	92.7	91.9	97.1
2	66.8	65.8	71.5	72.8	73.9	77.5	99.1	101.0	101.2	104.0	104.0	115.5
3	67.6	66.3	71.5	79.1	81.6	87.8	95.9	96.7	101.2	104.0	104.0	115.5
4	66.8	65.6	71.5	72.5	74.2	77.5	87.8	87.2	83.5	102.1	101.0	115.5

Table 1: The comparison among TTAS, IR, and THERMIC, power 12 watts, power pulse length 200 milliseconds. The surface of the object is covered with high emissivity black spray, the emissivity is 0.94.

thermal test chip CMOS2 assumed four heat sources (symmetric to the center of the silicon chip) are uniformly distributed on the active surface of silicon chip. The model consists of a silicon chip attached with a die attach material (silver filled Polyimide) to a ceramic package (40 pin CERDIP). The heat transfer between the bottom of the package and temperature controlled heatsink is performed by conduction. The model assumes dimensions of heatsink infinite and ability of heat absorbing also infinite. Thermal conductivities of materials of the package is isotropic.

Table 1 also shows simulation results of finite deference method THERMIC against TTAS (**bold underline** number is the junction temperature above the voids). With the higher percentage voids, the difference between THERMIC and TTAS is getting bigger, this was due to the difference of the voids between the practical packages measured by TTAS and THERMIC models. The original value of thermal conductivity of the die attach material (silver filled Polyimide) offered by manufacturer is 0.38 watts/°C. This value leads a large error between the modeling and the experiment. Through a series of calculations and modeling, the value of thermal conductivity of the die attach material (silver filled Polyimide) is believed to be 2.6 watts/°C. With this value, the modeling and the experiment have a much better agreement.

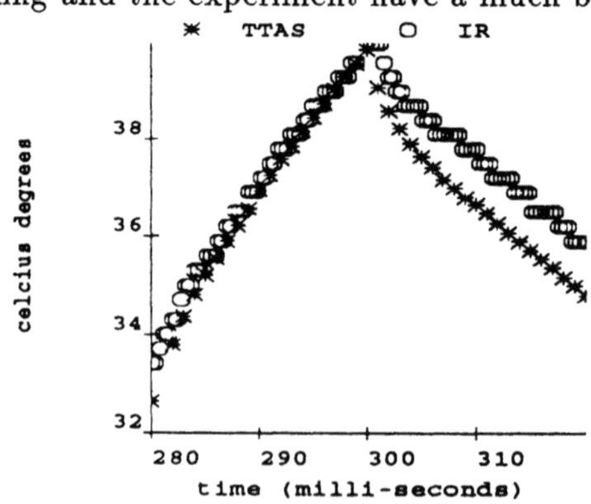

Figure 6: (a) The comparison of Infrared measurement IR6V50A.006 with TTAS (b) Sequential 'one line' images (IR6V50A.006) of junction temperature of CMOS2

Infrared Thermal Imaging System has a line measurement speed up to 400μsec. per line. Figure 6 shows the comparison of this measurement with TTAS. The time duration is from 280 milliseconds to 320 milliseconds, which is the sixth period of the repeat pulse. Both techniques get very similar measurement results. The difference in the cooling-down period is insignificant, about 1 °C.

Thermal Transient Measurement vs. RC Networks

Figure 7(a), (b), and (c) is the π type, L type, T type model of the silicon chip respectively. The thermal resistance is $R\Theta = \frac{x}{k \times A}$, x is the distance of the thermal path, k is the thermal conductivity of the material, A is the area normal to the thermal path. The thermal capacitance is $C\Theta = c \times d \times v$, where c is specific heat of the material, d is density of the material, v is volume of the material.

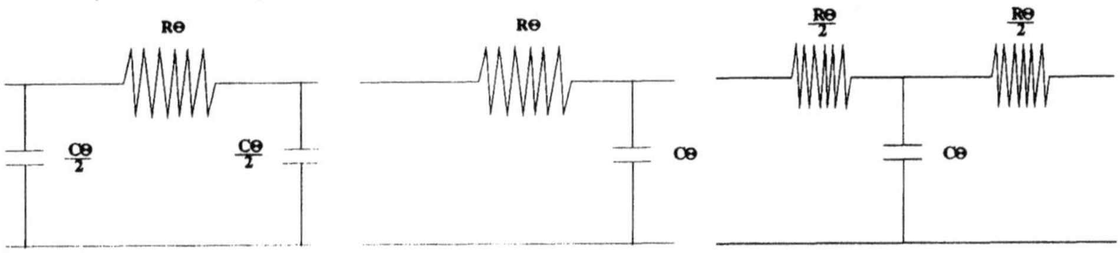

Figure 7: (a) π type lumped element equivalent circuit model of the silicon chip (b) L type lumped element equivalent circuit model of the silicon chip (c) T type lumped element equivalent circuit model of the silicon chip

Table 8(a) shows the thermal resistances and capacitances of the 40pin CERDIP. Figure 8(b) shows the full three-section π type model of the package 40pin CERDIP, which is run on HSPICE. Node-1 (N1) is modeled as surface of the silicon chip (CMOS2), Node-2 (N2) is modeled as surface of the die attach between silicon chip and die attach, Node-3 (N3) is modeled as surface of the ceramic package between die attach and ceramic package. Node-4 (N4) is modeled as bottom of the ceramic package , which is the contacting surface between ceramic and heatsink. V4 is modeled as the constant temperature of heatsink. V200 is modeled as initial temperature of all over the package, which is regarded as the same temperature as heatsink. I (current) is modeled as power source of the package, which is generated on silicon chip.

Though there are three models shown above, only one is agreed by the TTAS measurement. In Figure 9, the simulation results of the three HSPICE models (π type, L type, T type) are compared with the measurement of the TTAS under the same conditions. Also Figure 9(a) shows that the simulation result of the π type model best agrees with the measurement of TTAS. The accuracy is about 6.7%. The other two models are shown that the predicted temperatures are much higher than the measurement of TTAS. The accuracy of the π model decreases to maximum 10% after 600 milliseconds (see Figure 9(b)) due to the structure dependent character of the RC network analysis.

	silicon chip	die attach	ceramic package
thermal resistance (celius/watt)	0.1538	1.0054	2.727
thermal capacitance (Joule/celius)	2.136×10^{-2}	8.021×10^{-4}	1.571×10^{-1}

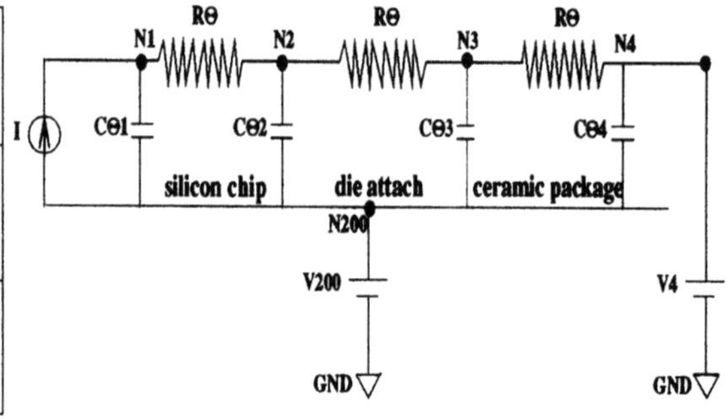

Figure 8: (a) The value of thermal resistances and capacitances of the 40pin CERDIP package (b) Full model of lumped elements of the 40pin CERDIP package

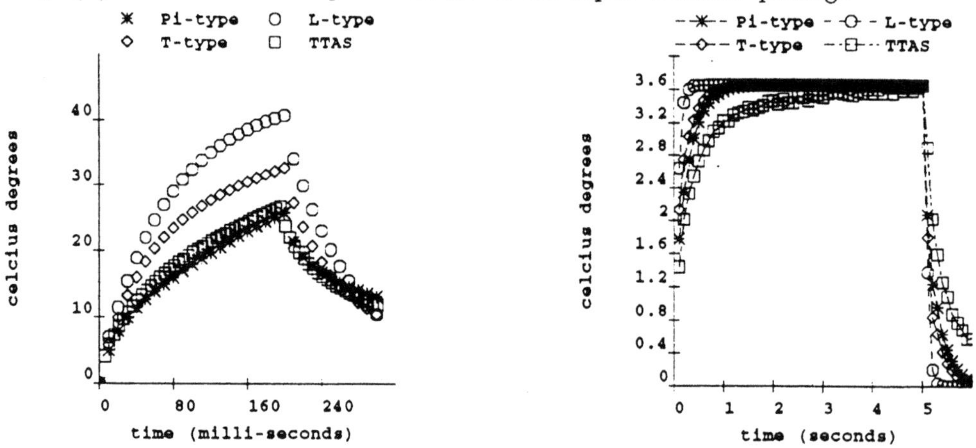

Figure 9: (a) The comparison between HSPICE with three RC models — π type, L type, T type and TTAS. Single pulse, length of power pulse is 200 milliseconds. Surface junction temperature of CMOS2 under single power pulse (12 watts). (b) The comparison between HSPICE with three RC models vs. TTAS. Single Pulse, length of power pulse is 5 seconds. Power is 1 watts.

IV. Conclusions

Thermal Transient Analysis System (TTAS) associated with CMOS2 thermal test chip has been used to investigate effects of voids in die attach.

With the symmetric temperature distribution of thermal test chip CMOS2, the integrity of die attach could be detected due to the asymmetric temperature distribution caused by the presence of voids in die attach.

The capability of non-destructive Infrared Thermal Imaging System is extended to record transient thermal imaging, and can be used to detect the voids in die attach. Both high speed images and line scans of CMOS2 die surface showed very good agreement with the TTAS measurement.

The THERMIC modeling shows that voids in die attach increases junction temperature, which shows excellent agreement between THERMIC and TTAS. The THERMIC model also modifies the value of the thermal conductivity of die attach material.

This also shows that sometimes there is a big difference on values of thermal conductivities between the manufacturers and the users. With THERMIC, the configuration of the model — the geometry, materials of the model and the pattern of the voids, is easy to modify.

With the π type equivalent electrical circuit model, HSPICE simulates thermal performance of the package with the accuracy 6.7%, while the L type and T type are far away from the experiment. After 600 milliseconds, the accuracy of the π model is about 10%. It is very easy to calculate the lumped thermal elements R and C, and to build the RC network of the package. Usually the model is quite simple, the time to run is short (a few seconds). The RC network modeling in this paper developed an optimized thermal model to fit the measurement data.

Acknowledgements

The work has been sponsored by Power Electronics Ireland and European Space Agency. Thanks are due to laboratory staff involved in various aspects of the work: Tony Compagno, Willy Lawton, Helene Shinners, Tom Hayes, Martin Wiseman.

References

1. J. Anderman, J. Tustaniwskyi, and R. Usell. "Die Attach Evaluation Using Test Chips Containing Localized Temperature Measurement Diodes". *IEEE Transactions CHMT*, Vol. 9, No. 4, pp. 410 - 415, Dec. 1986.
2. K.W. Awkward and R.D. Lewis. *Proceedings 7th IEEE Semi-Therm Symposium,* 1991, pages 86-90.
3. J. Barrett, T. Hayes, R. Doyle, C O Mathuna, T. Yamada and A. Boetti. "Reliability Evaluation of High-Pincount Hermetic Ceramic IC Packages for Space Applications". *Proceedings Electronic Components Technology Conference*, 1992, pages 190-199.
4. J.A. Cooke and S.W. Lee. "Finite Element Thermal Analysis of 144 Pin Plastic Flat Packs". *Proceedings 5th IEEE Semi-Therm Symposium*, pages 59-62, 1989.
5. J.S. Fitch "A Comparison of Acoustic and Infrared Inspection Techniques for Die Attach" *Proceedngs IEEE Intersociety Conference on Thermal Phenomena*, 1992, pages 159-164.
6. G.B. Gao and J.J. Zhu. "Monitoring and Analyzing the Dynamic Junction Temperature Distribution of RF Power Transistor by Using RM-50 Infrared Micro Imaging", *Proceedings 4th IEEE SemiTherm Symposium*, 1988, pp 84-87.
7. T. Hopkins, C. Cognetti, and R. Tiziani. "Designing with Thermal Impedance". *Proceedings 4th IEEE Semi-Therm Symposium,* 1988, pages 55-61.
8. Meta-Software Inc. "HSPICE User's Manual". 1990.
9. J.M. Kallis, G.S. Egan and M.P. Wirick. "Nondestructive Infrared Inspection of Hybrid Microcircuit Substrate-to-Package Thermal Adhesive Bonds". *IEEE Transactions CHMT*, Vol. 4, No. 3, 257-260, Sept. 1981.
10. U. Kumar. "Modelling of Distributed Lossless and Lossy Structures - A Review". *IEEE Circuit and System Magazine*, 2(3):12-16, 1980.
11. M. Mahalingam and J.A. Andrews. "Thermal Design Studies of a High-Current Transistor Package". *IEEE Transactions CHMT*, Vol. 5, No.4, 463-469, Dec. 1982.
12. M. Mahalingam, M. Nsgarkar, and L. Lofgran, "Thermal Effects of Die Bond Voids in Metal, Ceramic and Plastic Package". *IEEE Transactions CHMT*, pp. 469-477, 1984.
13. C. O Mathuna. "Thest Chips for the Evaluation of the Performance of IC Packaging and Interconnection Technologies". *Proceedings IEMT Japan*, pages 291-294, 1989.

14. C. O Mathuna. "Development and Analysis of an Automated Test System for the Thermal Characterisation of IC Packaging Technologies". *IEEE Transactions CHMT*, Vol. 15, No. 5, pp. 615-624, 1992.
15. Cian O Mathuna and Daguang Liu. "Development and Application of Test Chips and Test Systems for the Thermal Characterisation of IC Packaging Technologies". *Proceedings 8th IEEE Semi-Therm Symposium*, pp. 47-54, March 1992.
16. W.E. Newell. "Transient Thermal Analysis of Solid-State Power Devices - Making a Dreaded Process Easy". *Transactions on Industry Applications*, 1A-12(4):405 - 420, Aug. 1976.
17. Swedish Institute of Production Engineering Research. "Thermic Manual 5.1". 1992.
18. H.M. Olson. "Temperature Transients in IMPATT Diodes". *Transactions on Electron Devices*, 23(5):494-503, May 1976.
19. B. Ozmat. "Interconnect Technologies and the Thermal Performance of MCM". *Proceedings Inter-Society Conference on Thermal Phenomena*, 1992, pp. 226-245.
20. R. Pendse and B.J. Shanker. "A Study of Thermal Performance of Package Using a New Test Die". *IEEE Transactions CHMT*, pages 50-52, 1988.
21. Al Pshaenich. "300A Delta Vf Tester". *Proceedings 5th IEEE Semi-Thermal Symposium*, 1989, pp. 140-145.
22. T. Sakurai. "Approximation of Wiring Delay in MOSFET LST". *IEEE Journal of Solid-State Circuit*, 18(4):418-426, Aug. 19483.
23. G.L. Skibinski and W.A. Sethares. "Thermal Parameter Estimation Using Recursive Identification". *Proceedings of Industrial Applications*, pages 1581-1588, 1990.

8. THERMAL STRESS AND DIE ATTACH DEFECTS

Thermal Stress In Convectively-Cooled Plastic-Encapsulated Chip Packages

Sorin Witzman[1], Devin E. Mix[2] and Avram Bar-Cohen[2]
[1]Bell Northern Research Ottawa, Ontario, K1Y-4H7, Canada
[2]Department of Mechanical Engineering University of Minnesota, Minneapolis, Minnesota 55455, USA

ABSTRACT

The present study focuses on the analytical determination of the thermally-induced stresses in a convectively-cooled PDIP package. A model of thermal conduction inside the plastic package is combined with a methodology for the computation of the thermal resistance in the convective boundary layer to yield the internal temperature field. The component temperature distribution, obtained in this manner, is then used, via first-order elastic thermal stress relations, to determine the stresses both within the constituent materials and along the interfaces among them.

The analytical results provide insight into the potential failure mechanisms of this package and are found to compare well with a detailed finite-element solution of the temperature and strain/stress fields.

INTRODUCTION

The successful design of plastic integrated circuit (IC) packages for high performance, and relatively high power, VLSI chips is a crucial element in the development of cost-effective packaging technology. Unfortunately, however, most common plastic encapsulating and die-bonding materials provide relatively low thermal conductivities, as well as low mechanical strengths and a limited operating temperature range. As a consequence, plastic IC packages are susceptible to a large number of thermally-induced package failure modes. The package degradation that can occur due to the thermal strain and resulting stress may not produce a package failure directly. But it is still undesirable, since even a small amount of degradation, for instance partial delamination at an interface, may be sufficient to trigger other failure mechanisms such as wire bond fatigue, corrosion, adhesive chemical bond breakdown, or even secondary voltage breakdown (latchup).

In what follows, previously developed approximate analytical thermo-structural relations for the temperatures, strains, and stresses in a typical plastic IC package [1, 2] will be briefly discussed, with emphasis on those relations for predicting the quantities of interest at steady state. These relations will then be modified to include an analytical determination of the convective heat transfer coefficient to facilitate the purely analytical first order prediction of the thermo-structural behavior of a plastic IC package as well as the selection of materials for design purposes. Finally, the results obtained from these relations for a typical plastic PDIP containing a telecommunication chip will be presented and discussed.

THERMAL BEHAVIOR

Qualitative Description

The transient development of the temperature field in a typical plastic IC package (see Fig. 1) can be divided into four key time intervals as discussed in [1, 2] and depicted in Fig. 2. These intervals are described as an early, or chip, period - when the thermal effects are essentially confined to the chip; an intermediate, or package, period - when the

C. J. Hoogendoorn et al. (eds.), Thermal Management of Electronic Systems, 309–323.

die-bond and local encapsulant are involved; a quasi-steady cooling period - when external cooling begins; and finally a steady-state period - when the temperatures everywhere have stabilized. Of particular interest is the case where a "pulse" occurs when the package is operating at steady state conditions. The temperature fields that arise as a result of these operating conditions, steady power dissipation and steady power dissipation with a "pulse" will be examined in what follows.

While the chip (or "junction") temperature is a frequently used metric for chip and package reliability [3], "key" temperature differences such as across the die-bond and between the chip and encapsulant, are believed to better characterize the potential for thermally-induced failures in plastic IC packages. Relationships to predict these "key" temperature differences were derived in a previous study [1] and validated by comparison to the detailed temperature field obtained by exercising an axisymmetric FE model of a typical plastic IC package. Implementation of the analytical determination of the heat transfer coefficient into these expressions will be addressed in the following.

Thermal Analytical Models

Convection Heat Transfer Coefficient (HTC) Determination

In order to understand the development of an expression for the prediction of the HTC it is appropriate to review some definitions that are commonly encountered. The "component adiabatic temperature" is the temperature achieved when no convective heat transfer occurs from the component. In the absence of conduction and radiation from other elements, this is the temperature achieved when no power is applied to the component and the rest of the system is activated [4]. This is not to be confused with the "component adiabatic rise," which is the powered component temperature increase above the adiabatic temperature, i.e. $T_D = T_{AD} + \Delta T_{AD}$.

The term R_{conv} will also be encountered and is defined as the heat passing through the external surface of the component for each degree Kelvin (Celsius) difference between the component external surface and the adiabatic temperature, or $R_{conv} = \Delta T_{AD}/Q$.

The experimental study in [5] identified the diffusion of heat in the air boundary layer, surrounding the component, as the dominant factor for convectively-cooled elements. This conclusion justifies the creation of a thermal transfer equation that concentrates on thermal phenomena in the vicinity of the component, rather than considering the flow in the entire channel. The influence of the upstream (or downstream) channel conditions on the component temperature can be superimposed via the component adiabatic temperature. The latter represents, in fact, the temperature of the environment in which the component is located. Due to the relatively large space (large relative to the dimensions of the components) between components, it is postulated that very little interaction takes place between the thermal boundary layers of different components mounted on the same board. For boards with thermal conductivity in the range of 0.3 to 2 W/m^oC, this interaction is mainly in the form of a thermal wake produced by the high power components mounted upstream from the component under consideration. The results are similar for forced and free convection cooling of electronic cabinets, and to a certain extent they are valid for hermetic (re-circulation) enclosures (for cases where the distance between the component and the outside enclosure is larger than 15 mm).

It is found that the appropriate convective heat transfer coefficient can be obtained from measurements of the heat transfer coefficients for small suspended bodies in free and forced convection cooling, obtained at BNR and at the University of Waterloo [6], as well as from the literature related to the cooling of small bodies, e.g. [7]. These temperature and velocity data are presented in Fig. 3 in non-dimensionalized form. In this diagram the characteristic dimension is the square root of the entire area of the heated body. The results were corrected to account for radiation between the body and the environment. The local average convection heat transfer coefficient, which acts on the external surface of the package, can be computed from Eq. 1 [7].

$$Nu_{MC} = Nu_D + Nu_{FC} + Nu_{NC} + CR \tag{1}$$

In this expression Nu_D is the diffusive limit, $Nu_D = 2\sqrt{\pi}$, Nu_{FC} is the pure forced convection limit, $Nu_{FC} = 0.825\ Re^{0.5}\ Pr^{0.33}$, Nu_{NC} is the pure natural convection limit, $Nu_{NC} = 0.694\ Ra^{0.25}$, and CR is the mixed convection cross flow correction, $CR = 0.86 - 3.54\ [Ra/Re^2]^{0.25}$.

This equation covers the entire interval between completely static air to the beginning of the turbulent flow regime. In the case of vertical flow the CR term is equal to zero. The characteristic length, for isothermal bodies, is the square root of the external surface area of the heated body. The correlation is valid to within ± 5% [7] for any isothermal body geometry except for hollow bodies through which air can flow.

For bodies that are not isothermal, for example a real electronic component mounted on a printed circuit board, corrections that account for the nonuniformity of the thermal field on the surface of the body must be made. These corrections are usually made by applying the "surface effectiveness" concept. For bodies with nonuniform temperature fields, the physical surface area of the body can be replaced by the effective heat transfer area defined as $A_{eff} = \phi_c A_{comp} + \phi_b A_{bd}$. Considering the characteristics of the problem - a small heat source on a relatively large plate - a good approximation of the temperature field created by the component on the board is the temperature field created by a cylindrical hot body attached to a circumferential fin. The surface effectiveness variation for such a fin was studied in depth by Gardner [8].

It is interesting to note that when the effectiveness of the surface is computed for geometries and materials normally found in electronic circuit packs (10 to 20 mm plastic encapsulated silicon chips, components, on printed circuit boards with equivalent thermal conductivities between 0.7 to 2 W/m°C) the effective area of the component itself is approximately 83% of the geometrical area. At the same time, it is found that the equivalent effective area of the board is from 15 to 20% of the geometrical area of the component. Thus, use of the total surface area of the component as the reference surface in Eq. 1 (the equation used to predict the local average heat transfer coefficient for the component) provides a very good first approximation.

For a component mounted immediately after a component with large power dissipation, the replacement of the adiabatic temperature with the bulk air temperature at the component level is not valid and could generate relatively large errors. These components are immersed in the thermal wake produced downstream, and in this situation the device temperature becomes, $T_D = T_{BA} + \Delta T_{film} + \Delta T_{TW}$, where T_{BA} is the bulk air temperature at the component level (computed from a heat balance equation in which all of the heat dissipated upstream is considered to be absorbed by the moving air), and ΔT_{TW} is the temperature supplement created by the thermal wake, as defined by Ortega and Moffat 1985, 1986 [4]. To find ΔT_{TW} for a specific component the following empirical relation is proposed

$$\Delta T_{TW} = \frac{C(Q - 0.5)}{D} \tag{2}$$

where D (dimensionless) is the number of "rows" the component of interest lies downstream of the high power dissipation component, Q is the power dissipation of the upstream component which is influencing the downstream components, and C (dimensionless) is an experimental constant. The value of C decreases with increasing local air velocity with typical values of 3 for 1 m/sec, 2 for 2 m/sec, and 1 for air velocities higher than 3 m/sec.

A pragmatic and very efficient approach, especially for designers who deal with a relatively small number of similar products, is to approximate ΔT_{film} by the relation [9]

$$\Delta T_{film} = \frac{Q^T v^n}{Ci\ a^x\ \lambda^y\ b^z\ L^t} \tag{3}$$

where all of the values of the parameters in this equation are determined experimentally. The determined values can be extrapolated to a very large number of applications and the expected errors are very small, unless major changes in the component substrate technology are made.

Steady State Behavior

To determine the steady-state temperature on the active surface of the chip, the temperature differences across the chip, the die-bond, the "heat spreading" element (half-thickness), and the case temperature difference are summed. Using the Fourier equation for 1-D heat conduction [10], this summation can expressed as in Eq. 4, where A_i is the

$$T_{s,a} = Q[(\frac{\Delta x}{kA})_s + (\frac{\Delta x}{kA})_{d/a} + (\frac{\Delta x}{2kA})_{hs} + R_{conv}] + T_\infty, \tag{4}$$

thickness direction surface area of the ith component, and T_∞ is the ambient temperature.

Because of the complicated configuration in an electronic package and since the solids surrounding the heat source (device), including the board, have relatively high thermal conductivities compared to the air around the component, the flow of heat from the device (where the power is dissipated) to the passing air occurs via a complex thermal path. A strongly nonuniform temperature rise, at any point, is a function of a multitude of factors such as material conductivity and geometry, distance between the components which dissipate power, local air velocity, internal configuration of the device package, and the type of mounting scheme utilized, etc.

As a general rule, one should consider that part of the heat is transferred from the component surface to the surrounding air and part of the heat is passed first to the board and then from there to the air stream. Thus, the heat balance written for any specific component becomes

$$Q = Q_B + Q_D, \tag{5}$$

where Q is the total power dissipated by the device, Q_D is the heat transferred directly from the package, and Q_B is the heat transferred through the substrate. By dividing each term of this equation by $\Delta T_{c\text{-}a}$, the temperature difference between the case and ambient, Eq. 5 can be expressed in terms of thermal resistances, namely,

$$R_{conv} = \frac{R_B\ R_D}{R_B + R_D} \tag{6}$$

where R_{conv} represents the overall thermal resistance of the component package, R_B the resistance for the heat transferred to the air through the board, and R_D the resistance for the heat transferred directly to the air stream.

This formulation is advantageous because if allows separate approximations for each resistance, without having to solve the differential equations describing the entire temperature field at once. An approximate, but sufficiently accurate analytical solution can be obtained by treating both the package and the substrate as circular fins and arriving at the temperature field by using a Bessel functions solution [9]. The local average convection heat transfer coefficient needed in this analytical solution can be estimated with Eq. 1.

The steady state temperature, at the active surface of the chip, is just one of the temperatures of importance at steady state. The average temperature of the silicon and encapsulant are also of interest, since the differential expansion that occurs between the two depends on their average temperatures. The average steady state temperature of the silicon can be approximated by the half-thickness silicon temperature, since the gradient through the silicon is approximately linear. The expression for this temperature can be obtained by replacing the Δx in the silicon term of Eq. 4 by, $\Delta x/2$.

A first-order approximation of the average temperature of the encapsulant is the half-thickness temperature of the encapsulant. For a package that is uniformly cooled by an equivalent effective heat transfer coefficient, where it may be assumed that the heat flow off the upper and lower surfaces of the leadframe is equal and uniform, it is possible to express the half-thickness temperature of the encapsulant as

$$T_{half\text{-}enc} = \frac{Q}{2}\left[\left(\frac{\Delta x}{2kA}\right)_{enc}\right] + Q[R_{conv}] + T_{\infty}. \tag{7}$$

Finally, the die-bond through-thickness temperature gradient is also of interest. Since the steady state analytical model, for the active surface of the silicon, considers the case where all of the heat dissipated in the chip flows through the thickness of the chip and then the die-bond before spreading, across the entire plane of the package, it is possible to express the die-bond temperature gradient as

$$\Delta T_{d/a} = Q\left[\left(\frac{\Delta x}{kA}\right)_{d/a}\right]. \tag{8}$$

It is interesting to note that in convectively-cooled packages where a high thermal conductance leadframe, substrate, or heat spreader is present, the thermal resistances in the silicon and die-attach materials are approximately two orders of magnitude smaller than the effective convective resistance (R_{conv}), while the resistance of the leadframe, or heat spreader is found to be about four orders of magnitude smaller.

Transient Behavior

In the interest of being able to predict the peak temperature rise in a package that experiences a power pulse when operating at steady state conditions (as a result of a constant power dissipation as shown in Fig. 2) it is necessary to consider the transient heat-up that is a result of a power pulse. In [1] the formulation for predicting the temperature increase at the active surface of the chip, due to a power pulse, was developed using the semi-∞ solid formulation where the active silicon surface was treated as having a uniform heat flux incident upon it. Superposition of the ΔT predicted using the relationship in [1] on the ΔT found using Eq. 4 yields a reasonable estimate of this peak temperature rise.

Likewise in [1] it was shown that once the heat has reached the external surfaces of the package, it is the convective time constant, for the entire package, which determines the gross thermal behavior of the package. The composite convective thermal time constant developed in [1] can be modified to contain the analytically determined HTC thru replacement of $1/hA$ by R_{conv}.

Finally, the temperature at any location within the package, after the time at which the heat has spread to the external surfaces of the package, can be predicted using the relationship in [1] with the thermal time constant being replaced by the modified composite package time constant.

STRUCTURAL BEHAVIOR

Qualitative Description

The natural divisions that were present for the temporal thermal behavior of the plastic IC package are also present in the structural response of the package. During the "chip period," the expansion of the silicon is constrained by the surrounding encapsulant. Thus, the chip is momentarily loaded in a state of compression. As time progresses to the quasi-steady interval the signs of stresses reverse and the magnitudes of the strain and stress experience a marked increase with maximum values reached at steady state. In fact, at steady state, stress values are found to be 2 to 3 orders of magnitude larger than those during the early time frame (those occurring shortly after the pulse has terminated). Finally, the effect of an occurrence of a power pulse at steady operating conditions would serve to reduce the differential expansion between the chip and the surrounding materials resulting in a slight relaxation of the stress field.

Structural Analytical Models

Configuration

The configuration that was modelled consisted of a chip adhesively bonded to a substrate composed of a leadframe and underlying encapsulant. This arrangement provides a conservative estimate of the stresses that develop in the adhesive layer. The encapsulant that encapsulates the top and edge or perimeter of the chip is assumed to have no interaction with the chip. The absence of interaction between the chip and the encapsulant constitutes a worst case shear condition for the adhesive layer since the encapsulant is not able to alter the natural thermal expansion of the chip by "compressing" or "stretching" the chip. It is appropriate to consider the worst case shear condition in the adhesive as previous studies [2,11] have revealed that the die-attach is the most vulnerable.

Tri-material Plate Structure

Since the stiffness of the adhesive is much less than that of the chip or substrate (whose stiffnesses are comparable, see Tables 1, 2 [2]) the thermal expansion of the adhesive and bending in the package can be ignored. With these assumptions, satisfaction of the force balance requirement, and formulation of the appropriate normal and shear strain expressions it is possible to arrive at a second order differential equation for the shear stress in the adhesive layer as was done in [2]. The solution of this equation for non-isothermal conditions yields Eq. 9, an expression for predicting the shear stress in the adhesive layer as a function of distance from the line of symmetry of the plate configuration.

$$\tau = [(\alpha\Delta T)_1 - (\alpha\Delta T)_2] \frac{G(\mathrm{Sinh}\ \beta x)}{\beta(h_a)(\mathrm{Cosh}\ \beta l)} \tag{9}$$

where $\beta^2 = \frac{G}{h_a}[\frac{1}{(Eh)_1} + \frac{1}{(Eh)_2}]$, h_i is the thickness of the ith component, G is the adhesive shear modulus, E_i is the Young's modulus of the ith adherend, and α_i is the thermal expansion coefficient (TEC) of the ith adherend.

Incidentally, Eq. 9 reduces to the solution obtained in [12] when isothermal conditions are present. This expression is also similar in nature to the more sophisticated model that was developed by Suhir in [13].

In this formulation the maximum stress is encountered at the location x = l, which when substituted into Eq. 9 yields

$$\tau_{max} = [(\alpha\Delta T)_1 - (\alpha\Delta T)_2] [\frac{G(\tanh [\beta l])}{\beta h_a}]. \tag{10}$$

Once the shear stress is known it becomes possible to obtain expressions for the normal stresses in the adherends above and below the die-bond material by substituting Eq. 9 into the force balance equations, integrating, and solving for the necessary constants, using the boundary conditions, as was done in [2]. The expression in Eq. 11 can be used to predict the normal stress in the plane of the silicon. A similar expression can be used to predict the the normal stress in the plane of the substrate as was developed in [2].

$$\sigma_1 = \frac{F'_1}{h_1} = \left[\frac{(\alpha_1\Delta T_1 - \alpha_2\Delta T_2)G}{\beta h_a \mathrm{Cosh}\beta l}\right]\left[\frac{\mathrm{Cosh}\ \beta x}{h_1\beta}\right] + \frac{(\alpha_2\Delta T_2 - \alpha_1\Delta T_1)G}{h_1\beta^2 h_a} \quad (11)$$

The normal stress expressions predict stresses that are nearly constant, in the vicinity of the centerline of the adherend, which then tail off to zero as the free edge of the adherend is approached, as can be seen by substituting $x = l$ into Eq. 11 and the appropriate expression in [2]. This behavior is as expected with the maximum value of normal stress occurring at the centerline.

Furthermore, by applying the work of Mirman (et. al.) [14] to the tri-material analytical models that were developed, the 2-D strip relations can be used to generate axisymmetric analytical results. This can be done by simply replacing the α_j's by $\alpha_j(1-\nu_j)$ and the E_j's by $E_j/(1-\nu_j^2)$ where ν_j is the poisson's ratio of the ith adherend.

Thermo-Structural Analytical Models

To determine the shear stress, in the adhesive, and normal stresses that develop, in the chip and substrate, during operation of a typical PDIP it is possible to combine the temperature and stress relations derived above. This was done for all of the time frames of interest in [2] but since it was found in [2] that the steady state stresses dominate, by 2 to 3 orders of magnitude, only the steady thermo-structural relations will be reproduced here.

Steady State Period

The maximum shear stress in the adhesive can be determined using Eq. 12. The R_{conv} term in Eq. 12, as well as in the equation to follow, is as defined earlier in the thermal section of this paper.

$$\tau_{max} = \Big[\alpha_1\Big\{Q[(\frac{\Delta x}{2kA})_{si} + (\frac{\Delta x}{kA})_{d/a} + (\frac{\Delta x}{2kA})_{hs} + R_{conv}]\Big\} - \alpha_2\Big\{\frac{Q}{2}[(\frac{\Delta x}{2kA})_{enc} + R_{conv}]\Big\}\Big]\Big[\frac{G(\tanh[\beta l])}{\beta h_a}\Big] \quad (12)$$

Likewise, the normal stress in the chip can be predicted at steady state using Eq. 13. A similar relationship can be obtained for the normal stress in the substrate by substituting Eqs. 4 and 7 into an expression similar to Eq. 11 (see [2]).

$$\sigma_1 = \Big[\frac{\alpha_1\Big\{Q[(\frac{\Delta x}{2kA})_{si} + (\frac{\Delta x}{kA})_{d/a} + (\frac{\Delta x}{2kA})_{hs} + R_{conv}]\Big\}}{\beta h_a \mathrm{Cosh}\beta l} - \frac{\alpha_2\Big\{\frac{Q}{2}[(\frac{\Delta x}{2kA})_{enc} + R_{conv}]\Big\}}{\beta h_a \mathrm{Cosh}\beta l}\Big]\Big[\frac{G\ \mathrm{Cosh}\ \beta x}{h_1\beta}\Big] + \frac{G\ \alpha_2\Big\{\frac{Q}{2}[(\frac{\Delta x}{2kA})_{enc} + R_{conv}]\Big\}}{h_1\beta^2 h_a}$$

$$-\frac{G\,\alpha_1\left\{Q[(\frac{\Delta x}{2kA})_{si}+(\frac{\Delta x}{kA})_{d/a}+(\frac{\Delta x}{2kA})_{hs}+R_{conv}]\right\}}{h_1\beta^2 h_a} \tag{13}$$

FE MODEL DESCRIPTION

The FE models considered were 2-D and axisymmetric in nature. The 2-D model consisted of 276 nodes and 236 elements while the axisymmetric model contained 520 nodes and 475 elements as shown in Fig. 4. Half-symmetry arguments were utilized for both geometries.

The 2-D model dimensions were determined from package dimensions of length equal to 3.28 cm, depth equal to 0.81 cm, and thickness equal to 0.461 cm, and chip dimensions of length 0.2416 cm, depth 0.2438 cm, and thickness 0.0525 cm. The die-bond area was taken to be the same as the chip with a thickness of 0.00254 cm. Similarly, the leadframe area was taken to be the same as the package with a thickness of 0.0254 cm. The dimensions used in the axisymmetric model were effective dimensions based on the actual dimensions. In particular, effective radii which produced surface areas in the thickness direction that were equivalent to those in the actual package were used. With these actual dimensions the effective package and chip radii were determined to be 0.921 cm and 0.133 cm, respectively. The material properties used for the various components of the package are as outlined in Table 1.

The 2-D thermal B.C.'s consisted of a heat flux of 1000 W/cm^2 of duration 1 ms followed by a steady flux of 15.5 W/cm^2 incident at the top surface of the chip while the line of symmetry of the package was an adiabatic (thermally insulated) surface. The initial temperature throughout the package in both cases was 20 ^oC. In addition, convective cooling, with an h of 170 W/m^2 K (high to account for unit depth heat flux incident on chip), to ambient air at 20 ^oC took place at all external surfaces except the vertical edge and top surface of the chip.

The power generation in the axisymmetric model took place in the top layer of chip elements with volumetric power dissipations of 384,615 and 5962 W/cm^3 for the 1 ms duration pulse and steady power dissipation, respectively. Convective cooling to 20 ^oC ambient air with an h of 50 W/m^2 K, a typical value that Eq. 1 yields, was present at all external surfaces while the axis of symmetry acted as an adiabatic surface.

Structurally, identically meshed models were used while the thermal load input conditions came from the respective thermal model transient temperature distribution output. The structural boundary conditions restricted any lateral or radial displacement accordingly at the line of symmetry in the models while the lower left node in each model was constrained both axially and laterally (or radially), the bodies were freely suspended.

RESULTS

The temperatures predicted, analytically and numerically, at various "key" locations within the package during the early transient and steady state time intervals are tabulated in Fig. 1. A peak chip surface temperature rise of approximately 20 ^oC is observed at the termination of the power pulse while the peak rise at steady state is approximately 30 ^oC. The occurrence of a power pulse at steady state is found to produce a peak temperature rise of approximately 50 ^oC, a condition that would be most detrimental in terms of temperature.

While the overall temperature rises are higher at steady state it is during the initial transient that significant temperature differences between components of the package are observed. During the transient between component temperature differences on the order of 20 ^oC are present while a moderate temperature difference between components of approximately 5 ^oC is present in the package at steady state. However, the occurrence of a pulse at steady operating conditions produces the highest peak temperatures as well as the largest component temperature differences.

The above temperature distributions are responsible for the time-varying strain and stress states that develop in the PDIP package. The behavior of the adhesive shear stress at steady state, in the tri-material configuration, as predicted by the location dependent variation of Eq. 12, for the typical PDIP, is depicted in Fig. 5 along with the normal stress profiles for the adherends at steady state, as predicted by Eq. 13 and that in [2].

The peak stresses attained at steady state as predicted by analytical and numerical means are found tabulated in Table 2 [2]. Although the stresses during the initial time period (not shown) are 2 to 3 orders of magnitude lower than those occurring at steady state, the transient thermal behavior is responsible for a reversal in the shear and normal package stresses. It is also interesting to note that the occurrence of a pulse during the early transient results in the development of an undesirable stress field, while the occurrence of a pulse at steady state actually serves to reduce the differential expansion between the chip and the surrounding materials and thus results in a stress field of lower magnitude. However, the reduction is of little significance since the change in the stress level is from 2 to 3 orders of magnitude less than the nominal stress level at steady conditions.

The shear stress, of 1900 N/cm^2, predicted for the adhesive material is well above its expected failure strength which suggests that this material will be the most problematic. A comparison of the peak component stresses to the respective failure strengths (S_{max}/Sy) for these components confirms this conjecture as is demonstrated in Table 2 [2]. Although the magnitudes of the normal stresses that develop in the package are significantly lower than the respective failure strengths, the shear stress in the adhesive is substantial due to the large shear strain that develops through the adhesive layer.

Examination of the thermo-structural equations suggests that a low modulus adhesive with liberal thickness should be utilized in order to minimize the shear and normal stresses. Furthermore, the equations reveal that beyond a certain chip length the maximum adhesive shear stress will no longer continue to increase with increasing length of chip. Thus, if the conditions are such as this, use of a larger chip will not produce a more problematic stress state.

Comparison of Analytical and FE Thermo-Structural Results

During the initial or "chip" time period (on the order of 1 ms) the analytical prediction of the peak temperature rise in the silicon is seen to be somewhat conservative with a difference of approximately 0.5 °C (2.4%) between the analytical and FE results.

The prediction of package temperatures, for those quasi-steady times of interest, namely, for those times where the ΔT's from the initial temperature are approaching their largest values, $t \geq [(1)(\tau_{ICP})]$, were found to a differ from the numerical by 7% at $(1)(\tau_{ICP})$, which decreased to 2.1% as the steady state temperatures are approached. The analytical system time constant that was used was determined solely on the basis of plastic encapsulant properties. This value most closely agreed with the system time constant obtained from a FE analysis, with both predicting a value for τ_{ICP} of approximately 102 seconds.

The difference between the analytical and numerical prediction of the peak steady state temperature rise in the silicon, using Eq. 4, was found to be 2.1%. It was also possible to obtain the steady state temperatures at other locations within the package using variations of Eq. 4. For example, the analytical and numerical predictions of the ΔT across the die-attach material were both less than 1 °C with a difference of approximately 0.3 °C, while the difference between the analytical and numerical half-thickness encapsulant temperature was found to be 2.4%.

In Table 2 [2], the maximum stress values from the analytical relations and FE analysis are presented. Results are provided for both 2-D plane stress and axisymmetric geometries. The "tri-material" analytical model provides 2-D stress results that are within a factor of two of the 2-D FE results. The Mirman transformed 2-D strip relations yield analytical axisymmetric results that are in similar agreement to axisymmetric FE results. A comparison of these results is found in Fig. 6 where the analytical and FE predictions of the maximum shear stress in the adhesive are plotted. The shear stress values are in agreement within a factor of two while the predicted trends are in excellent agreement. In addition, for the FE models, the maximum shear stress at the top and bottom surface of the die-bond material is tabulated along with the average shear stress between the two surfaces.

It is important to note that the fully encapsulated model yields a normal stress in the silicon that is higher than that in the removed encapsulant case. This occurs since the higher thermal expansion coefficient encapsulant is now acting directly on the chip, with a no-slip condition at the surfaces in contact, and thus produces the higher normal stress. The encapsulant effect also serves to reduce the shear stress in the die-bond material by reducing the shear strain that occurs across the die-bond. Similarly, the introduction of the encapsulant above the chip serves to reduce the normal stress in the encapsulant in the vicinity of the leadframe since the encapsulant stiffness is now a factor of two larger than that of the leadframe. However, the normal stress in the encapsulant directly above the chip will be higher than that below the chip due to the absence of a compliant layer between the chip and the encapsulant.

It is also interesting to note that the shear stress predicted by the axisymmetric model is about a factor of 2.5 lower than the 2-D removed encapsulant model value. This reduction reflects the effect of geometry and the presence of encapsulant on the top and side of the chip.

CONCLUSIONS

Purely analytical first order relations were developed for predicting the thermo-structural behavior of a plastic IC package. These analytical stress models yielded limiting values for the stresses, in the various components of the tri-material configuration, for both 2-D and axisymmetric geometries. The analytical results were in agreement with the FE results within a factor of 2 while predicted trends were in excellent agreement. Another finding of significance was the identification of a stress reversal ("pseudo" cycling) event, that takes place during the early time development of the stress field in a plastic IC package.

ACKNOWLEDGEMENTS

The authors would like to extend their thanks to Bell Northern Research for supporting this research. We are also indebted K. Smith, T. Nicoletta, and T. Petermann (of Northern Telecom) for their constructive comments and for providing helpful background information.

REFERENCES

1. Bar-Cohen, A. and Mix, D. E., *Thermal Modeling of Plastic IC Packages*, Proc. Mat. Res. Soc. Symposium, Vol. 226, Materials Research Society, 1991.
2. Mix, D. E., and Bar-Cohen, A., *Transient and Steady State Thermo-Structural Modeling of a PDIP Package*, Presented at ASME WAM, Paper # 92-WA/EEP-10,1992.
3. *Reliability Prediction of Electronic Equipment*, Military Standardization Handbook, M-HDBK-217.
4. Ortega, A., and Moffat, R., *Heat Transfer from an Array of Simulated Electric Components: Experimental Results for Free Convection With and Without a Shrouding Wall*, Proc. 23rd AICHE/ASME National Heat Transfer Conference, Denver, Colorado, pp 5-15.
5. Witzman, S., Nicoletta, T., and Mack, B., *Enhanced Heat Transfer for Electronic Devices Operating Inside Enclosures*, SEMITHERM 1992, Austin, Texas.

6. Mack, B., *Natural Convection from an Isothermal Cube on a Vertical Plate*, University of Waterloo, Master's Thesis, 1991.
7. Yovanovich, M., and Vanoverbeke, C., *Combined Natural and Forced Convection from Isothermal Spheres*, Proc. 1988 AIAA Thermophysics Plasma and Dynamics Lasers Conference, AIAA-88-2618, San Antonio, Texas.
8. Gardner, K. A., Tran. ASME, *Efficiency of Extended Surface*, 67: 621-631, 1945.
9. Witzman, S., Newport, D., and Nicoletta, T., *An Alternative Methodology for Computing the Temperature Rise of an Electronic Device Inside a Direct Air-Cooled Electronic Cabinet*, Proc. The Ninth International Heat Transfer Conference, Jerusalem, Israel, Volume 2, pp 313-319, 1990.
10. Holman, J. P., Heat Transfer, McGraw-Hill Books Company, 6th Ed., 1986, pp 49.
11. Witzman, S., and Smith, K., and Metelski, G., *Silicon Interconnect-A Critical Factor in Device Thermal Management*, IEEE Transactions on Components, Hybrids, and Manufacturing Technology, Volume 113, Number 4, December 1990.
12. Johnson, E. A., Chen, W. T., and Lim, C. K., *Mechanical Design*, Principles of Electronic Packaging, McGraw-Hill Book Company, 1989, Ch. 6, pp 168-171.
13. Suhir, E., *Thermal Stress Failures in Microelectronic Components - Review and Extension*, Advances in Thermal Modeling of Electronic Components and Systems, edited by Bar-Cohen, A., and Kraus, A., (Hemisphere Publishing Corporation, 1988, Volume 1), pp 337-412.
14. Mirman, B., and Mirman, I., *Thermal Stresses in Axisymmetric Bimaterial Assemblies: Microelectronic Applications*, Advances in Electronic Packaging, ASME, 1992, pp 425-435.

TABLES

IC Package Component Thermal/Structural Properties							
Material	ρ	Cp	k	E	ν	α	Sy
	(Kg/m^3)	(J/Kg K)	(W/m K)	(10^10 N/m^2)		(1/deg. C)	(10^8 N/m^2)
Encapsulant	1820	1884	0.67	1.2753	0.25	0.000016	0.5
Die-Attach - Shear	3800	647	4.5	0.8274	0.25	0.000046	0.08274
Die-Attach - Normal	3800	647	4.5	0.8274	0.25	0.000046	.7-1.05
Chip (Silicon)	2330	703	153	16.55	0.25	2.3E-06	2.5-4.14
Leadframe	8920	400	196.6	11.93	0.34	1.69E-05	

Table 1

SUMMARY OF ANALYTICAL/FE STRESSES					
MATERIAL	E	h	Eh	Sy	Smax/Sy
	(10^6 N/cm^2)	(cm)	(N/cm)	(10^4 N/cm^2)	
Encapsulant (Epoxy: EME-6300H)	1.2753	0.188	315637	0.5	(.16**)
Die-Attach (SFE: 84-1LMIS) - Shear	0.8274	0.0025	2102	0.08274	7
Die-Attach (SFE: 84-1LMIS) - Normal				.7-1.05	.57-.86
Chip (Silicon)	16.55	0.052	868213	2.5-4.14	.10-.16
Leadframe (Copper: Alloy 195)	11.93	0.026	304215		
MODEL	SFE SHEAR STRESS	SFE NORMAL STRESS	SI NORMAL STRESS	ENC NORMAL STRESS	
	(N/cm^2)	(N/cm^2)	(N/cm^2)	(N/cm^2)	
Taumax Unconstrained expansion of adherends.	-5764.6				
Trimaterial (Johnson, et.al.; 2-D)*	-2137.8		1786.7	-498.6	
Trimaterial (Johnson, et. al. Transformed by Mirman, et. al.; Axisymmetric)	-2165.8		1820.3	-447.3	
	Tau1 Tau (mid) Tau2				
Removed Encapsulant 2-D (coarse mesh)	(-3500.) (3075.) (2650.)		3250	-700	
Removed Encapsulant 2-D (fine mesh)	(-4800.) (3800.) (2800.)	-6000	1620	-705	
Fully Encapsulated Axisymmetric	(-2000.) (1937.5) (1875.)		4000	-375	

* Curvature and stress variation through thickness neglected.
** Composite Stress/Epoxy Failure Strength
NOTE: Maximum stress values are tabulated, unless indicated otherwise.
Tau (mid) = (Tau1 + Tau2)/2

TABLE 2

FIGURES

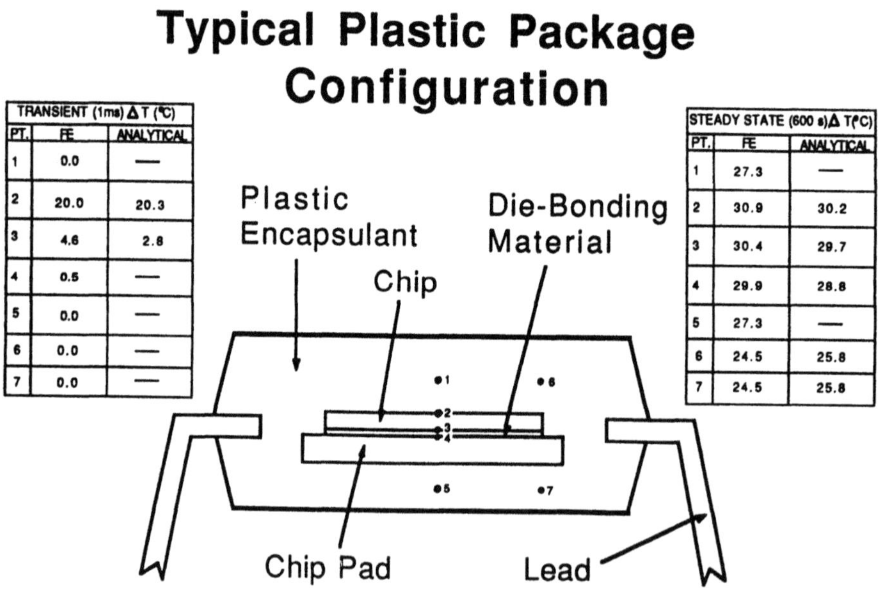

Figure 1

IC Thermal Response in a Typical Plastic Package. a) Constant Power Dissipation (1W). b) Initial Power Pulse Followed by Constant Power Dissipation (50W, 1W).

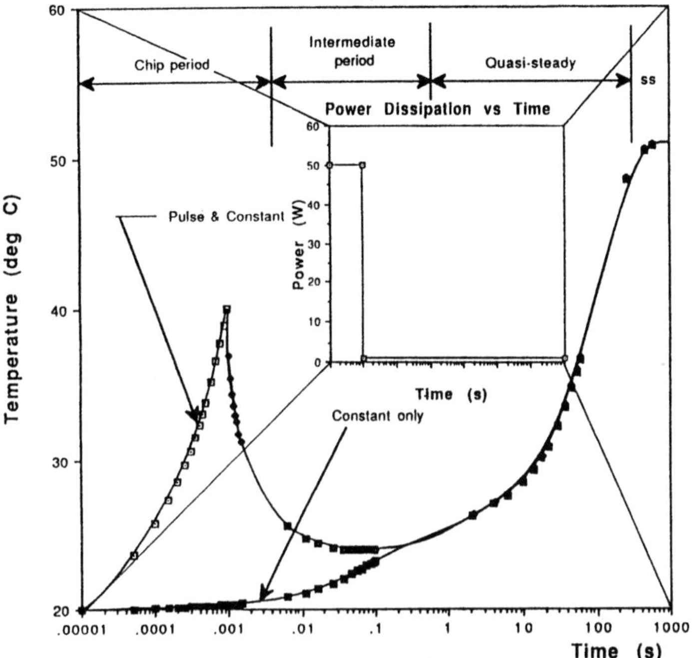

Figure 2

Nusselt Number vs Reynolds Number

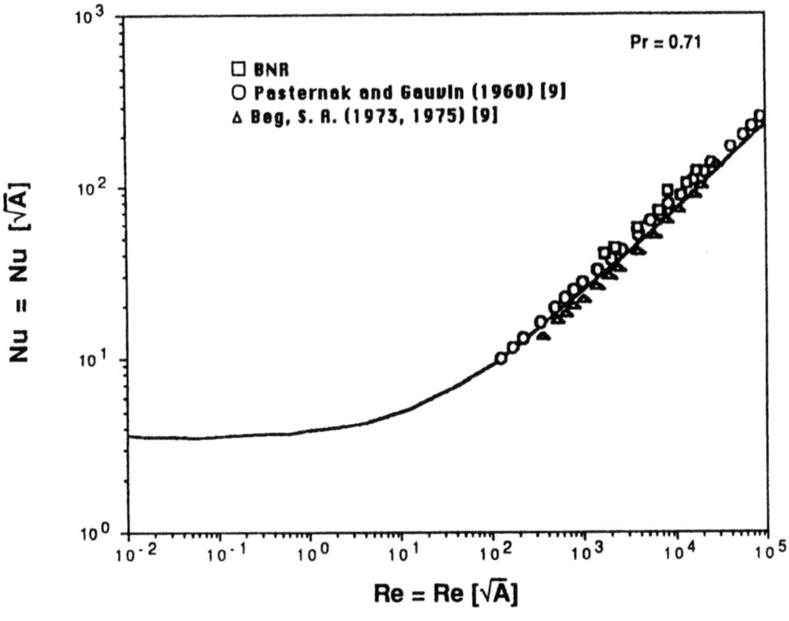

Figure 3

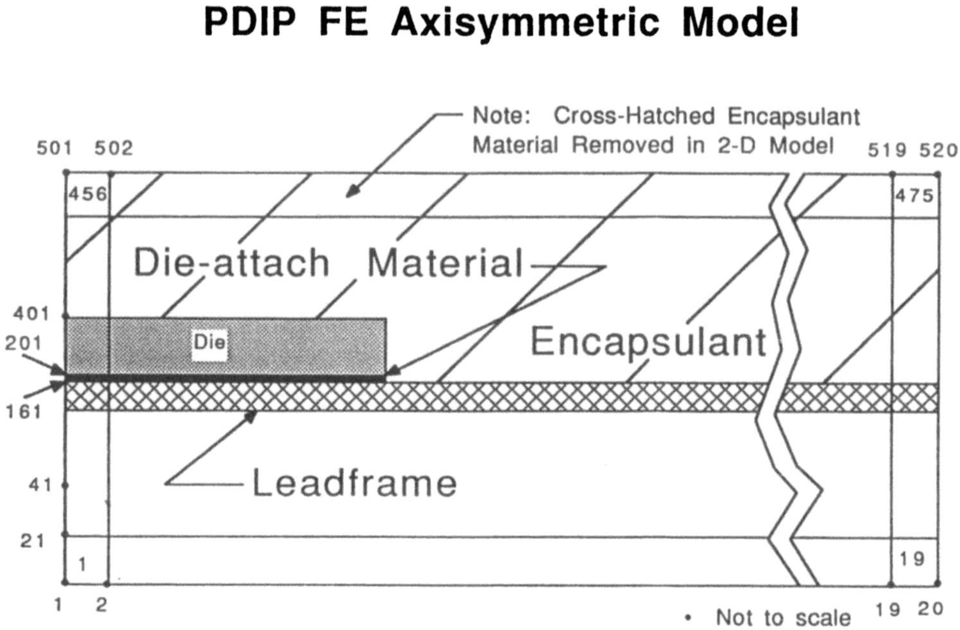

Figure 4

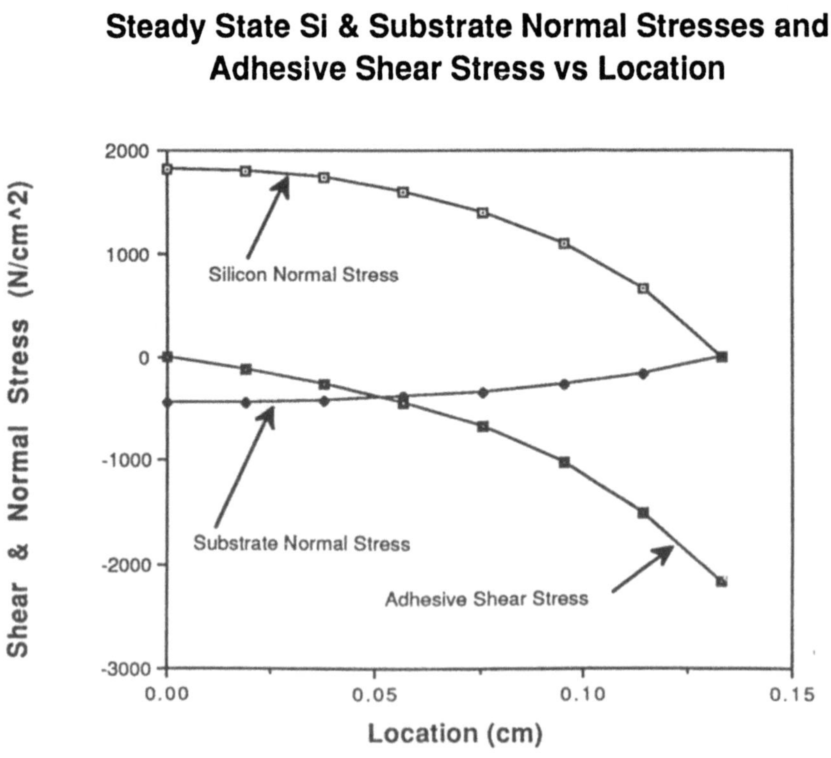

Figure 5

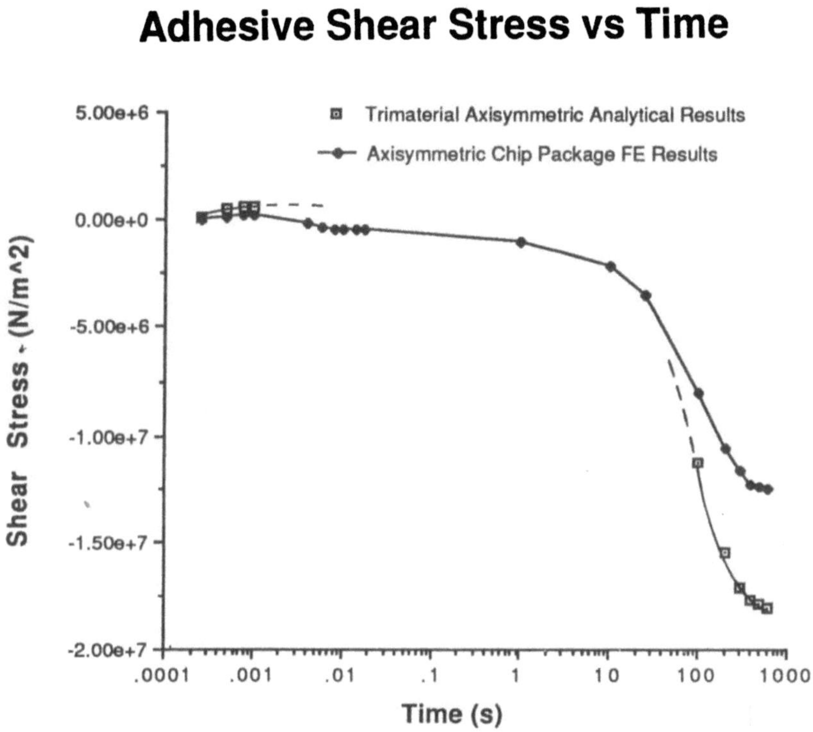
Adhesive Shear Stress vs Time
Trimaterial Axisymmetric Analytical Results
Axisymmetric Chip Package FE Results
5.00e+6
0.00e+0
-5.00e+6
-1.00e+7
-1.50e+7
-2.00e+7
Shear Stress - (N/m^2)
.0001
.001
.01
.1
1
10
100
1000
Time (s)

Figure 6

Impact of Die Attach Defects on Internal Temperature of Power Hybrids.

by:

F. Michard, J.M. Dupont, H. Ribot, M. Salagoity.

IBM France
International Procured Technologies
Test&Analysis
Department 1063
B.P. 100
33603 Pessac Cedex
FRANCE
Ph: (33) 56 07 53 76
Fax: (33) 56 07 50 16

Abstract. In this study, hybrid components from three different suppliers are first compared in terms of internal temperature using finite-element simulation softwares ANSYS[1] and CAEDS[2]. Then, the influence of die attach defects is investigated. Numerical models clearly prove that the position and shape of the major void is even more important than the total surface of the voids in die attach. This led us to define parameters that account for thermal flux distortion and can be correlated with the thermal resistance of the voided die attach.

Introduction.

Computer industry stimulated a lot of research on massive integration of components. Multi-chip modules and surface mounted technology have been breakthroughs towards highly integrated boards. This general trend results in an ever increasing dissipated power in systems and designers are now coming across severe thermal problems. Indeed, it is a challenge to maintain adequate temperature and improve reliability of machines while dissipating high power. Additionally, surface mounted components have to go through harsh assembly conditions especially when used on double-sided cards. Such demanding functional and environmental requirements make it necessary to develop a set of experimental and numerical techniques that constitute **Thermal Engineering**.

Computer-aided Thermomechanical modeling is one of these techniques. Based on finite elements/volumes methods, it gives access to temperature gradients and associated mechanical stresses for various thermal or mechanical loads. It is very flexible and can handle very complex models for which analytical approach is not practical. Such simulation software's have been primarily developed for automotive and aerospace industry, but can be readily used for electronic components. They are most efficient to compare components from different suppliers and to assess the impact of a defect or design change on the reliability of a given device. In this report, two software's are used: ANSYS for 3D temperature mapping in transient conditions and CAEDS for 3D thermal and mechanical mapping in steady states conditions. Temperature and stress in hybrids from three suppliers are compared in normal

C. J. Hoogendoorn et al. (eds.), Thermal Management of Electronic Systems, 325–334.

operating conditions. Finally, the influence of voids in die attach is investigated. Junction temperature is shown to depend on voids position and shape.

1) Hybrids description:

The hybrids under investigation are driving hard disk motors.
Components are made of 3 N-FET transistors, 3 P-FET transistors and a diode. In normal operating conditions, the N-FET transistor is dissipating 2.5 Watts, after transient regime (9 W, 1.2 seconds), and simultaneously the P-FET is dissipating 18 W for 10 seconds (start-up period), then 5 W continuously.

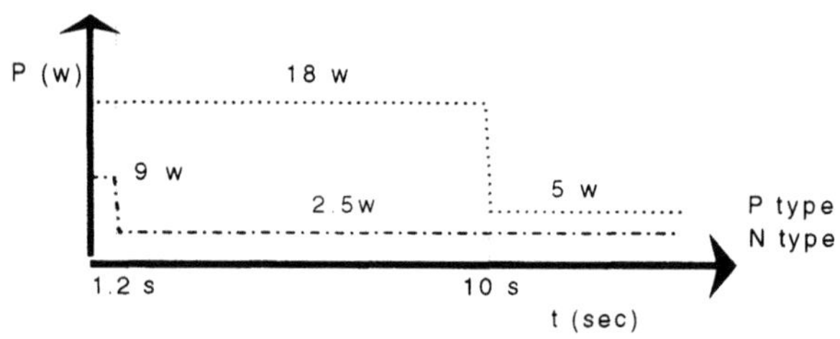

Fig. 1: Commutation timing for P and N transistors.

At manufacturing level, the hybrid is inserted in a metal housing filled with a non-conductive gel (Fig.2). The device is connected to an external heatsink (not shown here) by a conductive grease (20μm thick).

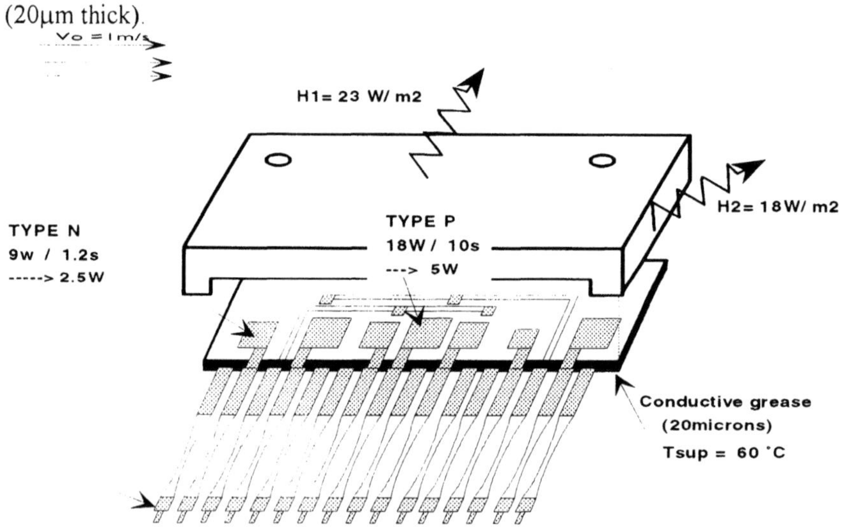

Fig. 2: Schematic of hybrid with boundary conditions.

Transistors are connected to the substrate by a stack of layers, which was designed differently by three manufacturers although the hybrids fulfill the same functional requirements (Fig.2). The simplest design is the hybrid Z, where the chip is attached to the Ag-plated alumina substrate by a $Pb_{88}Sn_{10}Ag_2$ solder.

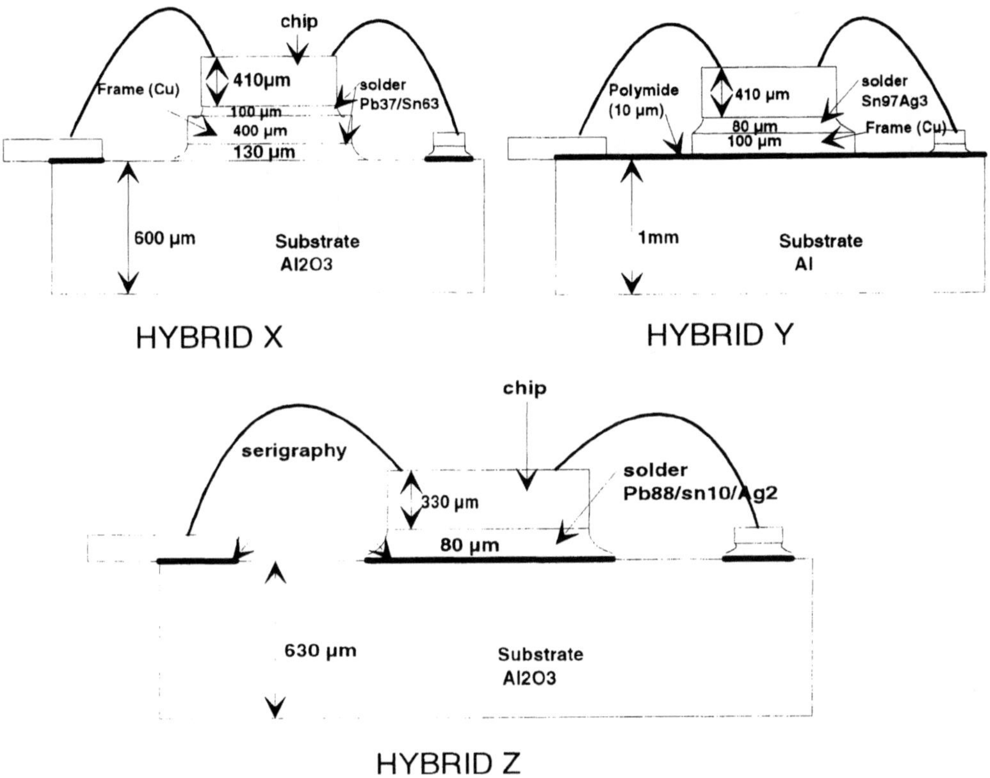

Fig. 3: Comparison of hybrids designs.

2) Comparison of the thermal behavior of hybrids using finite-element modeling:

First, a global transient 3D model is generated using ANSYS to visualize the thermal gradient and determine the critical areas. Boundary conditions are displayed on Fig. 1.
This first model clearly shows that thermal flux is mainly generated in the P-FET transistor which dissipates 18 Watts. Also, it appears that heat transfer through the non-conductive gel can be neglected.

Owing to these results, the size of the 3D model is significantly reduced by eliminating the regions where the heat flux is small (see Fig.4), but the model is also significantly refined in the die area:
- Housing is ignored since no heat dissipates through the gel.
- Wire Bonds and pins are included into the model.
- Only a **12x12 mm** area of the substrate is represented.
- Only the **Pfet** transistor is modeled.

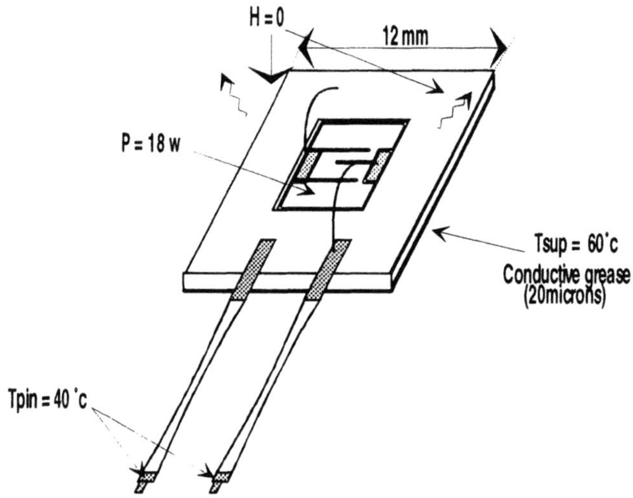

Fig.4: Boundary conditions on refined model.

Although the temperature gradients look similar on the three devices, component X exhibits a much higher maximum temperature in the chip, Tmax (121°C, see Fig.5). This is due to the larger thickness of the copper frame and solder in design X.

DEVICE	Tmax
X	121°C
Y	108°C
Z	110°C

Fig. 5: Comparison of maximum internal temperature in hybrids.

In view of such a difference in junction temperatures, component X is likely to be less reliable and designs Y,Z should be preferred. Hybrid Z design is used in the following calculations.

3) Influence of die attach quality on thermal field:

Various construction and failure analysis performed at IPT revealed defects in die attach. Our aim is to investigate the influence of such defects on the thermal properties of hybrids.
In current specifications, only the percentage of voids in the solder is taken into account[3]. In this study, we are also including defects distribution through five main defect types that were identified in construction and failure analysis. Typical patterns are shown on figure 6 and represent the voids distribution in the $Pb_{88}Sn_{10}Ag_2$ layer of hybrid Z.
Type 1 is basically an homogeneous distribution of voids.
In type 2, solder is missing in the center of the chip.
Type 3 is the opposite configuration where solder is confined to the center of the die.
Type 4 consists of a large void along one edge of the chip.
Type 5 is a lack of solder in one corner of the chip.

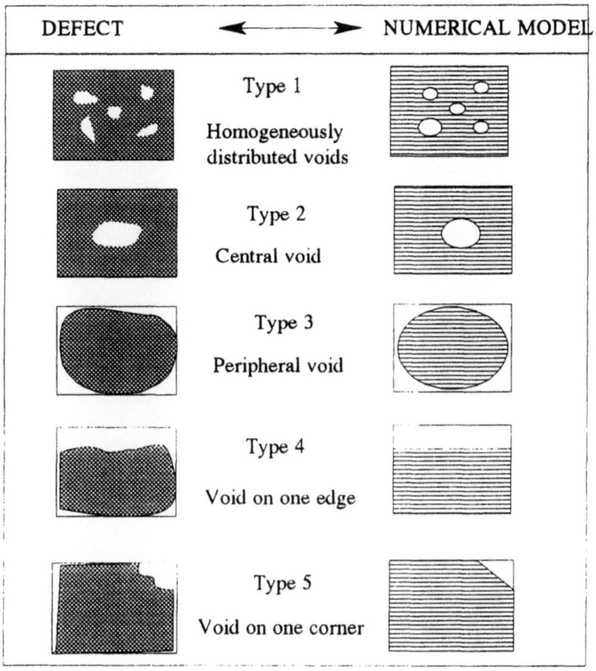

Fig. 6: Classification of defects in die attach (top view of solder).

Solder layer thermal resistance, Rint is related to Tmax, Tbase, temperature at the base of the substrate and P, dissipated power in the circuit:

Rint = (Tmax-Tbase)/P

Rint is plotted as a function of the percentage of voids, ε=Sair/Ssolder, on figure 7 for the above defects:

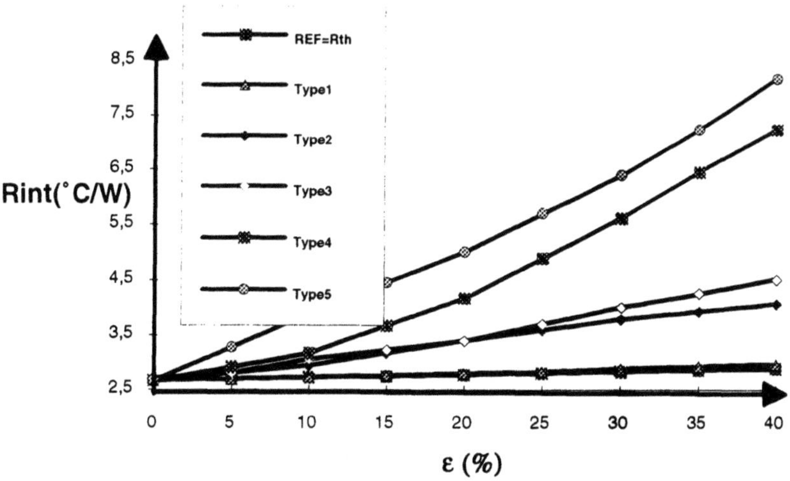

Fig. 7: Calculated junction thermal resistance as a function of defects surface and type.

For type 1, Rint does not increase much compared to a perfect die attach. A 40% voids rate results in a 5% increase of Rint, which should be considered as a **minor** concern.
For types 2 and 3, the increase is sharper and can be as high as 20% for ε=40%. Such a change would be a **major** concern during a Construction Analysis.
Even more dramatic is the degradation for type 4 and 5, when ε gets beyond 10%. Rint almost doubles when 40% of the solder is missing on one side of the chip, and the calculated temperature gets higher than the solder fusion temperature. Therefore, catastrophic failure is expected and such defects should be considered as **critical**. The temperature field for defect type 5 is shown on Fig. 8.

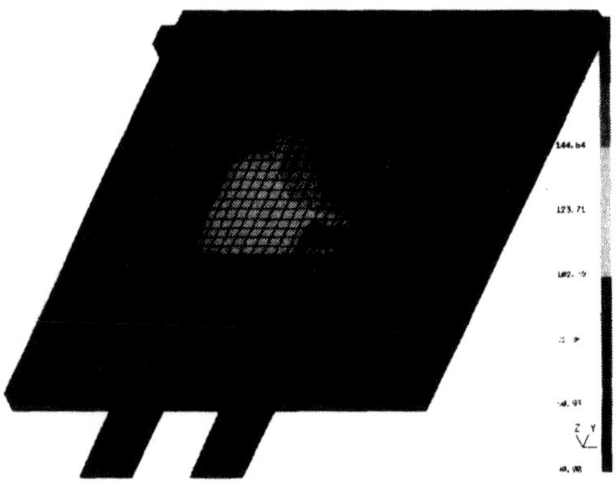

Fig. 8: Temperature field for defect type 5

Note that a classical approach using an equivalent series/parallel network of thermal resistances[4] cannot account for the observed discrepancies. In this case, the solder layer has the same thermal resistance, Rth, whatever the defect type is:

$$Rth = \frac{e}{((1-\varepsilon).\lambda_{sold} + \varepsilon\ .\ \lambda_{air})\ .\ Stot}$$

with: ε, previously defined, Stot, the solder layer total surface, e, its thickness and λ_{sold}, λ_{air}, respectively the solder and air conductivity.
Rth is always smaller than the value obtained by finite element modeling, except when defects are homogeneously distributed (defect type 1). The reason is that the analytical treatment cannot account for flux distortion[5], since it is basically a one-dimension approach. Yet, this is flux distortion which explains why defect types 4 and 5 lead to very high resistance and also why case 2 is slightly more favorable than case 3 for a given percentage of voids.

These results show that the total surface of voids is insufficient to account for junction temperature increase in electronic components. Indeed, the position of the major voids is even more important. Other parameters have to be derived which assess thermal flux distortion. One of them is the average path length of thermal flux around the void which is found to be closely related to the thermal resistance of die attach.

4) Assessment of heat flux distortion:

Let us consider the top view of a defective die attach as shown on figure 9, and define the following dimensionless parameter p_ϕ:

$$P_\varphi = \frac{1}{L \cdot S_d} \cdot \int_{S_d} p \cdot ds$$

With:

Sd, defect area, p, distance from a given point M to the edge of the defect, L characteristic length of die attach (L=a, in case of a square-shaped die attach).
p_ϕ accounts for the defect shape and position as well as for its area. It is easily calculated for simple shapes, such as defect types 2, 4 and 5, respectively, circular, rectangular or triangular voids.

4.a Case of a circular void in the center of die attach:

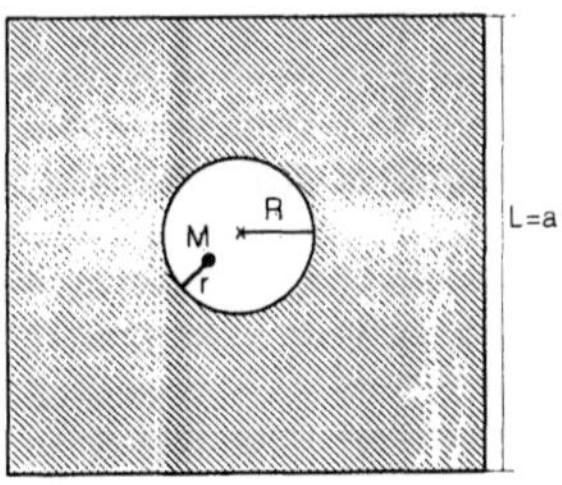

Fig. 9: Schematic of a central circular defect

$$p_\varphi = \frac{1}{\Pi . R^2 . a} \int_0^R (R - r) . 2 . \Pi . r . dr = \frac{R}{3 . a}$$

and since one has:

$\pi \,.\, R^2 = \varepsilon \,.\, a^2$

Therefore:

$$\boxed{p_\phi = \frac{\sqrt{\varepsilon}}{3 . \sqrt{\Pi}}}$$

4.b Case of a triangular void in a corner:

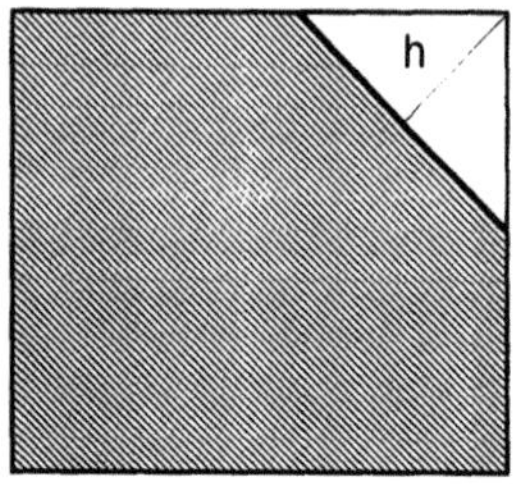

Fig. 10: Schematic of a triangular defect

h, being the height of the isosceles triangle, one has:

$h^2 = \varepsilon \,.\, a^2$

and:

$$p_\phi = \frac{2}{h^2.a}\int_0^h (h-x).x.dx = \frac{h}{3.a}$$

or:

$$\boxed{p_\phi = \frac{\sqrt{\varepsilon}}{3}}$$

4.c Case of a rectangular void:

Fig. 11: Schematic of rectangular defects

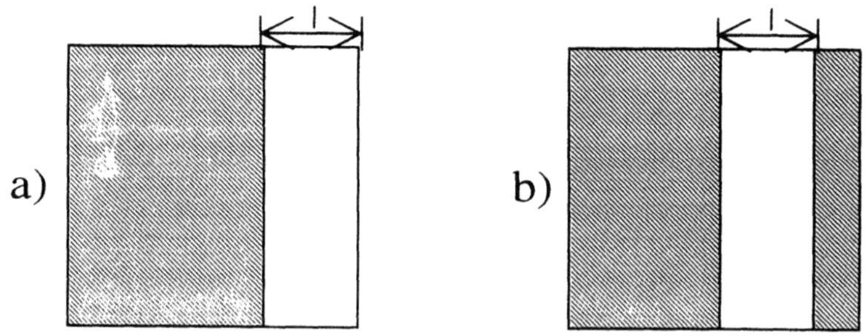

In the case of a rectangular void, p_ϕ is either:

$$\boxed{p_\phi = \frac{\varepsilon}{2}}$$

If the defect is along one edge of the die attach (Fig. 11a)

or:

$$\boxed{p_\phi = \frac{\varepsilon}{4}}$$

in any other case (Fig. 11b).

This abrupt change of the distortion parameter, p_ϕ, is very well correlated with our numerical results plotted on figure 12 where temperature is displayed as a function of a dimensionless parameter x.

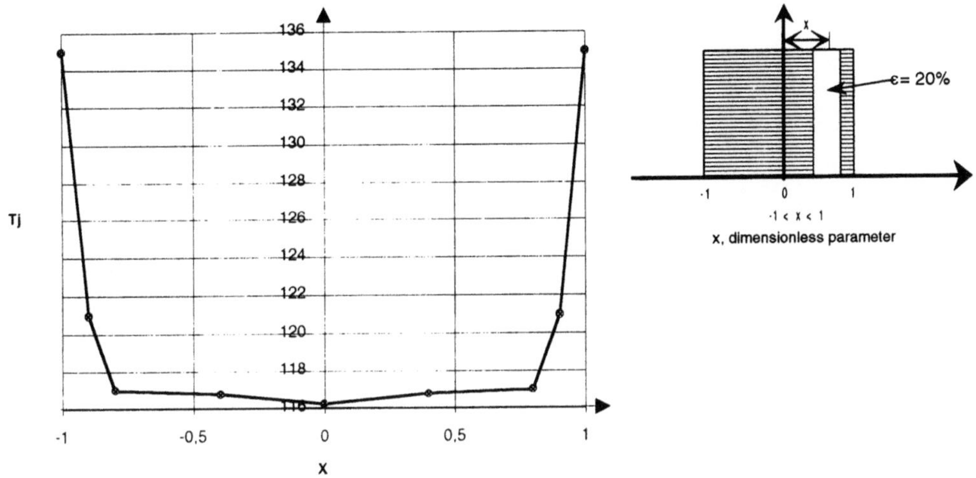

Fig. 12: Tj as a function of rectangular void position.

Clearly, the calculated temperature jumps as the void approaches the edge of the die attach.

4.d) Correlation between p_ϕ and the calculated deviation from Rth:

Deviation from Rth, expressed in (Rint - Rth)/Rint, and dimensionless parameter p_ϕ look very similar when plotted as a function of ε. This leads us to investigate the simple relation:

(Rint - Rth)/Rint = k(geom,mat) . p_ϕ

or:

$$\text{Rint} = \frac{Rth}{(1-k.p_\phi)}$$

where k is a function which depends on the hybrid geometry and materials properties while p_ϕ accounts for the flux distortion caused by a specific defect. To check the validity of this expression, one plots Rint calculated from the above equation (with experimental value k=3.1), together with the value of Rint derived from the finite element model (Fig.13).

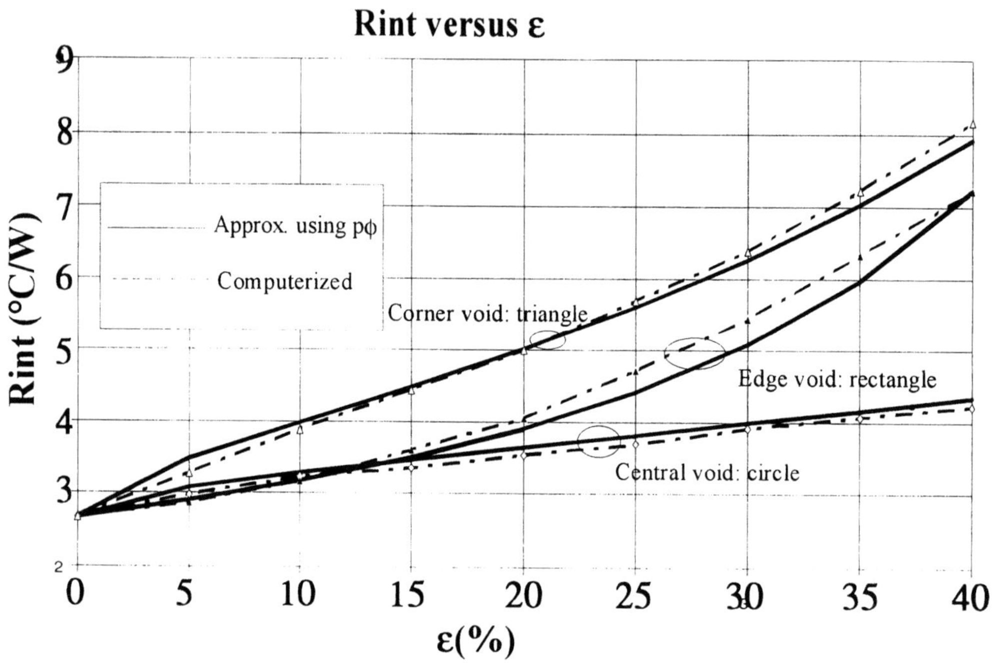

Fig.13: Thermal resistance as a function of ε: Approximate versus computerized solution.

Satisfactory correlation, validates our assumption that flux distortion in the die attach can be decoupled from geometrical and material properties by a simple dimensionless parameter. This parameter can be used very conveniently to estimate the resistance increase caused by any die attach defect.

Conclusion:

This study gives some useful insight of which deviations are acceptable in die attach process. A substantial lack of solder on one edge of the chip or in a corner is a very dangerous pattern which should trigger immediate corrective actions on the process.
Additionally, we were able to demonstrate that the chip temperature increase is primarily governed by the flux distortion caused by the major defect in the die attach. A dimensionless parameter is proposed, which makes it possible to decouple the distortion contribution from other geometrical and material parameters, in the total increase of the thermal resistance.

References:

[1] ANSYS® is a registered mark of Swanson Analysis Systems, Inc.

[2] CAEDS® is a registered trademark of International Business Machines Corporation.

[3] Military Standard 883C, method 2010.7 (internal visual) and method 2012.5 (radiography).

[4] Microelectronics Packaging Handbook. p.173, R.R. Tummala, E.J. Rymaszewski, Van Nostrand Reinhold, 1989.

[5] D.P. Kennedy, "Heat Conduction in a Homogeneous Solid Circular Cylinder of Isotropic Media", TR 00.15072. 699. IBM. December 1959.

AUTHOR INDEX

C. J. Hoogendoorn et al. (eds.), Thermal Management of Electronic Systems, 335.

MIX
Papier aus verantwortungsvollen Quellen
Paper from responsible sources
FSC® C105338

If you have any concerns about our products,
you can contact us on
ProductSafety@springernature.com

In case Publisher is established outside the EU,
the EU authorized representative is:
Springer Nature Customer Service Center GmbH
Europaplatz 3, 69115 Heidelberg, Germany

Printed by Libri Plureos GmbH
in Hamburg, Germany